Physical Aspects of Fracture

NATO Science Series

A Series presenting the results of scientific meetings supported under the NATO Science Programme.

The Series is published by IOS Press, Amsterdam, and Kluwer Academic Publishers in conjunction with the NATO Scientific Affairs Division

Sub-Series

I. **Life and Behavioural Sciences**	IOS Press
II. **Mathematics, Physics and Chemistry**	Kluwer Academic Publishers
III. **Computer and Systems Science**	IOS Press
IV. **Earth and Environmental Sciences**	Kluwer Academic Publishers

The NATO Science Series continues the series of books published formerly as the NATO ASI Series.

The NATO Science Programme offers support for collaboration in civil science between scientists of countries of the Euro-Atlantic Partnership Council. The types of scientific meeting generally supported are "Advanced Study Institutes" and "Advanced Research Workshops", and the NATO Science Series collects together the results of these meetings. The meetings are co-organized bij scientists from NATO countries and scientists from NATO's Partner countries – countries of the CIS and Central and Eastern Europe.

Advanced Study Institutes are high-level tutorial courses offering in-depth study of latest advances in a field.
Advanced Research Workshops are expert meetings aimed at critical assessment of a field, and identification of directions for future action.

As a consequence of the restructuring of the NATO Science Programme in 1999, the NATO Science Series was re-organized to the four sub-series noted above. Please consult the following web sites for information on previous volumes published in the Series.

http://www.nato.int/science
http://www.wkap.nl
http://www.iospress.nl
http://www.wtv-books.de/nato-pco.htm

Series II: Mathematics, Physics and Chemistry – Vol. 32

Physical Aspects of Fracture

edited by

Elisabeth Bouchaud

Service de Physique et Chimie des Surfaces et des Interfaces,
Direction des Sciences de la Matière, CEA,
Saclay, France

Dominique Jeulin

Centre de Morphologie Mathématique,
Unité Mixte de Recherche CNRS,
Ecole des Mines Paris, France

Claude Prioul

Université Paris 13 and Laboratoire Mécanique des Sols,
Structures et Matériaux,
Unité Mixte de Recherche CNRS,
Ecole Centrale Paris, France

and

Stéphane Roux

Laboratoire Surface du Verre et Interfaces,
Unité Mixte de Recherche CNRS,
St-Gobain, France

Kluwer Academic Publishers

Dordrecht / Boston / London

Published in cooperation with NATO Scientific Affairs Division

Proceedings of the NATO Advanced Study Institute on
Physical Aspects of Fracture
Cargèse, France
5–17 June 2000

A C.I.P. Catalogue record for this book is available from the Library of Congress.

ISBN 0-7923-7146-1 (HB)
ISBN 0-7923-7147-X (PB)

Published by Kluwer Academic Publishers,
P.O. Box 17, 3300 AA Dordrecht, The Netherlands.

Sold and distributed in North, Central and South America
by Kluwer Academic Publishers,
101 Philip Drive, Norwell, MA 02061, U.S.A.

In all other countries, sold and distributed
by Kluwer Academic Publishers,
P.O. Box 322, 3300 AH Dordrecht, The Netherlands.

Printed on acid-free paper

Printed in the Netherlands.

CONTENTS

vi

Fatigue and stress corrosion

Dynamics of fracture

PREFACE

The main scope of this Cargèse NATO Advanced Study Institute (June 5-17 2000) was to bring together a number of international experts, covering a large spectrum of the various Physical Aspects of Fracture. As a matter of fact, lecturers as well as participants were coming from various scientific communities: mechanics, physics, materials science, with the common objective of progressing towards a multi-scale description of fracture.

This volume includes papers on most materials of practical interest: from concrete to ceramics through metallic alloys, glasses, polymers and composite materials. The classical fields of damage and fracture mechanisms are addressed (critical and sub-critical quasi-static crack propagation, stress corrosion, fatigue, fatigue-corrosion, ... as well as dynamic fracture). Brittle and ductile fractures are considered and a balance has been carefully kept between experiments, simulations and theoretical models, and between the contributions of the various communities. New topics in damage and fracture mechanics - the effect of disorder and statistical aspects, dynamic fracture, friction and fracture of interfaces - were also explored. This large overview on the Physical Aspects of Fracture shows that the old barriers built between the different scales will soon "fracture". It is no more unrealistic to imagine that a crack initiated through a molecular dynamics description could be propagated at the grain level thanks to dislocation dynamics included in a crystal plasticity model, itself implemented in a finite element code. Linking what happens at the atomic scale to fracture of structures as large as a dam is the new emerging challenge.

This summer school has been particularly efficient in improving the communication and exchanges between the 90 scientists coming from 20 different countries, among which the large representation of Eastern Europe is to be emphasised. The richness of these exchanges is an essential success of this Summer School since the culture, the tools, the methods and even the objectives of the lecturers differed quite strongly. Hence we do believe that this volume addresses all the scientists interested in fracture, whatever their background may be. We are indebted to all the lecturers who contributed to this volume, for the special effort they made to produce particularly pedagogical papers, with a self-contained and explicit presentation.

Finally, we wish to acknowledge our sponsors, without whom this school, and the present volume, would have never existed: NATO, the Commissariat à l'Energie Atomique, the Centre National de la Recherche Scientifique, Saint-Gobain, the French Ministry of Research, the French Ministry of Foreign Affairs, and the Institut d'Etudes Scientifiques de Cargèse. We are particularly indebted to Elisabeth Dubois-Violette and her team in Cargèse, to Raymonde Boschiero for her help in the financial matters, and, last but not least, to Marcelle Martin, who has been of a great help in organising the meeting, and made the considerable effort to put everything together for the elaboration of the present volume.

Elisabeth Bouchaud, Dominique Jeulin, Claude Prioul, Stéphane Roux.

International Scientific Committee

G. Cailletaud	École des Mines de Paris, France
D. Fisher	Harvard University, USA
D. François	École Centrale Paris, France
A. Hansen	Trondheim University, Norway
F. Lund	Santiago University, Chile
M. Kardar	MIT, USA
B. Lawn	NIST, USA
G. Pijaudier-Cabot	École Normale Supérieure de Cachan, France
A. Pineau	École des Mines de Paris, France
J. Rice	Harvard University, USA.

Organizing Committee

E. Bouchaud	CEA, France
R. Goldstein	Russian Academy of Sciences
D. Jeulin	École des Mines de Paris, France
C. Prioul	École Centrale Paris, France
S. Roux	CNRS/Saint-Gobain, France.

Sponsors

Commissariat à l'Energie Atomique (CEA)
Centre National de la Recherche Scientifique (CNRS)
French Ministry for Education, Research and Technology (MNERT)
French Ministry of Foreign Affairs (MAE)
North Atlantic Treaty Organisation (NATO)
SAINT-GOBAIN.

LIST OF PARTICIPANTS

Prof. ALEXANDROV Serguei
Alcoa Technical Center, Bldg B,
100 Technical Drive, ALCOA CENTER, PA
15069-0001 – USA

Prof. AROUTIOUNIAN Vladimir
Yerevan State University-
Dpt of Physics of Semiconductors
1 Alex Manoukian St., 375049 YEREVAN
Rep. of Armenia

Dr. AURADOU Harold
CEA Saclay – SRIM
91191 GIF sur YVETTE CEDEX – France

Mr. BALAZSI Csaba
Institute of Technical Physics and Materials
Science, Ceramic Dpt
PO. Box 49, H-1525 BUDAPEST – Hungary

Dr.. BARTHELEMY Marc
CEA-Centre d'Etudes de Bruyères-le-Chatel
Service de Physique de la Matière Condensée
BP 12
91680 BRUYERE-LE-CHATEL - France

Dr. BELJAKOVA Tatiana
Moscow State University
Vernadsky prospect, 127-664
MOSCOW 117 571 – Russia

Prof. BERTHAUD Yves
ENS Cachan – LMT
61, ave du Président Wilson , F-94235
CACHAN CEDEX – France

Dr. BOUCHAUD Elisabeth
CEA Saclay – SPCSI
91191 GIF sur YVETTE CEDEX – France

Dr. BRONNIKOV Serguei
Institute of Macromolecular Compounds
Bolshoi Prospekt 31
199004 ST.PETERSBURG - Russia

Dr. CANTELAUBE Florence
Saint-Gobain – Mechanics Laboratory
CRIR - 19, rue Emile Zola, 60290
RANTIGNY – France

Dr. CHAUVOT Carine
Fraunhofer Institut fuer Werkstoffmechanik
Woehlerstrasse 11
D-79108 FREIBURG - Germany

Dr. CHILKO Evgueni
Institute of Strength Physics and Materials
Science,
2/1, Pr. Academicheskiy
TOMSK 634021 – Russia

Dr. CHLUP Zdenek
IPM AS CR
Zizkova 22 - 616 62 BRNO – Czech Republic

Dr. CORREIRA PEREIRA DA SILVA Filipe
Universidade de Minho – Dpt Eng. Mecanica
Azurem – 4800 GUIMARAES - Portugal

Dr. DANG VAN Ky
Ecole Polytechnique
Route de Saclay
91128 PALAISEAU CEDEX – France

Dr. DAVYDOVA Marina
Institute of Continuous Media Mechanics
1, Ak Korolev str.
614013 PERM - Russia

Prof. DEGALLAIX Suzanne
Ecole Centrale de Lille – LML
Cité Scientifique
59491 VILLENEUVE D'ASCQ CEDEX
France

Dr. DUCOURTHIAL Elodie
ONERA – 29, ave Division Leclerc
92322 CHATILLON CEDEX – France

Dr. EFROS Boris
Donetsk Physics & Technology Institute NASc
of Ukraine
72,R.Luxemburg str.
83114 DONETSK - Ukraine

Prof. FISCHER Daniel
Harvard University
Cambridge
MA 02138 – USA

Prof. FRANCOIS Dominique
Ecole Centrale Paris – LMSSMAT - Grande
Voie des Vignes,
F-92295 CHATENAY-MALABRY – France

Dr. GATELIER Nicolas
Université Joseph Fourier - Lab. Sols Solides
Structures (3S)
Domaine universitaire – BP 83
F-38041 GRENOBLE CEDEX 9 – France

Dr. KONEVA Helen
Institute of Theoretical and Applied Mechanics
Institutskaja str. 4/1
630090 NOVOSIBIRSK - Russia

Dr. GRABCO Daria
Institute of Applied Physics of Moldavian
Academy of Sciences
Academy str. 5
MD-2028, KISHINEV – Moldova

Dr. GROUZDKOV Alexey
St.Petersburg State University, RSD
191119 Russia
ST. PETERSBURG, Borovaya 24-9

Dr. GUILLOT Claude
CEA Saclay – DRECAM/SRSIM bât 462
91191 GIF sur YVETTE CEDEX – France

Dr. GY René
Saint-Gobain Recherche
39, quai Lucien Lefranc, BP 135
F-93303 AUBERVILLIERS CEDEX – France

Prof. HANSEN Alex
Institut for fyusikk – Norges teknisk-naturv.
University
N-7491 TRONDHEIM – Norway

Mr. HAUSILD Petr
Ecole Centrale Paris – LMSS-Mat
Grande Voie des Vignes
92295 CHATENAY-MALABRY CEDEX –
France

Dr. HILD François
Ecole Normale Supérieure Cachan – LMT
61, avenue du President Wilson
F-94235 CACHAN CEDEX – France

Dr. INOCHKINA Irina
St.Petersburg State University
Bibliotechnaya pl.2, 198904
ST.PETERSBURG – Russia

Dr. JABARI-LEE Ouma
Louisiana State University
Concurrent Computing Laboratory for
Materials Simulations
Department of Physics and Astronomy
BATON-ROUGE, LA 70803-4001 USA

Dr. JEULIN Dominique
Ecole des Mines de Paris – CMM
35, rue Saint Honoré, F-77300
FONTAINEBLEAU – France

Prof. KACHANOV Mark
Dept.of Mech.Eng-g,Tufts University,
Medford, MA 02155 - USA

Prof. KALIA Rajiv
Louisiana State University
Concurrent Computing Laboratory for
Materials Simulations
Department of Physics and Astronomy
BATON-ROUGE, LA 70803-4001 USA

Prof. KAMENOVA Tzanka
Institute of Metal Science
67, Shipchensky prohod str
1574 SOFIA – Bulgaria

Prof. KANAUN Sergey
Ecole des Mines de Paris – CMM
35, rue Saint Honoré, F-77300
FONTAINEBLEAU – France

Dr. KANDER Ladislav
Vitkovice, J.S.C., R&D Division
Pohranicni 31, 70602 OSTRAVA 6
Czech Republic

Prof. KIRCHNER Helmut
Université Paris-Sud
Lab. Métallurgie Structurale – Bât. 413
91405 ORSAY – France

Dr. KOHOUT Jan
Military Academy in Brno
Dpt. of Physics, Military Technology Faculty
Kounicova St. 65, CZ-61200 BRNO
Czech Republic

Dr. KOVACIK Jaroslav
Institute of Materials & Machine Mechanics
Racianska 75, 83812 BRATISLAVA
Slovak Republic

Prof. KRASOWSKY Arnold
Institute for Problems of Strength-National
Academy of Sciences
Timiryazevskaya St. 2, KIEV 01014 - Ukraine

Dr. KYSAR Jeffrey
Brown University Division of Engineering
Providence, RI 02912 – USA

Dr. LAMBERT Astrid
Ecole des Mines Paris – Centre des Matériaux
BP 87 – 91003 EVRY – France

Dr. LAVRYNENKO Sergiy
Kharkov State Polytechnic University
Saltivske Shose 145-6, kv. 54
KHARKOV 61112 – Ukraine

Prof. LUND Fernando
Universitad de Chile, Physics Dept.
Casilla 487-3, SANTIAGO, Chile

Prof. MAGNIN Thierry
Ecole des Mines de Saint Etienne - labo SMS
158, cours Fauriel
42023 SAINT-ETIENNE CEDEX – France

Dr. MAN Jiri
Institute of Physics of Materials - Academy of
Sciences Czech Rep.
Zizkova 22, 61669 BRNO – Czech Republic

Dr. MARLIERE Christian
Université de Montpellier II – Laboratoire des
Verres- CC.69
Place E . Bataillon
34095 MONTPELLIER CEDEX 5 – France

Dr. MARTINEZ-DIEZ Juan-Alberto
Facultad de Ciencias
Dpt Fisica de la Materia Condensada,
Cristalografia y Mineralogia
C/Prado de la Magdalena s/n, 47011
VALLADOLID – Spain

Dr. MINOZZI Manuela
Universita La Sapienza - NFM, Dipartimento di
Fisica, E. Fermi
P. le A. Moro 2 – 00185 ROMA – Italy

Dr. MOREL Stéphane
Université Bordeaux 1 – Lab. De Rhéologie du
Bois
Domaine de l'Hermitage
33610 CESTAS GAZINET – France

Prof. MUGHRABI Hael
UniversitaetErlangen-Nuernberg
Institut fuer Werkstoffwissenschaften
Lehrstuhl I – Martensstr. 5
D-910058 ERLANGEN, Germany

Prof. NAIMARK Oleg
Institute of Continuous Media Mechanics
1, Ak Korolev str., 614013 PERM, Russia

Prof. NEDBAL Ivan
Technical University in Prague
CVUT-FJFI-KMAT
Trojanova 13, 12000 PRAHA 2
Czech Republic

Dr. NOBILI Maurizio
Université de Montpellier II – place E.
Bataillon CC026
F-34095 MONTPELLIER CEDEX 5 – France

Dr. OLIVE Jean-Marc
Université Bordeaux 1 – Lab. de Mécanique
Physique
351, cours de la Libération
33405 TALENCE CEDEX – France

Dr. PETIT Jean
ENSMA – LMPM
Site du Futuroscope - BP. 109
86960 FUTUROSCOPE CEDEX - France

Dr. PFUFF Michael
GKSS Research Centre
Max Planck strasse
D-21502 GEESTHACHT – Germany

Prof. PIJAUDIER-CABOT Gilles
Ecole Centrale de Nantes
Lab. de Génie Civil de Nantes Saint-Nazaire
BP. 92101 – NANTES – France

Prof. PINEAU André
Ecole des Mines Paris – Centre des Matériaux
BP 87 – 91003 EVRY – France

Dr. PONIKAROV Nikolai
St.Petersburg State University, Elasticity Dept.
Bibliotechnaya pl.2
198904 ST.PETERSBURG – Russia

Dr. PRASILOVA Alice
Institute of Physics of Materials - Academy of
Sciences Czech Rep.
Zizkova 22, 61669 BRNO – Czech Republic

Prof. PRIOUL Claude
Ecole Centrale Paris – LMSS-Mat
Grande Voie des Vignes
92295 CHATENAY-MALABRY CEDEX –
France

Prof. RAVI-CHANDAR K.
University of Texas
Dept of Aerospace Eng. And Eng. Mech
AUSTIN, TX 78712, USA

Prof. REY Colette
Ecole Centrale Paris – LMSS-Mat
Grande Voie des Vignes
92295 CHATENAY-MALABRY CEDEX –
France

Prof. RICE James
Harvard University-Division of Engineering
and Applied Sciences
And Dpt of Earth and Planetary Sciences
224 Pierce Hall – 29 Oxford street,
CAMBRIDGE, MA 02138 USA

Prof. RITTEL Daniel
Faculty of Mechanical Eng. – Technion
32000 HAIFA – Israël

Dr. ROSSOLL Andreas
Ecole Polytechnique Fédérale de Lausanne
Département des Matériaux-
Lab. de Métallurgie Mécanique
CH-1015 LAUSANNE – Suisse

Dr. ROUNTREE Cindy
Louisiana State University
Concurrent Computing Laboratory for
Materials Simulations
Department of Physics and Astronomy
BATON-ROUGE, LA 70803-4001 USA

Dr. ROUX Stéphane
Unité mixte CNRS/Saint-Gobain – Surface du
Verre et Interfaces
39, quai Lucien Lefranc, BP 135
93303 AUBERVILLIERS CEDEX – France

Dr. SARRAZIN-BAUDOUX Christine
ENSMA – LMPM
Site du Futuroscope BP. 109
86960 FUTUROSCOPE CEDEX - France

Dr. SCHALLER Robert
Ecole Polytechnique Fédérale de Lausanne,
Inst. Génie Atomique
CH-1015 LAUSANNE – Suisse

Dr. SCHMITTBUHL Jean
Ecole Normale Supérieure de Paris, Dpt TAO
24, rue Lhomond
75231 PARIS CEDEX 05 – France

Dr. SHIKIMAKA Olga
Institute of Applied Physics of Moldavian
Academy of Sciences
Academy str. 5
MD-2028, KISHINEV – Moldova

Dr. SHISHKOVA Natalya
Donetsk Physics & Technology Institute NASc
of Ukraine
72,R.Luxemburg str.
83114 DONETSK – Ukraine

Dr. SMOLINE Igor
Institute of Strength Physics and Materials
Science,
Russian Academy of Sciences, Siberian Branch
2/1, Pr. Academicheskiy
TOMSK 634021 – Russia

Dr. TANGUY Benoît
Ecole des Mines Paris – Centre des Matériaux
BP 87 – 91003 EVRY – France

Dr. TARABAN Vladimir
St. Petersburg State University, RCD
Grazhdansky av., 77-2-13
195257 ST.PETERSBURG – Russia

Dr. TARASOVS Sergejs
Institute of Polymer Mechanics
Aizkraukles 23-211, LV 1006 RIGA – Latvia

Dr. TCHANKOV Dimitar
Technical University of Sofia – Dpt Strength of
Materials
1000 SOFIA – Bulgaria

Dr. TOUSSAINT Renaud
Géosciences Rennes, bureau 211
Campus de Beulieu, bât 15
ave du Général Leclerc
35042 RENNES CEDEX – France

Dr. UTKIN Andrey
Institute of Theoretical and Applied Mechanics
Institutskaja str. 4/1
630090 NOVOSIBIRSK – Russia

Dr. UVAROV Sergey
Institute of Continuous Media Mechanics
1, Ak Korolev str., 614013 PERM - Russia

Dr.VAN BRUTZEL Lauren
Louisiana State University
Concurrent Computing Laboratory for
Materials Simulations
Department of Physics and Astronomy
BATON-ROUGE, LA 70803-4001 USA

Prof. VAN DER GIESSEN Erik
Delft University of Technology –
Micromechanics of Materials
Mekelweg, 2
2628 CD DELFT – The Netherlands

Prof. VAN MIER Jan
Delft University of Technology –
Micromechanics of Materials
Mekelweg, 2
2628 CD DELFT – The Netherlands

Dr. VANDEMBROUCQ Damien
Unité mixte CNRS/Saint-Gobain – Surface du
Verre et Interfaces
39, quai Lucien Lefranc, BP 135
93303 AUBERVILLIERS CEDEX – France

Dr. VAVRIK Daniel
Institute of Theoretical and Applied Mechanics
Proseckà 76, 19000 PRAHA 9
Czech Republic

Dr. VIDAL Beatriz
Facultad de Ciencias
Dpt Fisica de la Materia Condensada
Cristalografia y Mineralogia
C/Prado de la Magdalena s/n
47011 VALLADOLID – Spain

Dr. ZAVOITCHINSKAIA Eleonora
Moscow State University - Mechanical and
Mathematical Dpt
Nametkina, 1.,32, 117393 MOSCOW – Russia

OPENING OVERVIEW

New trends in fracture mechanics

SOME STUDIES OF CRACK DYNAMICS

JAMES R. RICE
Harvard University
Cambridge, MA 02138 USA

Abstract

Recent developments in fracture dynamics include the discovery of elastic waves which propagate along moving crack fronts in three-dimensional solids, and the identification of possible sources of the roughening and low terminal speeds of tensile cracks in brittle amorphous solids. Those topics are discussed briefly here.

Keywords: fracture / crack dynamics / elastodynamics

1. Introduction

The focus in this brief report is on two areas of tensile crack dynamics in elastic-brittle solids that have received recent focus. These involve the crack front waves revealed in numerical and analytical studies of cracking along a plane in a 3D solid, and the problem of understanding the significantly sub-Rayleigh limiting speeds of tensile cracks in brittle amorphous solids, with associated clusters of microcracks and roughening of the fracture surface. A related discussion of recent studies on the dynamics of crack and fault rupture is given in [1]. That includes some additional topics that were covered in the oral version of this presentation, on dynamic frictional slip, especially the unstable form it takes along interfaces between elastically dissimilar materials, on slip-rupture along earthquake faults, and on combined tensile and shear failures at bimaterial interfaces.

Here tensile cracking is addressed in the framework of continuum elastodynamics. The governing equations are the equations of motion $\nabla \cdot \sigma = \rho \partial^2 \mathbf{u} / \partial t^2$ and the stress - displacement gradient relations $\sigma = \lambda(\nabla \cdot \mathbf{u})\mathbf{I} + \mu[(\nabla\mathbf{u}) + (\nabla\mathbf{u})^T]$. Two different fracture formulations are used. The first is a singular crack model, in which one sets $\sigma_{yy} = 0$ on the mathematical cut along $y = 0$ (see lower part of Figure 1) which is the crack surface. That leads to a well known singular field of structure

$$\lim_{r \to 0} \left(\sqrt{r}\sigma_{\alpha\beta} \right) = \sqrt{\mu G}\Sigma_{\alpha\beta}(\theta, v / c_s, c_p / c_s)$$

where r, θ are polar coordinates at the crack tip, v is the speed of crack propagation, and the $\Sigma_{\alpha\beta}$ are dimensionless universal functions [2,3]. Here c_p, c_s are the body wave speeds. The strength of the singularity has been normalized in terms of G, which is the

3

E. Bouchaud et al. (eds.), Physical Aspects of Fracture, 3–11.

energy release rate (energy flow to crack tip singularity, per unit of new crack area), expressed by

$$G = \lim_{\Gamma \to 0} \int_{\Gamma} \left[n_x (W + \tfrac{1}{2} \rho \, | \, \partial \mathbf{u} / \partial t \, |^2) - \mathbf{n} \cdot \boldsymbol{\sigma} \cdot \partial \mathbf{u} / \partial x \right] ds \ .$$

In that expression, W is the strain energy density, coordinate x points in the direction of crack growth, Γ is a circuit which loops around the crack tip at the place of interest, and s is arc length along Γ, whose outer normal is $\mathbf{n}$.

From the Freund [4] solution for the unsteady tensile crack motion it is known that G has the structure $G = g(v(t))G_{rest}$. Here $g(v)$ is a universal function of crack speed v. It satisfies $g(0^+) = 1$ and diminishes monotonically to $g(c_R) = 0$ at the Rayleigh speed c_R, which is therefore the theoretical limiting speed at least so long as the crack remains on a plane (see below). The term represented by G_{rest} is a complicated and generally untractable functional of the prior history of crack growth and of external loading, but is independent of the instantaneous crack speed $v(t)$. Owing to that structure for G, it is possible for cracks to instantaneously change v if the requisite energy supply, G_{crit}, for cracking changes discontinuously along the fracture path. For a solid loaded by a remotely applied tension, it will generally be the case that G_{rest} increases as the crack lengthens. It also increases, quadratically, with the intensity of the applied stress. Thus, if G_{crit} is bounded and if the energy flows into a single crack tip (e.g., single crack moving along a plane), then in a sufficiently large body, $G_{crit} / G_{rest} \to 0$ as the crack lengthens. Since that ratio is just $g(v)$, so also will $g(v) \to 0$, which means that v will accelerate towards the value c_R at which $g(v) = 0$. That is the sense in which c_R is the limiting crack speed. In the simplest model, which includes the classical Griffith model, we take G_{crit} as a material constant, although more realistically G_{crit} must be considered as a function of crack speed v, to be determined empirically or by suitable microscale modeling.

An alternative to the singular model is the Barenblatt-Dugdale cohesive zone fracture formulation, a displacement-weakening model with finite stresses. It is often more congenial for numerical simulation, even in cases for which the process zone is small enough that one would be happy to use the singular crack model. It provides for gradual decohesion by imposing a weakening relation between the tensile stress σ_{yy} and displacement-discontinuity δ_y on the crack plane $y = 0$. See the lower portion of Figure 1. The singularity at the crack tip is then spread into a displacement weakening zone. Its length R (measured in the direction of crack growth) scales [5,6] as, roughly, $\mu \delta_o / [\sigma_o f(v)]$ with σ_o being the maximum cohesive strength and δ_o the displacement at which cohesion is lost, and with $f(v)$ being a universal function [5,6] satisfying $f(0^+) = 1$ and $f(c_R) = \infty$. (The latter limit poses a challenge for numerical simulation of fracture at speeds very close to c_R.) When $R \ll$ all length scales in the problem (crack length, distance of wave travel, etc.), predictions of the displacement-weakening model agree with those of the singular crack model, with G_{crit} identified (Figure 1) as the area under the cohesive relation.

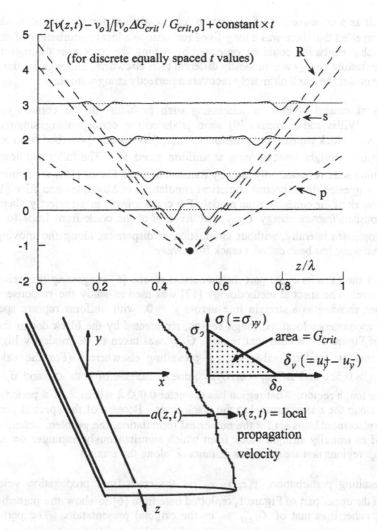

Figure 1. Cracking along a plane in a 3D elastic solid, and (at top)
results of numerical simulation [6] showing crack front wave.

2. Crack Front Waves

Figure 1 shows, in its lower part, a tensile crack growing along the plane $y = 0$ in a
3D solid. Earlier work by the author and coworkers [7,8] addressed a simplified version
of this problem for a model elastic theory. That involved a single displacement
component u satisfying a scalar wave equation, in a 3D solid with a planar crack of non-
straight front. It shows $1/\sqrt{r}$ crack tip singularities in ∇u, of a speed-dependent
angular structure somewhat similar to those of actual elasticity. An energy release rate

G, as well as a cohesive zone formulation, can be defined for that scalar model. Its solution revealed that there was a long-lived response to a local perturbation of the crack front. Such perturbation could be generated by having the crack pass through a region where the fracture energy was modestly different than elsewhere, although in the case of the scalar model the crack ultimately recovers a perfectly straight front.

That work created interest in addressing such problems in the context of actual elasticity. Willis and Movchan [9] soon produced the corresponding singular crack solution, for a crack whose front position $x = a(z,t)$ is linearly perturbed from $x = v_o t$, that is, from a straight front moving at uniform speed v_o. The fuller implications of their solution were revealed only later by Ramanathan and Fisher [10], confirming what had been suggested from spectral numerical simulations of Morrissey and Rice [11,6] in the framework of the cohesive zone model: For crack growth in a perfectly elastic solid with a constant fracture energy G_{crit}, perturbation of the crack front leads to a wave which propagates laterally, without attenuation or dispersion, along the moving crack front. That wave has been called a crack front wave.

Figure 1 shows in its upper part the numerical results [6] suggesting the existence of such a wave. The spectral methodology [12] was used to study the response when a crack front, moving as a straight line across $y = 0$, with uniform rupture speed v_o, suddenly encounters a localized tough region, represented by the black dot in the upper section of Figure 1. Within that region, G_{crit} was taken to be modestly higher, by ΔG_{crit}, than its uniform value $G_{crit,o}$ prevailing elsewhere. (For the calculation shown, $v_o \approx 0.5 c_R$ and $\Delta G_{crit} \approx 0.1 G_{crit,o}$ due to increase of both σ_o and δ_o by 5% within the tough region. That region has diameter $0.002\,\lambda$ where λ is a periodic repeat distance along the z axis, parallel to the crack front. Because of the spectral periodicity in the displacement basis set for the numerical formulation, the problem actually solved is that of an initially straight crack front which simultaneously impinges on a row of such tough regions that are spaced at distance λ along the z axis.)

The resulting perturbation, $v(z,t) - v_o$, in the crack front propagation velocity is shown in the upper part of Figure 1, replotted here from [6] to show the perturbation of velocity, rather than that of G_{rest} as in the original presentation. The perturbation seemed to propagate as a persistent wave, for as long as it was feasible to do the numerical calculation. The dashed lines show where p, s and R waves would intersect the moving crack front, so that the crack front wave speed c_f, measured relative to the position of the tough region where it nucleated, is seen to be a high fraction of c_R. The motivation in [6,11] to examine the response to such a small localized perturbation arose from understanding of the scalar model [7], for which the form of response to an isolated excitation was critical to understanding the response to sustained (small) random excitation [8] generated by a spatial fluctuation of $\Delta G_{crit} = \Delta G_{crit}(x,z)$ along the crack plane.

For the singular model, Ramanathan and Fisher [10] further developed the Willis-Movchan [9] analysis and showed that the equations of elastodynamics do indeed imply a persistent wave, at least when G_{crit} is independent of crack speed v. They also noted that increase of G_{crit} with v damps the wave (viscoelastic properties of the material would be expected to do also); that has not yet been fully quantified. Further, they showed that the speed is a high fraction of c_R, consistent with the wave as seen in the simulation. Figure 2 shows results [13], obtained by evaluating the analytically determined expression [10], for the crack-front-parallel wave speed, $\sqrt{c_f^2 - v_o^2}$, as a function of crack speed v_o for different values of the Poisson ratio v. It is seen that typically $\sqrt{c_f^2 - v_o^2} \approx 0.96\sqrt{c_R^2 - v_o^2}$ at modest speeds, thus showing that $c_f \approx 0.96 c_R \sqrt{1 + 0.1 v_o^2 / c_R^2}$.

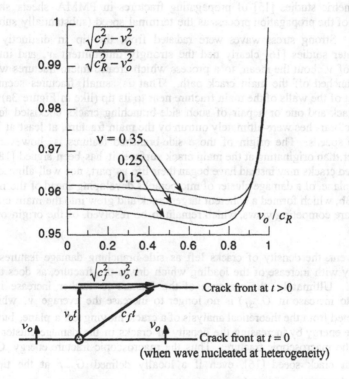

Figure 2 Crack front wave speed, from [13] based on solution in [10].

Many aspects concerning these waves remain to be understood. These include damping due to $dG_{crit} / dv > 0$ and to material viscoelasticity and, most especially, to such generalizations of crack front waves as may exist when there is non-planarity of growth involved in perturbations of the crack front. E. g., such non-planarity must be a part of the Wallner line process [14] since those features are observed optically on what are, otherwise, mirror smooth surfaces.

3. Limiting Fracture Speed and Fracture Roughening

As explained above, according to the singular crack model, for ruptures along a plane we expect that under typical remote loading conditions v will accelerate towards the limiting speed c_R. However, to summarize observations [3,15,16]: v in brittle amorphous solids (glass, PMMA) has an upper limit of order $0.50 - 0.60 c_s$ (i.e., $0.55 - 0.65 c_R$). The fracture surface is mirror-smooth only for $v < 0.3 - 0.4 c_R$. The crack surface roughens severely, with profuse micro-crack formation off of the main crack plane, and v becomes strongly oscillatory, at higher (average) speeds. The crack forks at the highest speeds. There are exceptions for which v approaches c_R. These involve cases for which the fracture does remain confined to a plane: brittle single crystals and incompletely sintered solids [17].

Interferometric studies [15] of propagating fractures in PMMA sheets showed the irregularity of the propagation process as the terminal speed (substantially sub- c_R) was approached. Strong stress waves were radiated from the tip in distinctly separated pulses. Later studies [16] clearly tied the strongly intermittent v, and intensity of fluctuation of v about the mean, to a process which created micro-fractures whose path was side-branched off the main crack path. That is, small fractures seem to have emerged out of the walls of the main fracture near to its tip (like in Figure 3a) such that the main crack and one or a pair of such side-branching cracks coexisted for a time, although the branches were ultimately outrun by the main fracture, at least at low mean propagation speeds. The origin of those side-branched features is, however, disputed [18]. Rather than originating at the main crack surface, it has been argued [18] that the side branched cracks may instead have began their life as part, not well aligned with the main crack plane, of a damage cluster of microcracks developing ahead of the main crack tip, Figure 3b, which formed a coherent larger crack and grew into the main crack walls. Thus there are competing views, which remain to be resolved, of the origin of dynamic roughening.

In any event, the density of cracks left as side-branching damage features increases substantially with increase of the loading which drives the fracture, as does the surface roughening. Ultimately, the response of the fracture to further increase in loading (precisely, to increase in G_{rest}) is no longer to increase the average v, which is the result expected from the theoretical analysis of a crack moving on a plane, but rather to absorb more energy by increasing the density of cracks in the damage cluster formed at the tip of the macroscopic fracture. Thus the macroscopic fracture energy G_{crit} rises steeply with crack speed [16], even if a locally defined G_{crit} at the tip of each microcrack does not.

The deviation of the fracture process from a plane is tied yet more definitively to the intermittency and low limiting crack speeds by studies [17] of weakly sintered plates of PMMA. The resistance to fracturing along the joint was much smaller than for the adjoining material. That kept the crack confined to a plane, which led to smooth propagation of the fracture (no evidence of jaggedness of interferograms by stress wave emission), with v reaching $0.92\,c_R$.

The attempts to explain low limiting speeds and fracture surface roughening begin with the first paper, by Yoffe [19], in which the elastodynamic equations were solved for a moving crack. That was done for a crack which grew at one end and healed at the other, so that the field depended on $x - vt$ and y only (2D case). Nevertheless, it sufficed to reveal the structure of the near tip singular field, including the v dependence of the $\Sigma_{\alpha\beta}$ functions above. The hoop stress component $\sigma_{\theta\theta}$, at fixed small r within the singularity-dominated zone, reaches a maximum with respect to θ for $\theta \neq 0$ when $v > 0.65\,c_R$. That gives a plausible explanation of limiting speeds and macroscopic branching, but does not explain the side-branching and roughening which set in at much smaller speeds.

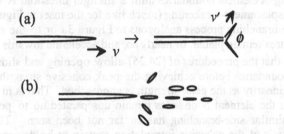

Figure 3. Two views of the origin of surface roughening.

Eshelby [20] took a different approach in posing the question of what is the v at which a fracture must be moving so that, by slowing down to a speed v', it can provide enough energy to drive two fractures. The extreme is to slow down to $v' = 0^+$. That question still has no precise answer when the angle ϕ (Figure 3a) is non-zero, but Eshelby answered it in the limit $\phi \to 0$ for mode III (the Freund [4] mode I solution was not yet available). That $\phi \to 0$ version reduces to the solved problem of sudden change of speed of a fracture moving along the plane. The condition is $g(v) < 1/2$ (assuming no v dependence of G_{crit}), which for mode III gave $v > 0.60\,c_S \approx 0.65\,c_R$. Freund [4] addressed the same $\phi \to 0$ limit for mode I; with re-calculation [21] that leads to $v > 0.53\,c_R$. By working out the static version of the $\phi \neq 0$ problem, Adda-Bedia and Sharon [21] optimize with respect to ϕ to estimate that the lowest v for branching is $\approx 0.50\,c_R$. Material response with $dG_{crit}/dv > 0$ will reduce that threshold but, at present, it seems unlikely that the inferred onset of roughening at speeds in the range $0.3 - 0.4\,c_R$ can be explained by a branching instability at a crack tip in a solid which is modeled, otherwise, as a linear elastic continuum. That argues that perhaps the scenario sketched in Figure 3a should be replaced by another, possibly like that discussed in connection with Figure 3b.

Nevertheless, it is interesting that some discrete numerical models of cracking do give surface roughening, and in some cases side-branching from the tip like in Figure 3a, at roughly realistic speeds. A transition to zig-zag growth was shown to set in around $0.33\,c_R$ in molecular dynamics simulations [22]. It is not yet clear how the v at its onset, or whether roughening sets in at all, depends on details of the force law. Cohesive finite elements [23], or cohesive element interfaces [24-27] that fully model the separation process allow, within constraints of the mesh, for self-chosen fracture paths. They have shown off-plane fracturing by a process similar to what is shown in Figure 3a. In particular, Xu and Needleman [24] reproduce the observations that when the crack is confined to a weak plane, it accelerates towards c_R, whereas for cracking in a uniform material, side branches form at $0.45\,c_R$ or at a realistic slightly lower speed [25] if statistical variation in cohesive properties is introduced.

Recent work [26] suggests that such propensity for low-speed side branching is not universal for all cohesive finite element models. In particular, models which involve no opening or sliding at element boundaries until a strength threshold is reached [27], after which there is displacement-weakening (much like for the inset in Figure 1), have so far not shown a side-branching process analogous to Figure 3a or to the results of [24,25]. Rather, the fractures remain planar, or nearly so, and accelerate towards c_R [26]. It now seems important that the procedures of [24,25] allow opening and sliding displacements at the element boundaries before achieving the peak cohesive strength and that there is substantial non-linearity as the peak strength is approached. That is in contrast with the models in which the element boundaries remain unseparated up to peak strength [26], and for which similar side-branching has so far not been seen. The origin of this sensitivity to details of the cohesive formulation remain to be determined. The results may be confirming a proposal that local nonlinear features of the pre-peak deformation response are critical to understanding the onset of roughening [28].

Nearly all the theoretical work on crack roughening has been carried out in the context of a process like in Figure 3a, in which conditions for a branching or other instability are sought at the tip of the main fracture. More attention would be appropriate to understanding the effects of the opening, and sometime joining together, of multiple microcracks or microcavities ahead of the main fracture [18], like in Figure 3b. That is likely to be an important case not only for brittle amorphous solids under dynamic stressing as discussed here, but also for surface roughening in broad classes of materials [29], whether subjected to quasi-static or dynamic failure conditions.

4. Acknowledgment

The studies reported were supported by the Office of Naval Research, Solid Mechanics Program.

5. References

[1] Rice, J. R. (2001) New perspectives in crack and fault dynamics, in *Mechanics for a New Millennium* (Proceedings of the 20th International Congress of Theoretical and Applied Mechanics, 27 Aug - 2 Sept 2000, Chicago), eds. Aref, H., and Phillips, J. W., Kluwer Academic Publishers, Dordrecht, pp. 1-23, in press.

[2] Freund, L. B. (1989) *Dynamic fracture mechanics,* Cambridge University Press, Cambridge.

[3] Broberg, K. B. (1999) *Cracks and Fracture,* Academic Press, San Diego.

[4] Freund, L. B. (1972) Crack propagation in an elastic solid subject to general loading, I, Constant rate of extension, II, Non-uniform rate of extension, *J. Mech. Phys. Solids,* 20, 129-140, 141-152.

[5] Rice, J. R. (1980) The mechanics of earthquake rupture, in *Physics of the Earth's Interior* (Proc. International School of Physics 'Enrico Fermi', ed. Dziewonski, A. M. and Boschi, E.), Italian Physical Society and North-Holland Publ. Co., Amsterdam, pp. 555-649.

[6] Morrissey, J. W., and Rice, J. R. (1998) Crack front waves, *J. Mech. Phys. Solids,* 46, 467-487.

[7] Rice, J. R., Ben-Zion, Y., and Kim, K. S. (1994) Three-dimensional perturbation solution for a dynamic planar crack moving unsteadily in a model elastic solid, *J. Mech. Phys. Solids,* 42, 813-843.

[8] Perrin, G., and Rice, J. R. (1994) Disordering of a dynamic planar crack front in a model elastic medium of randomly variable toughness, *J. Mech. Phys. Solids,* 42, 1047-1964.

[9] Willis, J. R., and Movchan, A. B. (1995) Dynamic weight functions for a moving crack. I. Mode I loading, *J. Mech. Phys. Solids,* 43, 319-341.

[10] Ramanathan, S., and Fisher, D. (1997) Dynamics and instabilities of planar tensile cracks in heterogeneous media, *Phys. Rev. Lett.,* 79, 877-880.

[11] Morrissey, J. W., and Rice, J. R. (1996) 3D elastodynamics of cracking through heterogeneous solids: Crack front waves and growth of fluctuations (abstract), *EOS Trans. Am. Geophys. Un.,* 77, F485.

[12] Geubelle, P., and Rice, J. R. (1995) A spectral method for 3D elastodynamic fracture problems, *J. Mech. Phys. Solids,* 43, 1791-1824.

[13] Morrissey, J. W., and Rice, J. R. (2000) Perturbative simulations of crack front waves, *J. Mech. Phys. Solids,* 48, 1229-1251.

[14] Sharon, E., Cohen, G., and Fineberg, J. (2001) Propagating solitary waves along a rapidly moving crack front, *Nature,* 410, 68-71.

[15] Ravi-Chandar, K., and Knauss, W. G. (1984) An experimental investigation into dynamic fracture, I, Crack initiation and crack arrest, II, Microstructural aspects, III, Steady state crack propagation and crack branching, IV, On the interaction of stress waves with propagating cracks, *Int. J. Fracture,* 25, 247-262, 26, 65-80, 141-154, 189-200.

[16] Sharon, E., Gross, S. P., and Fineberg, J. (1995) Local crack branching as a mechanism for instability in dynamic fracture, *Phys. Rev. Lett.,* 74, 5096-5099. Also, Sharon, E., and Fineberg, J. (1996) Microbranching instability and the dynamic fracture of brittle materials, *Phys. Rev. B,* 54, 7128-7139.

[17] Washabaugh, P. D., and Knauss, W. G. (1994) A reconciliation of dynamic crack velocity and Rayleigh wave speed in isotropic brittle solids, *Int. J. Fract.,* 65, 97-114.

[18] Ravi-Chandar, K., and Yang, B. (1997) On the role of microcracks in the dynamic fracture of brittle materials, *J. Mech. Phys. Solids,* 45, 535-563.

[19] Yoffe, E. H. (1951) The moving Griffith crack, *Phil. Mag.,* 42, 739-750.

[20] Eshelby, J. D. (1970) Energy relations and the energy-momentum tensor in continuum mechanics, *Inelastic Behavior of Solids,* ed. Kanninen, M. F., Adler, W. F., Rosenfield, A. R., and Jaffe, R. I., McGraw-Hill, New York, 77-115.

[21] Adda-Bedia, M., and Sharon, E. (2000) private communication.

[22] Abraham, F. F., Brodbeck, D., Rafey, R. A., and Rudge, W. E. (1994) Instability dynamics of fracture: a computer simulation investigation, *Phys. Rev. Lett.,* 73, 272-275.

[23] Johnson, E. (1992) Process region changes for rapidly propagating cracks, *Int. J. Fracture,* 55, 47-63.

[24] Xu, X.-P., and Needleman, A. (1994) Numerical simulations of fast crack growth in brittle solids, *J. Mech. Phys. Solids,* 42, 1397-1434.

[25] Xu, X.-P., Needleman, A., and Abraham, F. F. (1997) Effect of inhomogeneties on dynamic crack growth in an elastic solid, *Model. Simul. Mat. Sci. Engin.,* 5, 489-516.

[26] Falk, M. L., Needleman, A., and Rice, J. R. (2001) A critical evaluation of dynamic fracture simulations using cohesive surfaces, submitted to *Proceedings of 5th European Mechanics of Materials Conference* (Delft, 5-9 March 2001), *Journal de Physique IV - Proceedings.*

[27] Camacho, G. T., and Ortiz, M. (1996) Computational modelling of impact damage in brittle materials, *Int. J. Solids Structures,* 33, 2899-2938.

[28] Gao, H. (1996) A theory of local limiting speed in dynamic fracture, *J. Mech. Phys. Solids,* 44, 1453-1474.

[29] Bouchaud, E. (1997) Scaling properties of cracks, *J. Phys.: Condens. Matter,* 9, 4319-4344.

BRITTLE FRACTURE

FRACTURE OF METALS
Part I : Cleavage Fracture

Dominique FRANCOIS
Laboratoire Matériaux – Ecole Centrale de Paris
UMR CNRS 8579
Grande Voie des Vignes
92295 Châtenay Malabry Cedex (France)

André PINEAU
Centre des Matériaux – École des Mines
UMR CNRS 7633
BP 87 – 91003 Evry Cedex (France)

Abstract

The structural integrity of any flawed mechanical structure is usually assessed by global and/or local methodologies which can be used under any kind of loading conditions. In this assessment brittle cleavage fracture is the most dangerous fracture mode. Ductile dimple fracture mode is the most common one. This paper only concerned with metallic materials is divided in two parts: the first one is devoted to cleavage fracture and the second one to ductile fracture. In this first paper the main characteristics of cleavage fracture in metals and alloys, in particular in BCC steels, including scatter, size effect, shallow crack effect, are firstly reviewed in the light of a conventional global approach based on fracture mechanics (K, J). Then in the second part of the paper, a number of attempts made to account for the loss of constraint effect with large scale yielding, by using a second parameter (T, Q) in addition to K(J) parameter are briefly mentioned. The third part of the paper is devoted to a review of cleavage micromechanisms and to the application of the local approach to cleavage fracture. The Weibull weakest link hypothesis is shown to account reasonably well for a number of the characteristics associated with cleavage fracture. Applications of the theory to a number of results obtained on laboratory specimens and on large components are shown.

Keywords

Cleavage fracture / fracture mechanics / local approach / ferritic steels / nucleation / growth / statistics / weakest link theory

15

E. Bouchaud et al. (eds.), Physical Aspects of Fracture, 15–33.
© 2001 *Kluwer Academic Publishers. Printed in the Netherlands.*

Notations (Part I and II)

a: axial dimension of an inclusion or radius of a spheroidal inclusion

A_n : nucleation rate

b: radial dimension of an inclusion or Bürgers vector

B: specimen thickness

C: plastic constraint factor

C_m : numerical constant

D: damage parameter

d: grain size

E: Young's modulus

E_{eq}^p : plastic equivalent deformation

E_{ij}: remote strain tensor

E_p : linear hardening slope

f: volume fraction of voids

f*: Tvergaard porosity parameter

f_0: initial porosity

f_c: critical porosity

H_{ijkl}: tensor of anisotropy

J: characterizing fracture parameter of Rice

k: shape parameter and parameter in crack nucleation from inclusions

k_y: parameter in Hall-Petch equation

k'_y : parameter in fracture stress equation

K : stress intensity factor

$K_{I\ min}$: minimum fracture toughness

K_{IC} : fracture toughness

L : distance between voids

m : shape factor of the Weibull law

N : work hardening exponent

P_R : probability of failure

Q : stress to take account of constraint effects at the crack tip

q_1, q_2 : parameters of the Gurson, Tvergaard, Needleman model

q_w : parameter of the model of Garajeu

r : ratio of radial dimension of a void to the distance between voids

R : size of a void

R_0 : initial size of voids

R_p : yield strength

R_x : radial dimension of a void

S : shape parameter = log(s)

s = a/b excentricity of an inclusion

S_{ij} : stress deviator

S_{ijkl} : tensor of Eshelby

T : stress parallel to the crack tip

V : volume of a void

Vo : characteristic volume in Weibull theory

Σ_{eq}: Von Mises equivalent stress

Σ_h: equivalent stress in the model of Garajeu

Σ_{ij} : applied stress

Σ_m, σm : hydrostatic stress

Σ_r : radial stress

Σ_R : remote principal stress initiating fracture at an inclusion

Σ_z : axial stress

Σ_I maximum principal stress

β : factor describing defect size distribution

γ_s : surface energy

δ :accelerating factor

ε : deformation

μ :shear modulus of the matrix

μ^I :•shear modulus of an inclusion

ν : Poisson ratio

σ_d : fracture stress of an inclusion or of its interface

σ_f : fracture stress

σ_{ij}^I : stress tensor in an inclusion

σ_0 : flow stress

σ_u : normalizing stress in Weibull law

σ_{th} : theoretical cleavage stress

σ_{xx} : local radial stress

σ_y : yield stress

τ : stress triaxiality

Ψ : plastic potential

•a : crack growth

1. Introduction

Scientists and engineers are very much concerned with brittle fracture modes observed in metals, ie intergranular fracture and cleavage fracture. Brittle intergranular fracture which is not considered in this paper is usually associated to the segregation of impurities along the grain boundaries, for instance, P, S, etc … in ferritic steels. Cleavage fracture occurs preferentially over crystallographic atomic planes : {100} in BCC metals, like ferritic steels, {0001} in metals with a HCP structure, like Mg, Zn, Ti, Be, etc … It can easily be shown that the theoretical cleavage stress in a crystalline solid is given by:

$$\sigma_{th} = (E\gamma_s / b)^{1/2} \approx E/10 \qquad (1a)$$

where E is Young's modulus, b, Burgers vector and γ_s the surface energy. This theoretical cleavage stress is much higher than what is usually found (about 10^3 MPa for steels). The reasons for this large difference will be examined in the following.

Whatever the damage mechanism, its initiation requires the breaking of atomic bonds. The expansion of a small cavity corresponds to an increase of the surface energy whereas the hydrostatic stress produces a work to increase the volume. The classical theory of nucleation shows that the critical radius of a cavity nucleus is given by the relation:

$$r^* = 2\gamma_s / \sigma_m \qquad (1b)$$

γ_s being the surface energy, related to the energy of cohesion and σ_m the hydrostatic stress.

For the smallest possible nucleus, without thermal activation, r^* being of the order of b the atomic distance, the corresponding stress is the atomic bonds fracture stress:

$$\sigma_{th} \approx 2\gamma_s / b \qquad (1c)$$

On the other hand, the calculation of the stress needed to pull apart two halves of a crystal yields Eq.1a. The two expressions are compatible with $E = 4\gamma_s$, which is the right order of magnitude. The same is found again by a cohesive zone model at the tip of a crack, in a purely linear elastic calculation. This atomic bond strength is usually high compared with the fracture strength of usual materials which means that fracture is due to large stress concentrations. They are the result of heterogeneous deformations and, in many cases, of the presence of inclusions.

Once a small fracture nucleus created, it will develop either as a cleavage crack or as a cavity, depending on the relative values of the surface energy of the cleavage plane and of the energy of other surfaces, and on the mobility of dislocations. Such a problem was studied by Rice and Thomson (1974) who examined the conditions for the nucleation of dislocation loops at the tip of a crack, unabling its blunting. According to their model, it would be expected cleavage to appear in rhodium, iridium, molybdenum, tungsten and chromium. The fact that it occurs in other metals must be related to the effect of impurities which might lower the cleavage surface energy or the mobility of dislocations.

Another great concern in brittle cleavage fracture, in particular for the engineers, is the assessment of the mechanical integrity of any flawed mechanical structure in the presence of this fracture mode. There exist two ways of approaching this problem. The first way is essentially based on the extensive development of linear elastic fracture mechanics (LEFM) or elastic-plastic fracture mechanics (EPFM). In this approach it is assumed that the fracture resistance can be described in terms of a single parameter, such as K_{IC}, J_{IC} or CTOD. Rules uniquely based on the mechanical conditions of test specimens have been established for « valid » fracture toughness measurements, without paying attention to the fracture micromechanisms. This approach which could be named the « global » approach is extremely useful, but it bears also a number of limitations, in particular when dealing with large-scale conditions or with non-isothermal loading. A number of these limitations are illustrated below. Another limitation is the size effect which is usually observed when steels are tested in the brittle domain or in the ductile-brittle transition regime. This raises the important problem related to the transferability of laboratory test results to components.

More recently, another way of approaching this problem has been developed. This is the so-called « local » approach in which the modelling of fracture toughness is based on local fracture criteria applied to the crack tip situation [1,2]. The development of this approach requires that, at least, two conditions are fulfilled: (i) micro mechanistically based models must be established and validated ; (ii) a perfect knowledge of the stress-strain field ahead of a stationary and propagating crack is required. This has been made possible thanks to the advent of analytical and numerical solutions. Actually the global and the local approaches to fracture are more complementary than contradictory, as illustrated in this paper.

The presentation of the paper is divided into three parts. After a brief review of a number of salient features related to brittle fracture toughness of ferritic steels, the basis of the global approach to fracture are presented. In particular we present briefly the results of recent studies in which a second load-geometry parameter was introduced to accurately describe the crack tip stress-strain field. Then the local approach is applied to brittle cleavage fracture in ferritic steels.

2. Key features of cleavage fracture in steels

The first characteristic of brittle fracture observed, in particular, in ferritic steels, is the scatter in test results, as illustrated in Fig.1 which shows the variation of fracture toughness results obtained on a pressure vessel steel (A508) [3].

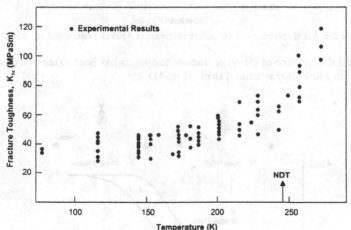

Figure 1. Fracture toughness versus temperature in pressurised water reactor steel

Another important feature associated to brittle fracture is the size effect illustrated in Fig.2 [4]. These results show that the fracture toughness decreases when the specimen size, in particular the thickness, is increased. These observations indicate that a statistical analysis of cleavage fracture is necessary.

The third important factor is the so-called « short-crack » effect which has been investigated more recently (see eg. [5]). This effect is illustrated in Fig. 3. Although there is still some debate about the interpretation of this effect, it is usually found that the ductile-brittle transition temperature is higher for deep cracks than for short cracks, as indicated schematically in Fig. 3.

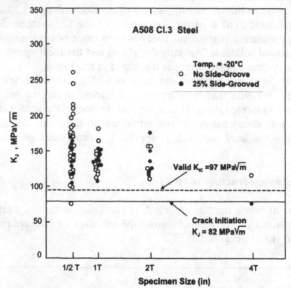

Figure 2. Effect of specimen size on fracture toughness of A508 cl.3 steel tested at – 20 °C

This geometrical dependence of cleavage fracture toughness has been extensively investigated by Sumpter [6] in a low alloy structural steel   (Fig. 4).

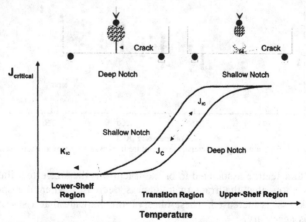

Figure 3. Sketch showing the effect of crack depth on fracture toughness transition temperature curves.

A fourth factor associated with brittle cleavage fracture is the effect of metallurgical variables: carbide distribution in steels, second phase in welds, inclusions, etc ... The effect of these metallurgical parameters will also be briefly discussed.

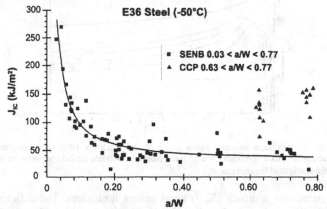

Figure 4. Variation of fracture toughness with specimen geometry (SENB and CCP specimens) and crack length in E36 steel tested at – 50 °C

3. Global approach – Crack tip stress-strain field

The stress field ahead of the tip of a stationary crack is shown schematically in Fig. 5. Three domains can be distinguished: (i) a zone associated with the crack blunting effect; (ii) a zone where the Hutchinson-Rice-Rosengren (HRR) [7-8] is prevailing, and (iii) a domain with the elastic field. It is recalled that the stress singularity in the HRR field is given by $r^{-1/N+1}$, where N is the work-hardening exponent of the stress-strain law for the material:

$$\varepsilon / \varepsilon_o = \alpha \sigma / \sigma_o^N \qquad (2)$$

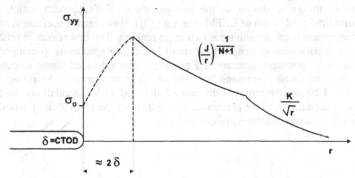

Figure 5. Sketch showing the variation of maximum principal stress ahead of the crack-tip in the crack plane. Three domains are defined.

22

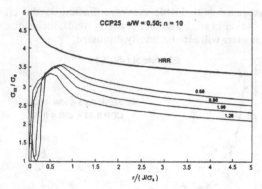

Figure 6. Variation of maximum principal stress ahead of the crack-tip. HRR field compared to numerical calculations on a CCP specimen (W = 25 mm). Numbers indicate the ratio between the applied stress and the yield strength of the material, σ_o.

This single parameter approach (K, J) bears severe limitations. The deficiencies of such one-parameter idealisation become increasingly clear examining the evolution of the crack-tip stress field for different ranges of near tip constraint induced by shallow crack configuration and/or remote tensile loading (see e.g [9, 10]). This is also illustrated in Fig. 6 where the results of FEM calculations applied to a CCP specimen with N=10 are compared to the asymptotic HRR field. It is observed that the calculated stress field largely deviates from the theoretical HRR field, more especially as the load is increased. In order to account for this difference a second parameter, Q, has been introduced, such as the crack-tip stress field is written as:

$$\sigma_{ij} = \sigma_o \, \dot{J} / \alpha \sigma_o I_N r^{1/N+1} \tilde{\sigma}_{ij} \, \dot{\theta}, N + Q \delta_{ij} \qquad (3)$$

where I_N is a tabulated function and $\tilde{\sigma}_{ij}$ is a nondimentional function (see e.g [11]. It is recommended to determine the Q parameter at a distance of the crack-tip such as $r/(J/\sigma_o)=2$.

The J-Q methodology is similar to the two-parameter K-T approach which has been proposed to extend the application of LEFM (see e.g [12]). However the usefulness of the two-parameter J-Q approach is questionable for two main reasons. The first reason is related to the transferability of Q parameter determined on small laboratory specimens to components. No solutions can easily be tabulated, contrary to T parameter since detailed plastic calculations are necessary not only on small specimens but also on large components. Moreover, accurate determination of Q parameter requires the use of detailed FEM calculations, in particular refined meshes. It appears therefore preferable to use directly the methodology based on local criteria to assess the fracture toughness of components.

4. Local approach to cleavage fracture

In this part an attempt is made to review briefly the micromechanisms of cleavage fracture, then to model brittle cleavage fracture toughness before showing a number of applications of these models.

4.1. MICROMECHANISMS OF CLEAVAGE FRACTURE

4.1.1. *Nature of cleavage initiation sites.*

In pure metals the large difference observed between the theoretical cleavage stress (σ_{th}) given by Eq. 1 and the measured fracture stress, σ_f, is usually explained by the existence of local stress concentration effects produced either by dislocation pile-up at grain boundaries (Stroh theory [13]) or at the intersection between slip band or mechanical twins generated during plastic deformation (Cottrell theory [14]). In the Cottrell theory, which is not detailed here, it is well to remember that fracture is assumed to be controlled by the propagation of a cleavage microcrack along a {001} plane initiated at the intersection of two {110} slip planes. The yield strength, σ_y, is given by Hall-Petch equation:

$$\sigma_y = \sigma_1 + K_y d^{-1/2} \qquad (4)$$

while the fracture stress, is expressed as:

$$\sigma_f = k_y' d^{-1/2} \qquad (5)$$

In these expressions σ_i is the friction stress which is strongly temperature and strain-rate dependent in BCC and HCP metals, d is the grain size. while the proportionality factor k_y' is related to k_y by :

$$k_y' = E\gamma_s / \pi(1 - v^2)k_y \qquad (6)$$

In ferritic steels, a number of studies have shown that cleavage cracks initiate from carbides, in particular those located along grain boundaries (Smith theory [15]). In these materials the fracture stress is given by a Griffith type criterion:

$$\sigma_f = (E\gamma_s / \alpha a)^{1/2} \qquad (7)$$

where α is a numerical constant while a is the size of the coarsest observed second phase particles equated to the size of the microcrack nuclei.

In structural steels there is now a growing body of evidence showing that cleavage cracks nucleate from microstructural defects such as MnS inclusions or martensite-austenite zones in welds. This situation is most frequently observed when cleavage fracture occurs in the ductile-brittle transition regime (see e.g [16]). In this case the fracture stress is given by an equation similar to Eq. 7 where a is the size of the inclusions or that of the inclusion clusters.

4.1.2 Nucleation or growth controlled cleavage fracture

The proof that fracture is controlled by the initiation of cleavage microcracks or by the propagation of arrested microcracks initiated from nucleation sites is most often difficult to be made, except in few circumstances (see e.g the studies devoted to cleavage fracture of silicon [17]). Theoretical studies have examined how the stress ahead of the tip of a cleavage crack can be relaxed by dislocation nucleation (see e.g [18,19]).

Many studies devoted to ferritic steels have concluded that, in these materials, the mechanism of cleavage fracture is growth controlled (see e.g [20]). This conclusion is consistent with the temperature independence of the cleavage fracture stress, as shown by Eq. 7.

4.1.3 Statistical aspects of cleavage fracture

Rather surprisingly, although the scatter in cleavage stress measurement is well established, it is only rather recently that models have been proposed to account for the scatter (for a review, see e.g [21]). For a homogeneous material in which there is no statistical spatial correlation between two adjacent areas, the Weibull weakest link concept is largely used to account for brittle fracture. The Weibull statistical distribution is only a specific case, as shown by Jeulin (see e.g [22]) who introduced other « random field » models. The Weibull statistics was the basis of the model proposed by Beremin (see e.g [23]) and Wallin et al (see e.g [24]). The fracture probability, P_R, of a specimen of volume V submitted to a homogeneous stress state, σ is given by:

$$P_R = 1 - \exp - \sigma^m V / \sigma_u^m V_o \qquad (8)$$

where V_o is an arbitrary unit volume, σ_u is the average strength of that unit volume, while m is the shape factor of the Weibull distribution. Eq. 8 is a simplified expression since it contains no threshold parameter.

In three dimensions this equation can be expressed as:

$$P_R = 1 - \exp(-\sigma_w / \sigma_u)^m \quad \text{where} \quad \sigma_w = \left[\int_{PZ} \frac{\sum_1^m dV}{V_0} \right]^{1/m} \qquad (9)$$

σ_w is the Weibull stress calculated over the plastic zone, while Σ_1 is the maximum principal stress.

Tests on notched specimens with various notch radii (to change the values of Σ_1) and various diameters (to test the size effect) are usually performed to determine the two parameters, ie m and the product $\sigma_u^m V_o$ introduced in Eq. 9. Typical values of 20 are found for m in pressure vessel steels while V_o, the elementary volume element statistically representative of cleavage fracture process is typically of the order of few grains. With these values for m and V_o, σ is found to be of the order of 2500MPa in pressure vessel steels with a tempered bainitic microstructure (see e.g [23]). It is worth adding that for a power law distribution of defect size:

$$P(a) = \alpha / a\beta \tag{10}$$

it can easily be shown that m is simply related to β by [23, 25]:

$$m = 2\beta - 2 \tag{11}$$

More recently, as previously stated, a number of studies have shown that in structural steels cleavage cracks are preferentially initiated from inclusion clusters. The size distribution of these clusters has been described by an exponential function. Under these conditions the probability to fracture can be written as:

$$P_R = 1 - \exp\left\{ \int_{dec} -\exp\left[-\left(\frac{\sigma_c}{\Sigma_1}\right)^4 \right] \frac{dV}{V_0} \right\} \tag{12}$$

where the integral is calculated over the volume in which the inclusion-matrix decohesion has taken place and σ_c is a normalizing stress similar to σ_u in Eq. 8 [26,27].

4.1.4 Microstructural effects

In steels with a BCC structure, many metallurgical factors influence the cleavage strength. This includes the grain size in ferritic steels [28], the packet size in bainitic/martensitic steels [29, 31], the presence of local brittle zones (Martensite/Austenite constituents in welded microstructures). As a rule, the dislocation mobility controls the fracture strength. Factors affecting this mobility, such as impurity pinning or irradiation embrittlement may strongly affect the cleavage stress [32]. It has also been reported that, when cleavage fracture occurs after large plastic deformation, a strain induced anisotropy affect is observed, the cleavage stress being increased in the direction of the maximum principal strain [23].

The criterion for nucleation-controlled cleavage fracture must obviously include other terms of the stress or strain tensor than the maximum principal stress. Modelling of cleavage fracture under these circumstance bears to some extent, a strong analogy with cavity nucleation in ductile rupture. Recently, an attempt in this direction has been presented to modify the Beremin model (Eq. 9) [33].

4.2 MODELLING CLEAVAGE FRACTURE TOUGHNESS

4.2.1. Cleavage under small-scale yielding conditions

The application of the Beremin model (Eq. 9) to the crack-tip situation is straightforward if it is assumed that the stress-strain field ahead of the crack-tip under small-scale yielding conditions is simply scaled by the ratio $x/(K/\sigma_o)^2$ or $x/(J/\sigma_o)$, where σ_o is the yield strength and x is the distance to the crack-tip [34, 36]. This means that the T stress or O stress effects introduced in section III are neglected. The probability distribution for the fracture toughness of a specimen containing a 2D crack can simply be expressed as:

$$P_R = 1 - \exp - \left[K_{IC}^4 B \sigma_0^{m-4} C_m / V_0 \sigma_u^m \right] \qquad (13)$$

where B is the specimen thickness and C_m is a numerical factor.

The above expression can simply be extended to a 3D crack of length, ℓ, in which the K factor is function of the curvilinear abscissa, s. It can thus be predicted that, for a given probability to fracture, the following expression is satisfied:

$$\int K_I^4 (s) \, ds = K_{IC}^4 \ell \text{ with } \ell = \int ds \qquad (14)$$

As shown earlier [37-39] and as illustrated below, the predicted variations of K_{IC} with specimen thickness, temperature and strain rate-via the temperature dependence of the yield strength-are in good agreement with the experimental results. Eq. 8 and 9 contain only two parameters. However in this theory there is a hidden condition for the initiation of cleavage fracture, which is the initiation of plastic deformation as a prerequisite for cleavage fracture. This condition does not explicitly appear in Eq. 13.

A similar model but including a third parameter was developed by Wallin [40]. Moreover, this author proposed a model able to take into account ductile crack growth, which eventually precedes cleavage fracture. In the absence of ductile crack extension, the fracture probability, P_R, is expressed as:

$$P_R = 1 - \exp - \left[B / B_0 \left(K_{IC} - K_{min} \right)^4 / \left(K_0 - K_{min} \right)^4 \right] \qquad (15)$$

where B_0 is an arbitrary normalizing thickness, K_0 is a parameter depending on temperature, while K_{min} is a limiting value below which fracture is impossible $\left(K_{min} \approx 20 MPa\sqrt{m} \right)$.

To conclude this part, it is worth remembering that the HRR field used to establish Eq. 13 is only an approximation, as already stated. An analytical expression as simple as Eq. 13 cannot be derived when Q parameter is introduced in the crack-tip stress (Eq. 3). Only numerical calculations can be made. Such calculations made on a CT specimen (a/w ≈ 0.60) showed that the size effect predicted by Eq. 13 was slightly different. An exponent close to 6, instead of 4, was found to be more appropriate [41].

4.2.2 Cleavage under large scale yielding conditions

In many situations, the analytical expression given by Eq. 13 cannot be directly applied to small laboratory specimens, in particular when they are tested in the ductile-brittle transition regime. The influence of the loss of constraint effect for large scale yielding (LSY) conditions is schematically indicated in Fig. 7. In this figure, the apparent increase in fracture toughness of specimens tested at various temperatures (T_1, T_2, T_3) is indicated when LSY conditions are reached. This diagram can qualitatively be used to explain the geometrical effect observed with shallow cracks (Fig. 3).

4.2.3. *Cleavage fracture after ductile tearing*

The fracture resistance of ferritic steels in the ductile-brittle transition regime is controlled by the competition between ductile tearing and cleavage fracture. Under typical conditions, a crack initiates and grows by ductile tearing but ultimate fracture occurs by catastrophic cleavage fracture. A typical example obtained on A 508 pressure vessel steel is shown in Fig. 8 where a large scatter in results obtained at room temperature is observed. This scatter which corresponds to noticeable crack advances cannot be modelled using the Beremin model which is valid only for a stationary crack. This model is only able to account for the results obtained at lower temperature in the absence of significant crack extension.

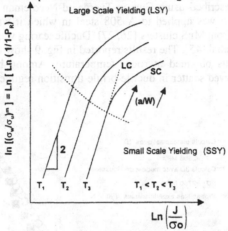

Figure 7. Sketch showing the variation of the probability to fracture with loading (J/σ_o) at various temperature T_1, T_2, T_3. The transition from small scale to large scale yielding conditions is indicated. The difference between long cracks (LC) and short cracks (SC) is shown.

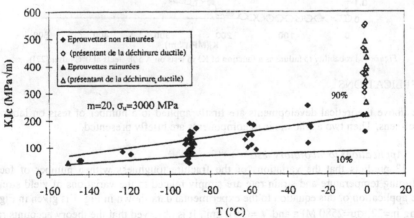

Figure 8. Variation of fracture toughness in A 508 steel with temperature. Comparison between experiments and Beremin model for stationary cracks.

Two factors contribute to the increase in the probability to cleavage fracture and to the increase in the scatter for a growing crack. The first factor is related to the stress distribution ahead of a propagating crack. It has been shown that under these conditions the maximum principal stress increases with crack length [42]. The second factor is simply related to the statistics of brittle fracture. More material is being sampled as the crack advances by ductile tearing, and eventually the advancing zone of high stress samples a critical cleavage crack nucleus. This produces unstable cleavage fracture [40]. Both factors have been taken into account in a cell element simulation which was shown to predict trends in ductile-brittle transition consistent with experiments [42]. In this simulation, ductile tearing was described using Gurson-Tvergaard-Needleman model [43, 44]. More recently, this approach was applied to A 508 steel in which it was shown that cleavage cracks were initiated from MnS clusters [26, 27]. Ductile tearing was numerically simulated using Rousselier potential [45]. The results reported in Fig. 9 show that the model is able to account for the results obtained at room temperature. Among both factors explaining qualitatively the observed scatter in ductile-brittle transition regime, the statistical factor is prevailing.

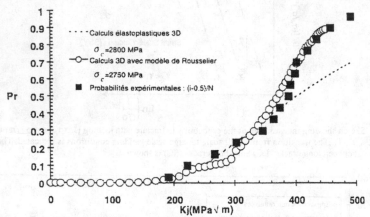

Figure 9. Probability to failure as a function of KJ in tests on A 508 steels at 0°C (after [27]).

4.3. APPLICATIONS

The above theoretical developments are firstly applied to a number of tests on laboratory specimens. Then two applications to components are briefly presented.

4.3.1 *Applications to laboratory tests*

Eq. 13 predicts that the variations of the fracture toughness with a number of factors, including temperature and strain rate are simply related to the variations of yield strength. The application of this equation to the experimental data shown in Fig. 1 is given in Fig. 10, using m=22, σu=2580 MPa and V_u=$(50)^3$ μm^3. It is observed that the theory accounts rather well for the variation of K_{IC} with temperature and for the scatter in test results measured in this pressure vessel steel (A 508). In particular, it is noted that the predicted ratio K_{IC} (P_R = 0.90) / K_{IC} (P_R = 0.10) = 2.20 is a good estimate of the observed scatter. Similarly the theory was applied to interpret the variations observed for K_{IC} with loading rate [46].

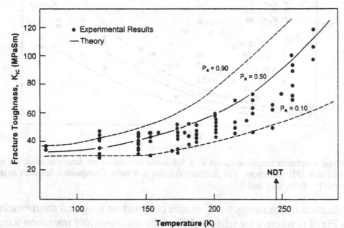

Figure 10.   Fracture toughness versus temperature in HSST program (see Fig. 1). Comparison between experiments and the Beremin model (Ref. [23]).

Another application was made for the assessment of irradiation embrittlement on A 508 steel [32]. It was shown that the theory was able to account for the decrease in K_{IC} observed in the lower shelf regime after irradiation in terms of increase in yield strength produced by neutron irridiation, without modifying the value of parameters σ_u, V_o and m.

As previously stated, the theory (Eqs. 13 and 15) implies a size effect in the results of fracture toughness measurements. A number of authors have confirmed this size effect already shown in Fig. 2 (see e.g [40]). One of the best example of the applicability of this size effect concept has been reported by Wallin [40]. This size effect has now been introduced in the ASTM standards.

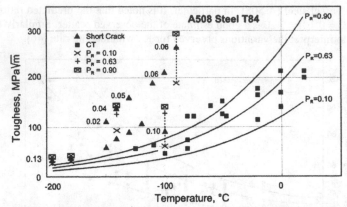

Figure 11. Fracture toughness transition curves in A 508 steels determined on deeply notched CT specimen and short crack 3PB specimens. The numbers indicate a/W ratio. Comparison between experiments and theory (P = 0.10, 0.63 and 0.91).

The application of the theory to K_{IC} results obtained on long and short cracks in A 508 steel is shown in Fig. 11, where a significant shift of the ductile-brittle transition temperature is observed ($\approx$ 100 °C) when shallow 3 point bend cracks are used to measure K_{IC} instead of long CT cracks. In Fig. 11, it is shown that the Beremin model (m = 22, σ_u = 2600 MPa, V_o = (50)^3 μm^3) describes rather well the experiments. Similar calculations were made to interpret the data published by Sumper [6] (Fig. 4). The results reported in Fig. 12 show that the model accounts also for these data obtained on another material provided that the parameters are changed (m=15, σ_u = 1810 MPa, σ_u = (100)^3 μm^3).

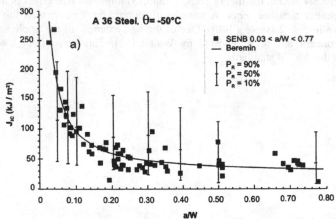

Figure 12. Effect of specimen geometry and crack length on fracture toughness of A36 steel tested at – 50 °C. Comparison between experiments (Fig.4) and theory (m=15, σ_u = 1810 MPa). a) SENB specimens

More recently the model has been applied to warm prestressing (WPS) effect with two loading paths, as indicated schematically in Fig. 13 [47]). In both cases a large increase in apparent fracture toughness is measured. Two main factors contribute to this beneficial effect of WPS effect, as already discussed [48]: (i) crack-tip blunting effect and (ii) residual stresses effect.

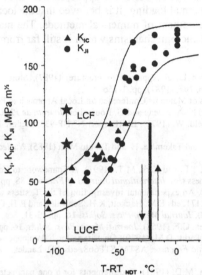

Figure 13.   Results of fracture toughness measurements in pressure vessel steel and sketch showing two loading cycles for warm prestressing.

4.3.2   *Application to components*

The first example relies on thermal shock experiments performed on large components of A 508 steel. A large cylinder containing inner crack was submitted to a severe liquid nitrogen thermal shock. Details are given elsewhere [49]. Crack initiation followed by crack arrest was observed. Numerical calculations showed a good agreement between K factor calculated at crack initiation and Beremin model provided that the full length of the crack (B= 1000 mm) was considered.

The second example is that of large mock-up of A 508 steel in which a 3D shallow fatigue crack was introduced before covering one face with a cladding of stainless steel. Details can be found elsewhere [50, 51]. This mock-up was tested at – 170 °C under 3P bending conditions. The Weibull stress was determined using FE calculations. The results showed that, using the values for m and σ_u parameters typical of those found in A 508 steel, the calculated fracture load reached a value equal to the observed load for a reasonable value of the calculated fracture probability. This example illustrates the advantages of the local approach to fracture when considering complex situations.

5. Concluding remarks

An attempt has been made to show a number of advantages provided by the analysis of the micromechanisms for the cleavage fracture and by the use of the local approach. As already indicated, the local approach and the global approach are more complementary than contradictory. The main advantage of the local approach lies in the application to complex situations, including non-isothermal loading. It is believed that the local approach will continue to develop thanks to the development of numerical methods. The main limitations may lie in the knowledge of the fracture micromechanisms which is still far from being complete.

References
1. International Seminar on Local Approach to Fracture, (1987), Moret sur Loing, June 3-5. Nuclear Engineering and Design, 105, (1987), pp. 1-156.
2. 1ᵉ European Mechanics of Materials Conference on Local Approach to Fracture, (1996), Fontainebleau, 9-11. September. Edited by A. Pineau and G. Rousselier. Journal de Physique IV, 6, (1996), pp. C6-558.
3. Server, W.L. and Oldfield, W. (1978), EPRI report N° NP-933, Electric Power Research Institute, Palo Alto, CA.
4. Iwadate, T., Tanaka, Y. and Takemata, H. and Tabuto Mori (1985), Nuclear Engineering and Design, 87, pp. 89-99.
5. Dodds, R.H. Anderson, T.L. and Kirk, M.T. (1991), A Framework to Correlate a/w Effects on Elastic-Plastic Fracture Toughness (Jc), International Journal of Fracture, 88, pp. 1-22.
6. Sumpter, J.D.G. (1993), An experimental investigation of the T. stress approach, Constraint Effects in Fracture, ASTM STP 1171, eds E.M. Hackett, K.H. Schwalbe and R.H. Doods, pp. 492-502.
7. Hutchinson, J.W. (1968), Journal Mech. Phys. Solids, 16, pp. 13-31.
8. Rice, J.R. and Rosengren, G.F. (1968), Journal Mech. Phys. Solids, 16, pp. 1-12.
9. McMeeking, R.M. and Parks, D.M. (1979), On criteria for J. dominance of crack-tip field in large-scale yielding in Elasto-Plastic Fracture, ASTM STP668, eds J.D. Landes, J.A. Begley and G.A. Clark, pp. 175-194.
10. Shih, C.F. and German, M.D. (1981), Requirements for a one parameter characterization of crack-tip fields by the HRR singularity, Int. Journal of Fracture, 17, pp. 27-43.
11. Dodds, R.H., Fong Shih, C. and Anderson, T.L. (1993), Continuum and micromechanics treatment of constraint in fracture, Int. Journal of Fracture, 64, pp. 101-133.
12. Hancock, J.W. (1992), Constraint and stress state effects in ductile fracture in « Topics in Fracture and Fatigue », eds. A.S. Argon, Springer Verlag, pp. 99-144.
13. Stroh, A.N. (1954), The formation of cracks as a result of plastic flow, Proc. Roy. Soc., 223, pp. 132-203.
14. Cottrell, A.H. (1958), Theory of brittle fracture in steel and similar metals, Trans AIME, 212, pp. 132-203.
15. Smith, E. (1966), The nucleation and growth of cleavage microcracks in mild steel, Physical Basis of Yield and Fracture, conf. Proceedings Inst. Phys. and Phys. Soc. London, pp. 36-46.
16. Rosenfield, A.R., Shetty, D.K. and Skidmore, A.J. (1983), Fractographic observations of cleavage initiation in the ductile –brittle transition region of a reactor-pressure-vessel steel. Met. Trans. A, 14A, pp. 1934-1937.
17. Hirsch, P.B. and Roberts, S.G. (1996), Acta Mater, 44, pp. 2361-2371.
18. Rice, J.R. and Thomson, R.M. (1974), Ductile vs brittle behavior of crystals, Phil. Mag., 29, pp. 73-97.
19. Rice, J.R., Beltz, G.E. and Sun, Y. (1992), Peierls framework for dislocation nucleation from a crack-tip. Topics in Fracture and Fatigue, eds A.S. Argon. Springer-Verlag.
20. Curry, D.A. and Knott, J.F. (1976), Metal Science, 10, pp. 1-10.
21. Wallin, K. (1991), Statistical modelling of fracture in the ductile to brittle transition region. Defect assessment in components, eds J.G. Blauel and K.H. Schwalbe, Fundamentals and Applications, ESIS/EGF9, MEP, pp. 415-445.
22. Jeulin, D. (1990), Random fields models for fracture statistics. Proceed. Colloque de Métallurgie, INSTN, Ed. Revue de Métallurgie N°4, pp. 99-113.
23. Beremin, F.M. (1983), A local criterion for cleavage fracture of a nuclear pressure vessel steel, Met. Trans., 14A, pp. 2277-2286.

24. Wallin, K., Saario, T. and Torronen, K. (1984), *Metal Science*, **18**, pp. 13-16.
25. Mudry, F. (1987), *Nuclear Engineering Design*, **105**, pp. 65-76.
26. Renevey, S. (1997). Approche globale et locale de la rupture dans le domaine de la transition fragile-ductile d'un acier faiblement allié. Thesis Univ. Paris-Sud XI.
27. Carassou, S., Renevey, S., Marini, B. and Pineau, A. (1998), Modelling of the ductile to brittle transition of a low alloy steel, ECF12, ESIS, **II**, pp. 691-696.
28. Curry, D.A. and Knott, J.F. (1978), *Metal Science*, **12**, pp. 511-514.
29. Naylor, J.P. and Krahe, P.R. (1974). The effect of the bainitic packet size on toughness. *Met. Trans.*, **6A**, pp. 1699-1701.
30. Brozzo, P., Buzzichelli, G. Mascanzoni, A. and Mirabile, M. (1977). Microstructure and cleavage resistance of low-carbon bainitic steels, **11**, pp. 123-129.
31. Bouyne, E., Flower, H.M., Lindley, T.C. and Pineau, A. (1998). Use of EBSD technique to examine microstructure and cracking in a bainitic steel. *Scripta Mater*, **39**, pp. 295-300.
32. Al Mundheri, M., Soulat, P. and Pineau, A. (1989). Fatigue and Fracture of Eng. *Mater. and Struct.*, **12**, pp. 19-30.
33. Bernauer, G., Brocks, W. and Schmitt, W. (1999). Modifications of the Beremin model for cleavage fracture in the transition region of a ferritic steel. *Eng. Fracture Mechanics*, **64**, pp. 305-325.
34. Pineau, A. (1981), Proc. 5ᵗʰ Int. Conf. On Fracture, Cannes (Edited by D. François et Al), *Pergamon Press*, **2**, pp. 533-577.
35. Mudry, F. (1987). A local approach to cleavage fracture. *Nuclear Engineering and Design*, **105**, pp. 65-76.
36. Pineau, A. (1992). Global and local approaches of fracture, Transferability of laboratory test results to components, in *Topics in Fracture Fatigue*, eds A.S. Argon, Springer, Verlag, pp. 197-234.
37. Pineau, A. and Joly, P. (1991). Defect Assessment in Components – *Fundamentals and Applications*, ESIS/EGF9 (edited by J.G. Blauel and K.H. Schwalbe), MEP London, pp. 381-414.
38. Wallin, K. (1984). The scatter in K_{Ic} results, *Engng. Fract. Mech.*, **19**, pp. 1085-1093.
39. Wallin, K. (1985). The size effect in K_{Ic} results, *Engng. Fract. Mech.*, **19**, pp. 1085-1093
40. Wallin, K. (1991). Statistical modelling of fracture in the ductile to brittle transition region, in Defect Assessment in Components - *Fundamentals and Applications*, ESIS/EGF9 (edited by J.G. Blauel and K.H. Schwalbe), MEP London, pp. 415-445.
41. Yahya, O.M., Piques, R. and Pineau A.(1998), Effet d'échelle et rupture fragile intergranulaire de l'acier 16MND5, Journal de Physique, **8**, pp. 4/175-4/180.
42. Xia, L. and Shih, C.F. (1996). Ductile crack growth –III Transition to cleavage fracture incorporating statistics, *J. Mech. Phys. Solids.*, **44**, pp. 603-639.
43. Gurson, A.L. (1975). *J. Eng. Mat. Techn.*, pp. 2-15.
44. Tvergaard, V. and Needleman, A. (1984). *Acta Metall.*, **32**, pp. 157-169.
45. Rousselier, G. (1987). *Nuclear Engineering Design.*, **105**, pp. 97-111.
46. Henry, M., Marandet, B., Mudry, F. and Pineau, A. (1986). Effets de la température et de la vitesse de chargement sur la ténacité à rupture d'un acier faiblement allié – Interprétation par des critères locaux, *Journal de Mécanique Théorique et Appliquée*, **4**, pp. 741-768.
47. Roos, E., Alsmann, U., Elsässer, K., Eisele, W ; and Seidenfuss, M. (1998). ECF 12 , eds M.W. Brown, E.R. de los Rios and K.J. Miller, EMAS, **II**, pp. 939-944.
48. Beremin, F.M. (1981). Numerical modelling of warm prestress effect using a damage function for cleavage fracture. ICF5, eds D. François et al., **II**, pp. 825-832.
49. Saillard, P., Devaux, J.C. and Pellissier-Tanon, A. (1987). Numerical prediction of the fracture of a thick pressurized shell subjected to a liquid nitrogen shock, *Nuclear Engng. Design.*, **105**, pp. 83-88.
50. Moinereau, D., Brochard, J., Vatela, M.P., Bhandari, S., Guichard, D., France, C., Sanderson, D.J. and Sherry, A. (1997). The application of local approach to fracture to the structural integrity assessment of cladded vessels under thermal loading, SMIRT 14, **4**, pp. 655-663.
51. Moinereau, D., Frund, J.M., Brochard, J., Marini, B., Joly, P., Guichard, D., Bhandari, S., Sherry, A., Sanderson, D.J., France, C. and Lidbury, P.D.G. (1996). Local approach to fracture applied to reactor pressure vessel : Synthesis of a cooperative programme between EDF, CEA, Framatome and AEA, *Journal de physique*, **6**, pp. C6/243-C6/257.

THE WEIBULL LAW: A MODEL OF WIDE APPLICABILITY

F. HILD
LMT-Cachan
ENS de Cachan / CNRS UMR 8535 / Université Paris 6
61 avenue du Président Wilson, F-94235 Cachan Cedex, France

Abstract − The Weibull model is often used to study the degradation and failure of brittle or quasi-brittle materials. The paper aims at reviewing some applications of the model dealing with single or multiple cracking conditions.

1. Introduction

We will study the degradation and failure properties of brittle or quasi-brittle materials. The overall behavior is either linear elastic (i.e., brittle materials) or nonlinear (i.e., quasi-brittle materials) within the framework of infinitesimal strains (i.e., the strains to failure are assumed to be very small compared to unity). All these materials are sensitive to initial heterogeneities (e.g., inclusions, porosities). These imperfections are created during the fabrication of the material and are usually randomly distributed within the material or at the surfaces. They lead to a scatter of the failure load that can be modeled by a Weibull law.

2. An Overview of the Weibull Model

The Weibull law is commonly used to model the failure of brittle materials (e.g., ceramics, glass, rocks, cement, concrete, some brittle- or ductile-matrix composites). The overall behavior is linear elastic and the toughness is at most on the order of a few MPa$\sqrt{\text{m}}$.

2.1. THE WEIBULL MODEL FROM DIFFERENT VIEWPOINTS

The (cumulative) failure probability P_F is determined for structures made of brittle materials. As soon as local crack propagation occurs at one location of the material, the whole structure fails. In other words, as soon

35

E. Bouchaud et al. (eds.), *Physical Aspects of Fracture*, 35–46.

as the weakest link is broken, the whole chain (i.e., the structure) fails. Within the framework of the weakest link statistics [1], the local failure probability P_{F0} within a domain Ω_0 of volume V_0 is related to the global failure probability P_F of a structure Ω

$$P_F = 1 - \exp\left[\frac{1}{V_0} \int_{\Omega} \ln\left(1 - P_{F0}\right) \, dV\right]. \tag{1}$$

The cumulative failure probability P_{F0} can be postulated as [2]

$$\ln\left(1 - P_{F0}\right) = -\left(\frac{\langle\sigma_1 - S_u\rangle}{S_0}\right)^m \tag{2}$$

where σ_1 denotes the maximum principal stress in the considered element, $\langle\star\rangle$ the Macauley brackets (i.e., positive part of $\star$), m the shape parameter (also called Weibull modulus), $V_0 S_0^m$ the scale parameter and S_u the threshold parameter. When $S_u = 0$, the model is referred to as a two-parameter Weibull model; otherwise, it is called a three-parameter Weibull model. The failure stress σ_F is defined as the maximum equivalent stress (here σ_1) within the loaded sample or structure Ω.

From a physical point of view, P_{F0} is the probability of finding a critical defect within a domain Ω_0 and refers to a defect distribution characterized by a probability density function f. The initial flaw distribution, f_0, depends upon morphological parameters of the defect: first a size a (e.g., length or radius), a defect orientation (described by a unit normal $\underline{n}$), and other parameters (denoted by w). For a given load level, the set of defects D splits into two subsets. The first one, D_c, is related to the defects that are critical (i.e., they lead to the failure of a link, and therefore of the whole structure). The second one, D_{nc}, contains the defects that are not critical. When the local propagation is unstable, the cumulative failure probability $P_{F0}(Q)$ for a given loading parameter Q is given by

$$P_{F0}(Q) = \int_{D_c(Q)} f_0(a, \underline{n}, w) \, da \, d\underline{n} \, dw. \tag{3}$$

This definition ensures that P_{F0} lies between 0 (i.e., $D_c = \emptyset$: no defect is critical) and 1 (i.e., $D_c = D$: all defects are critical). By using Fracture Mechanics results, the defects are modeled as cracks. Consequently, only two morphological parameters are needed, viz. a size a (i.e., a crack length or a crack radius for penny shaped cracks) and a normal $\underline{n}$. The following fracture criterion is used

$$Y\sigma_1 G\left(\frac{\sigma_2}{\sigma_1}, \frac{\sigma_3}{\sigma_1}, \underline{n}\right) \sqrt{a} \geq K_c \tag{4}$$

where σ_2 and σ_3 denote the second and third principal stresses ($\sigma_1 \geq \sigma_2 \geq \sigma_3$), K_c the material toughness, Y a dimensionless parameter and G a function describing the equivalent stress. By assuming that the orientation is uniformly distributed and that the flaw size distribution is approximated by a power law function for large sizes ($a \to +\infty$), a two-parameter Weibull law can be retrieved [3]. Conversely, for a bounded flaw size distribution ($0 \leq a \leq a_M$) for which the flaw size distribution is approximated by a power law function for large sizes ($a \to a_M$), a three-parameter Weibull law is obtained [4]. These results show that the Weibull parameters can be related to an initial flaw distribution *and* a failure criterion. In particular, the equivalent stress chosen by Weibull in his pioneering work [5] corresponds to an assumption of a mode I mechanism of failure (even though this concept was not yet discovered). It is worth noting that the Beremin model [6], which is dealing with brittle failure induced by plastic deformation, is very close to the previous model. The main difference is that the integration volume no longer covers the whole structure Ω but only the part Ω_p where plasticity occurs, since slip nucleated cleavage can be dominant for steels at low temperatures.

From a mathematical perspective, it can be noted that the Weibull model can be described by an underlying Poisson point process of intensity $\lambda_t(\sigma_1)$

$$\lambda_t(\sigma_1) = \lambda_0 \left(\frac{\langle \sigma_1 - S_u \rangle}{S_0} \right)^m \tag{5}$$

where $\lambda_0 = 1/V_0$ is the normalizing defect density ($\lambda_t(\sigma_1)V_0$ is the average number of broken defects within a domain Ω_0 of volume V_0). The probability P of finding $N = n$ broken defects within a uniformly loaded domain Ω of volume V is written as

$$P(N = n) = \frac{[\lambda_t(\sigma_1)V]^n}{n!} \exp\left[-\lambda_t(\sigma_1)V\right]. \tag{6}$$

Within the weakest link framework, the failure probability P_F is the probability of finding at least one broken defect in a domain Ω of volume V

$$P_F = P(N \geq 1) = 1 - P(N - 0) - 1 - \exp\left[\frac{V}{V_0} \left(\frac{\langle \sigma_1 - S_u \rangle}{S_0} \right)^m \right]. \tag{7}$$

Equation (7) is identical to Eqns. (1) and (2) for a uniformly loaded domain Ω. In the present formulation, the failure sites are considered as point defects. It can be noted that this assumption is not necessary [7] but leads to simpler derivations.

Two different routes can be followed to identify the parameters of the Weibull model. The first one uses data obtained at the structural or sample level and deduces the behavior at the level of one volume element (i.e., link). The second one uses microscopic observations of the flaw distribution. If the only available data are failure probabilities P_F vs. failure stress σ_F, one can rewrite the cumulative failure probability as

$$P_F = 1 - \exp\left[-\frac{VH_m}{V_0}\left(\frac{\langle\sigma_F - S_u\rangle}{S_0}\right)^m\right], H_m = \int_\Omega \left(\frac{\langle\sigma_1 - S_u\rangle}{\langle\sigma_F - S_u\rangle}\right)^m dV \quad (8)$$

where VH_m denotes the effective volume [8]. For a two-parameter Weibull law, the stress heterogeneity factor H_m is independent of the failure stress and Eqn. (8) can be recast in the following form

$$\ln\left[\ln\left(\frac{1}{1-P_F}\right)\right] = m\ln\langle\sigma_F\rangle - \ln\left(\frac{V_0 S_0^m}{VH_m}\right). \quad (9)$$

Therefore, in a Weibull plot [5], it is expected that the material data follow a straight line whose slope corresponds to the Weibull modulus. Once the Weibull modulus is known, the stress heterogeneity factor H_m can be computed by using Eqn. (8) and by knowing the intercept, the shape parameter $V_0 S_0^m$ can be determined. One can note that a conventional least squares method can be utilized to identify the Weibull parameters or a maximum likelihood approach.

The second procedure is more time consuming, but its predictive capability is usually higher than in the previous case. Systematic observations of polished surfaces of samples by using a Scanning Electron Microscope (SEM) are performed to determine the flaw size distribution f_0 [7]. The value of V_0 is usually representative of the gauge volume of the sample. The choice of a failure criterion (see Eqn. (4)) is the last step of the identification.

2.2. PERSPECTIVES

Even though the Weibull law is well understood, the approximation of a defect by a crack [9, 10, 11] or by an initial damage variable [4] is not always satisfactory. In particular, the failure criterion (see Eqn. (4)) may be improved under multiaxial loading conditions. Furthermore, the flaw size distribution is very sensitive to the fabrication process. This dependence requires a significant number of experiments for each batch of material. Lastly, the weakest link statistics inherently contains a particular size effect that can be described by the dependence of the average failure stress $\overline{\sigma}_F$ on the volume V, the stress heterogeneity factor H_m and the Weibull parameters $(m, S_0 V_0^{1/m}, S_u)$

$$\overline{\sigma}_F = S_u + S_0 \left(\frac{V_0}{V H_m}\right)^{\frac{1}{m}} \Gamma\left(1 + \frac{1}{m}\right) \tag{10}$$

where Γ is the Euler function of the second kind. Furthermore, the weakest link theory assumes that when the crack propagation is unstable at a given point, it will remain unstable throughout its propagation. This may not be always the case in brittle / ductile multimaterials [12]. In that case a deterministic criterion [13] or a probabilistic treatment [14, 12] can be used to describe crack propagation and arrest.

3. Extension of the Weibull Model to Stable Crack Propagation

The previous section addressed unstable crack propagation at *any* scale. In this section, stable propagation is assumed to occur at the microscopic level. It follows that the framework of the weakest link statistics (see Eqn. (1)) can be used under the assumption of a gradual change of the flaw distribution with time.

3.1. GENERAL FRAMEWORK

When propagation is unstable, the failure probability $P_{F0}(Q)$ is expressed by Eqn. (3). When stable propagation occurs, the initial morphological parameters $(a, \underline{n}, w)$ change to become $(A, \underline{N}, W)$ after an instant τ (i.e., N cycles or time t). The morphological parameters $(A, \underline{N}, W)$ are assumed to be uniquely related to their initial values $(a, \underline{n}, w)$ through deterministic functions of C^1 class that model the crack propagation law. For a given instant τ and a fixed load Q, the failure probability $P_{F0}(Q, \tau)$ is linked with the flaw density function f_τ

$$P_{F0}(Q, \tau) = \int_{D_c(Q)} f_\tau(A, \underline{N}, W) \mathrm{d}A \mathrm{d}\underline{N} \mathrm{d}W \tag{11}$$

If no new cracks initiate during the load history, Eqn. (11) can be expressed as [15, 16]

$$P_{F0}(Q, \tau) = \int_{D_c^*(Q,\prime)} f_0(a, \underline{n}, w) \mathrm{d}a \mathrm{d}\underline{n} \mathrm{d}w \tag{12}$$

where D_c^* denotes the set of defects that become critical after an instant τ and a load level Q. Equation (12) constitutes a unified expression for the failure probability with or without stable propagation. In *both* cases, it relates the failure probability to the initial flaw distribution f_0. If we neglect the interaction between flaws, Eqn. (1) can be used.

3.2. APPLICATIONS

Fracture data obtained in high cycle fatigue (HCF) are usually scattered for many different materials. In HCF, most materials are loaded in their elastic domain. Their failure is often catastrophic with no warning signs such as non-linearities. Therefore the basis of the probabilistic treatment of fracture in HCF follows the afore-mentioned one. The crack propagation can be modeled by an extended Paris' law [17]. The identification procedure can be purely macroscopic when only Woehler curves are available. When the fatigue limits are known, the flaw size distribution can be identified. The crack propagation law is then obtained by considering one constant failure probability (say 50%) [18]. On the other hand, when possible, the flaw size distribution can be obtained by systematic observations of fractured surfaces of fatigued samples by using an SEM. The parameters of the crack propagation law are determined by considering one constant failure probability [19]. These results can be used to evaluate the reliability of structures in (very)high cycle fatigue [20].

The same framework can be used to model the effect of subcritical crack growth (e.g., by using the crack propagation law proposed by Evans and Wiederhorn [21, 22]) on failure probabilities of brittle materials such as ceramics or glasses. The same identification procedure as in HCF can be used [23].

3.3. FUTURE WORK

In all the previous applications, the applied load was assumed to be proportional (i.e., the principal stress directions are constant during the whole load history). In many fatigue applications, this hypothesis is too strong and may require a specific treatment. Furthermore, the load level itself may be random. The crack propagation cannot be described by a conventional Paris' law. The effect of (isolated) overloadings are known under plane stress conditions mainly. Are they still important for real defects at the surface, close to the surface of within the bulk of a material?

4. Damage Models: Multiple Cracking

When multiple cracking occurs, Continuum Damage Mechanics (CDM) is an appropriate means of describing degradation since changes in elastic moduli measured on a macroscopic level [24] provide a simpler and more robust way of measuring damage than does microscopic measurement of crack density, which requires the average of many readings before reliable values are established. The Weibull model becomes useful in deriving the kinetics of variables modeling gradual degradations as will be shown in the three following examples.

4.1. FIBER BUNDLE

Let us consider a fiber bundle consisting of an infinity of parallel fibers of length L subjected to the same displacement u [25, 26] and whose failure stress is obeying a two-parameter Weibull law ($m, L_0 S_0^m$ where L_0 is the gauge length). Because of the probabilistic nature of failure, all the fibers do not fail for the same load level. Therefore, the behavior becomes non-linear (i.e., quasi-brittle) and the overall stress / strain response is expressed as

$$\Sigma = E_f (1 - D) \frac{u}{L} \tag{13}$$

where E_f is the Young's modulus of the unbroken fibers and Σ the macroscopic stress. The damage variable D measures the percentage of broken fibers [27, 28] and is expressed as

$$D \equiv P_F = 1 - \exp\left[-\frac{L}{L_0}\left(\frac{\langle\sigma\rangle}{S_0}\right)^m\right]. \tag{14}$$

It can be noted that the damage parameter D is not driven by the *macroscopic* stress Σ but by the *effective* (i.e., microscopic) stress $\sigma = \Sigma/(1-D) = Eu/L$, since the stress acting on the unbroken fibers is equal to σ. The ultimate strength Σ_U of the fiber bundle is different from the average failure strength $\overline{\sigma}_F$ of the fibers (see Eqn. (10))

$$\Sigma_U = S_0 \left(\frac{L_0}{Lme}\right)^{\frac{1}{m}}. \tag{15}$$

However, the same size effect holds when the fibers are considered individually or in a bundle.

4.2. FIBER BREAKAGE

A unit cell in a fiber reinforced brittle-matrix composite of length L_R (Fig. 1-a) is considered where the matrix cracks are saturated with spacing L_m. The length L_R is the *recovery* length and refers to twice the longest fiber that can be pulled out and cause a reduction in the load carrying capacity. Away from a fiber break, as in the case of matrix-cracking, the fiber stress builds up through the stress transfer across the sliding fiber-matrix interface. If the interfacial shear stress τ is assumed to be constant, the recovery length is related to the maximum stress in the fiber by

$$L_R = \frac{RT}{\tau} \tag{16}$$

42

where the reference stress T is the fiber stress in the plane of the matrix crack, R is the fiber radius. Generally, $L_m \ll L_R$ and the stress field in the intact fibers [29, 30] for $0 \leq x \leq L_m/2$ is (Fig. 1-b)

$$\sigma_F(T, x) = T - \frac{2\tau}{R} x. \tag{17}$$

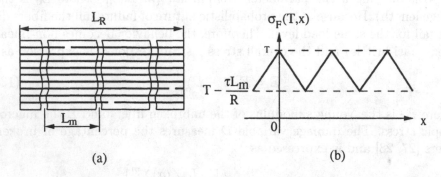

(a)   (b)

Figure 1. a-Depiction of the recovery length L_R when the density of matrix cracks (characterized by L_m) reaches saturation. b-Fiber stress field $\sigma_F(T,x)$ along a length L_R for a reference stress T when the fibers are intact.

If the fibers exhibit a statistical variation of strength that obeys a two-parameter Weibull law, then the probability that a fiber would break anywhere within the recovery length L_R at or below a reference stress T is given by (when $L_m \ll L_R$) [31]

$$P_F = 1 - \exp\left[-\left(\frac{\langle T \rangle}{S_c}\right)^{m+1}\right] \tag{18}$$

where S_c denotes the *characteristic* strength [32]

$$S_c^{m+1} = \frac{L_0 S_0^m \tau}{R}. \tag{19}$$

The cumulative failure probability is thus independent of the total length of the composite, provided the total length of the composite is greater than the recovery length [31]. Equation (18) allows one to define a damage parameter measuring the number of broken fibers within a length L_R. The kinetic law of the damage parameter $D \equiv P_F$ is dependent on the Weibull parameters m, $L_0 S_0^m$ and the interface shear strength τ. The average stress Σ applied to the composite is related to the reference stress T by [33]

$$\Sigma = fT\left(1 - \frac{P_F}{2}\right)$$

where f is the fiber volume fraction. The ultimate tensile strength of the composite is defined by the condition

$$\frac{d\Sigma}{dT} = 0$$

because the reference stress T is proportional to the average strain on the composite. This equation cannot be solved analytically. A first order solution of the ultimate tensile strength Σ_{UTS} is predicted to be [33]

$$\Sigma_{UTS} = fS_c \left(\frac{2}{m+2}\right)^{\frac{1}{m+1}} \frac{m+1}{m+2}. \tag{20}$$

Equation (20) shows that the characteristic strength S_c is the scaling parameter instead of S_0 for the fiber bundle (see Eqn. (15)). Furthermore, the ultimate tensile strength is no longer length-dependent, provided the composite length is greater than the characteristic length $\delta_c = RS_c/\tau$ [31]. These results can be used to predict pure flexural data as well as three-point flexure [31, 34].

The results derived so far can be used to describe the degradation of one- and two-directional composites by using an anisotropic damage model [35, 36]. Similarly, matrix-cracking can be described by introducing two other (anisotropic) damage variables [37]. The kinetic law of the damage variables depend upon *in-situ* Weibull parameters of the fibers or the matrix.

4.3. DYNAMIC FRAGMENTATION OF BRITTLE MATERIALS

In the bulk of an impacted ceramic, damage in tension is observed when the hoop stress induced by the radial motion is sufficiently large to generate fracture in mode I initiating on micro-defects [38]. It will be assumed that the defect population leading to damage and failure is *identical* when the material is subjected to quasi-static and dynamic loading conditions. The crack velocity is about $k = 40\%$ of the longitudinal stress wave celerity C_0 [39, 13]. Therefore, one may define a relaxation or obscuration domain of volume V_o around a crack (i.e., a volume in which the stresses are less than the applied stresses, therefore do not cause new crack initiations)

$$V_o = S\left[kC_0\left(T - t\right)\right]^3 \tag{21}$$

which is a function of a shape parameter S, the present time T and the time to nucleation $t < T$. The shape parameter S may depend on the Poisson's ratio ν but it is independent of time, i.e., the relaxed zones are self-similar. The flaw nucleation and crack propagation can be represented on a space-time graph (Fig. 2). The space locations of the defects are represented in a simple abscissa (instead of a three-dimensional representation) of an x-y graph where the y-axis represents the time (or stress) to failure of a given defect. The first crack initiation occurs for time T_1 (corresponding to a stress $\sigma(T_1)$) at the space location M_1 and produces an 'obscured volume' $V_o(T - T_1)$ increasing with time. For time T_2 (corresponding to a stress $\sigma(T_2) > \sigma(T_1)$), a second crack initiates in a non-affected zone and produces its own obscured volume. The third and fourth defects do not nucleate because they are obscured by the first and both first and second defects, respectively.

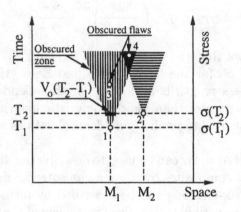

Figure 2. Fragmentation and obscuration phenomena.

Let us assume that the distribution of total flaws in a domain of volume V is modeled by a Poisson point process of intensity λ_t in accordance with Eqn. (5). New cracks will initiate only if the defect exists in the considered zone and if the flaw does not belong to the relaxed zone. It follows that the probability of obscuration P_o can be written as [40]

$$P_o\{\sigma_1(t)\} = 1 - \exp\left[-\widehat{V}_o(T)\lambda_t\{\sigma_1(T)\}\right] \qquad (22)$$

where $\widehat{V}_o$ is the mean obscuration volume

$$\widehat{V}_o(T)\lambda_t\{\sigma_1(T)\} = \int_0^T V_o(T-t)\frac{d\lambda_t}{dt}\{\sigma_1(t)\}\,dt. \qquad (23)$$

Equation (22) is yet another extension of the Weibull law. In particular, when the stress rate is small enough, only one defect leads to failure and $\widehat{V}_o = V$ (Eqn. (6)). Furthermore, Eqn. (22) is also the result obtained for the *boolean islands* model [41, 42, 43]. The fraction of relaxed zones P_o is a good approximation for a damage variable D_1 defined in the framework of CDM [44], with $D_1 = 0$ for the virgin material and $D_1 = 1$ for the fully broken one. It is worth noting that the damage variable is not *isotropic* even though it measures a volume ratio. Since the relaxation zones are relative to a cracking direction, an *anisotropic* damage description is needed. If one assumes that the damage directions coincide with the principal stress directions, the principal strain (E_1, E_2, E_3) / principal stress $(\Sigma_1, \Sigma_2, \Sigma_3)$ relationship of a damaged material can be written as

$$
\begin{bmatrix} E_1 \\ E_2 \\ E_3 \end{bmatrix} = \frac{1}{E} \begin{bmatrix} \frac{1}{1-D_1} & -\nu & -\nu \\ -\nu & \frac{1}{1-D_2} & -\nu \\ -\nu & -\nu & \frac{1}{1-D_3} \end{bmatrix} \begin{bmatrix} \Sigma_1 \\ \Sigma_2 \\ \Sigma_3 \end{bmatrix}
\tag{24}
$$

where E is the Young's modulus of the virgin material. The kinetic law of each damage variable $D_i \equiv P_o(\sigma_i)$ is obtained by derivation of Eqn. (22)

$$
\frac{d^2}{dt^2} \left(\frac{1}{1-D_i} \frac{dD_i}{dt} \right) = 6S \left(kC_0 \right)^3 \frac{d\lambda_t}{dt} (\sigma_i) \qquad \text{if } \sigma_i > 0 \text{ and } \dot{\sigma}_i > 0.
\tag{25}
$$

with $\Sigma_i = \sigma_i/(1 - D_i)$. Equation (25) shows that the kinetic law mainly depends on the Weibull parameters of the material that can be identified when a single fragmentation regime is observed (e.g., under quasi-static loading conditions [40]). In three-dimensional situations, the shape parameter S is equal to $4\pi/3$. The present model is valid as long as the principle stress directions do not change during the load history. Otherwise, more sophisticated analyses have to be carried out to account for non-proportionality of the load and crack closure phenomena (e.g., Refs. [45, 46]). These effects need to be considered and validated under dynamic loading conditions.

5. Summary

The Weibull model was used in different situations discussed herein. First, the weakest link hypothesis was made. Under this assumption, a single fracture event led to the complete failure of a structure. Then multiple cracking was discussed. A deterministic formulation can be used within the framework of CDM. However, strain-softening may lead to strain localization (i.e., macrocrack initiation) which is a discrete phenomenon again. Some solutions have been proposed, viz. higher order (i.e., non-local) damage models or discrete models. Consequently, one may wonder which approaches whether probabilistic or deterministic, continuous or discrete are able to predict the degradation and failure of (brittle) materials. This question is still debated nowadays.

References

1. A. M. Freudenthal, in Fracture, Academic Press, New York (USA), 2, (1968), 591.

46

2. W. Weibull, ASME J. Appl. Mech. 18 [3] (1951) 293.
3. A. de S. Jayatilaka and K. Trustrum, J. Mater. Sci. 12 (1977) 1426.
4. F. Hild and D. Marquis, Eur. J. Mech., A/Solids 11 [6] (1992) 753.
5. W. Weibull, Roy. Swed. Inst. Eng. Res., 151, 1939.
6. F. M. Beremin, Metallurgical Transactions A 14A (1983) 2277.
7. C. Berdin, PhD dissertation, Ecole Nationale Supérieure des Mines de Paris, 1993.
8. D. G. S. Davies, Proc. Brit. Ceram. Soc. 22 (1973) 429.
9. S. B. Batdorf and J. G. Crose, ASME J. Appl. Mech. 41 (1974) 459.
10. S. B. Batdorf, in Fracture Mechanics of Ceramics, 3, (1977), 1.
11. S. B. Batdorf and H. L. Neinisch Jr., J. Am. Ceram. Soc. 61 [7-8] (1978) 355.
12. Y. Charles and F. Hild, work in progress, LMT-Cachan, 2000.
13. M. F. Kanninen and C. H. Popelar, Advanced Fracture Mechanics, Oxford University Press, 1985.
14. D. Jeulin, thèse d'Etat, Université de Caen, 1991.
15. F. Hild and S. Roux, Mech. Res. Comm. 18 [6] (1991) 409.
16. F. Hild and D. Marquis, C. R. Acad. Sci. Paris t. 320 [Série IIb] (1995) 57.
17. P. C. Paris, M. P. Gomez and W. P. Anderson, The Trend in Engineering 13 (1961) 9.
18. F. Hild, A.-S. Béranger and R. Billardon, Mech. Mater. 22 (1996) 11.
19. H. Yaacoub Agha, A.-S. Béranger, R. Billardon and F. Hild, Fat. Fract. Eng. Mater. Struct. 21 (1998) 287.
20. I. Chantier, V. Bobet, R. Billardon and F. Hild, Fat. Fract. Eng. Mater. Struct. 23 (2000) 173.
21. A. G. Evans, J. Mater. Sci. 7 (1972) 1137.
22. A. G. Evans and S. M. Wiederhorn, J. Mater. Sci. 9 (1974) 270.
23. F. Hild, O. Kadouch, J.-P. Lambelin and D. Marquis, ASME J. Eng. Mater. Tech. 118 [3] (1996) 343.
24. J. Lemaitre and J. Dufailly, Eng. Fract. Mech. 28 [5-6] (1987) 643.
25. H. E. Daniels, Proc. R. Soc. London. A 183 (1944) 405.
26. B. D. Coleman, J. Mech. Phys. Solids 7 (1958) 60.
27. D. Krajcinovic and M. A. G. Silva, Int. J. Solids Struct. 18 [7] (1982) 551.
28. J. Hult and L. Travnicek, J. Méc. Théor. Appl. 2 [2] (1983) 643.
29. H. L. Cox, Br. J. Appl. Phys. 3 (1952) 72.
30. J. Aveston and A. Kelly, J. Mater. Sci. 8 (1973) 352.
31. F. Hild, J.-M. Domergue, A. G. Evans and F. A. Leckie, Int. J. Solids Struct. 31 [7] (1994) 1035.
32. R. B. Henstenburg and S. L. Phoenix, Polym. Comp. 10 [5] (1989) 389.
33. W. A. Curtin, J. Am. Ceram. Soc. 74 [11] (1991) 2837.
34. F. Hild and P. Feillard, Rel. Eng. Sys. Saf. 56 [3] (1997) 225.
35. F. Hild, P.-L. Larsson and F. A. Leckie, Int. J. Solids Struct. 29 [24] (1992) 3221.
36. F. Hild, P.-L. Larsson and F. A. Leckie, ASME J. Appl. Mech. 63 [2] (1996) 321.
37. A. Burr, F. Hild and F. A. Leckie, Eur. J. Mech. A/Solids 16 [1] (1997) 53.
38. P. Riou, C. Denoual and C. E. Cottenot, Int. J. Impact Eng. 21 [4] (1998) 225.
39. L. B. Freund, J. Mech. Phys. Solids 20 (1972) 129.
40. C. Denoual and F. Hild, Comp. Meth. Appl. Mech. Eng. 183 (2000) 247.
41. J. Serra, Image Analysis and Mathematical Morphology, Academic Press, 1982.
42. D. Jeulin, in Proceedings 4th Symp. on Stereology, Gotheburg (Sweden), (1985).
43. D. Jeulin, Acta Stereol. 6 (1987) 183.
44. J. Lemaitre, A Course on Damage Mechanics, Springer-Verlag, 1992.
45. J.-L. Chaboche, C. R. Acad. Sci. Paris t. 314 [Série II] (1992) 1395.
46. D. Halm and A. Dragon, Eur. J. Mech. A/Solids 17 [3] (1998) 439.

BRITTLE FRACTURE OF SNOW

H.O.K. KIRCHNER
Institut de Science des Matériaux. Bât. 410, Université Paris-Sud
F-91405 Orsay, FRANCE

ABSTRACT - Snow is a foam of ice. It shows a brittle to ductile transition (activation energy 0.6eV) as function of strain rate and temperature. Measured values of the fracture toughness K_{Ic} in the brittle regime are used in a theory of slab avalanches.

1. Introduction

Snow mechanics is considered a somewhat esoteric field by the Materials Science and Mechanics communities. This is surprising, because about as many people (a few hundred per year) die because of structural failure of the snow cover in avalanches, as because of structural failure of metals in airplane accidents. A single avalanche may kill many more people (4.000 in 1962 and 18.000 in 1970, both in Peru).

There are far fewer publications on snow than on ice, hundreds compared with thousands, but snow is catching up. The reason for the popularity of ice research was the cold war and military interest in the polar cap. There are books and reviews on ice [1 -3], but no book on snow has been on the market for 50 years [4]. The unpublished lecture notes of Gubler [5] are the best source available, the engineering properties of snow have been reviewed by Mellor [6].

Fig. 1: Snowflakes. From Bentley and Humphreys, 1963

47

E. Bouchaud et al. (eds.), Physical Aspects of Fracture, 47–57.

2. From ice to snow: structure

When water solidifies in air, it forms hexagonal single crystals of ice, called snow flakes. Some of the 6000 flakes photographed by the artist Bentley [7] are shown in Fig. 1. No two flakes are ever the same. Such a flake is perhaps formed at 3000 meters altitude and falls at a speed of 1 m/sec. During its hourlong journey it encounters different conditions of humidity and temperature, which provide variable conditions of nucleation and growth rarely encountered elsewhere. This results in a great richness of growth and forms.

Bentley [7] took his pictures of white snowflakes on a white background, they were hardly visible. To make them outstanding pictures he painted in a black background, these retouched photographs can therefore not be considered scientific. The first scientific book on snow appears to be by Dobrowolski [8], but copies are impossible to find. The first scientist to leave an impact was Nakaya [4], working in Hokkaido. Apparently his volume, which still is a standard reference, was the first scientific book to be translated from Japanese into English after World War II.

Fig.2 Metamorphism of snow flakes (a) Mass transport from convex to concave parts of a flake. (b) Rounding of a single flake by metamorphism at constant temperature. The numbers give the time in days, after Bader [12].

Once it touches the ground, a flake like shown in Fig. 1 does not keep its shape for long. The convex and concave parts have different chemical potential, there is mass transfer from the former to the latter and the appearance changes from dendritic to rounded. Glaciologist call this phenomenon metamorphism, for Material scientists it is the Gibbs-Thomson effect (the Thomson in question is not Lord Kelvin, but his older brother James [9]. As in sintering, the transport process are evaporation and condensation and surface or volume diffusion [10]. Temperature gradients arising during the day/night cycle also cause gradients in chemical potential and mass transport, the result is surface hoar [11]. In the language of materials scientists this is a strong texture of the surface layer. If later this layer is covered by fresh snow, it provides a mechanically weak interface and facilitates avalanches.

By the same mechanism the individual flakes sinter together and form snow. The individual single grains of 0.1 to 1 mm size retain the density of ice (0.917 $Mg.m^{-3}$), but the density of the agglomerate can vary widely: anything above 0.82 $Mg.m^{-3}$ is porous ice, firn (neve, Firn) lies between 0.82 and 0.55 Mg.m^{-3}. Typical snow on the ground has a density of 0.3 Mg.m^{-3}, and powder snow (poudreuse, Puderschnee) can have as little as 0.05 Mg.m^{-3} [13]. Unlike in sands, the individual grains (flakes, single crystals) are welded together.

Snow is not a granular material. Snow must be identified as a foam of ice (glaciologists do not use this terminology), and the foam theories of Gibson and Ashby [14] apply. For snow the structure shown in Fig. 3 has been proposed [15], and has effectively been confirmed by synchrotron tomography [16]. It is very difficult to photograph snow. A large depth of focus is required to get the three-dimensional structure, and the pictures end up white in white. Maybe a pinhole camera is the answer?

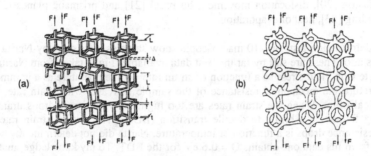

Fig.3: Snow as a foam of ice, of course schematical. (a) before metamorphose, (b) after metamorphose. From known mechanical data of the solid (ice) the properties of the foam (snow) can be derived by the strength of materials approach of foam theory [14, 15].

3. From ice to snow: mechanics

In principle, if the structure of the foam is known, one can, by foam theory [14,15], derive the mechanical properties of the foam (snow) from the mechanical properties of the solid (ice). If the structure of snow were really as simple and regular as in Fig. 3a, and consisted of quadratic beams and struts of thickness t and length l , only one geometric parameter, the aspect ratio (t/l) would be involved and the relative density (ρ_{snow}/ρ_{ice}) would be proportional to $(t/l)^2$. For this case foam theory predicts that Young's modulus and the yield strength should be proportional to ρ^2, and the fracture toughness K_{Ic} proportional to $\rho^{3/2}$, over the whole range of relative densities from 0 to 1. For any structure that is characterized by only one geometric parameter, such scaling laws hold, but with different numerical coefficients. For example, if the beams were tapered from thickness t to 2t between joints, the same power laws would hold. Metamorphism intervenes and rounds off corners, as in Fig. 3b. That structure is not characterized by one geometric parameter any more, and simple power laws do not hold over the whole density range, but only over parts of it with variable exponents. Nevertheless, the foam model provides a qualitative understanding.

Creep curves of polycrystalline fresh ice have been measured in the laboratory [17, 18]. Ice core samples from Greenland ice dated to the Holocene age, about the last 12 000 years

50

behaves the same way, but older ice creeps ten times faster, and nobody knows why [19]. This is of geophysical consequences. If ice were creeping faster, the ice sheets would have changed. For strain rates 10^{-10} s^{-1} < dε/dt < 10^{-5} s^{-1}, stresses 100kPa < σ < 3MPa, and temperatures between - 173 and -2°C, deformation follows the law

$$d\varepsilon/dt = A \sigma^n \exp(-Q/kT) \qquad (1)$$

with an activation energy of about Q = 0.7 eV and a stress exponent n = 3. The prefactor is about A = 10^8-10^9 s^{-1} (MPa)$^{-n}$. Concerning the activation energy it must be said that apparently all deformation and transport processes in ice have an activation energy of about 0.6 eV: volume diffusion [20], dislocation movement on basal [21] and prismatic planes [22], grain boundary sliding [23], heat of evaporation.

Pioneering deformation tests on 210 macroscopic snow samples were done by Narita [24, 25] twenty years ago. These are still by far the best data available! Fig.4, taken from Narita's thesis [25], shows tensile yield stress as a function of strain rate with temperature as a parameter. For each temperature the mechanical resistance of the sample increases with strain rate, but only up to a critical point. When the strain rates are too high, the yield stress drops dramatically. This striking feature is a brittle to ductile transition (BDT). The critical strain rate beyond which the resistance drops is a function of temperature. Narita did not determine the activation energy, but from his data one obtains Q = 0.6 eV for the BDT. To my knowledge, that was the first evidence ever of a ductile/brittle transition of a structural foam. This activation energy implies, for example, that tourists in the Antarctic (at say −75° C ski on very brittle snow. To the contrary, snowmen in the sun or snow on warm roofs can slowly creep hundreds of percent. The 0.6 eV tally more or less with the activation energy 0.7 eV of the BDT in ice [18]. As in other materials, the activation energy is the activation energy of dislocation mobility in ice. This allows the conclusion that the deformation mechanism of snow is the same as the deformation mechanism of ice, but does not allow to identify one or the other, because all transport mechanisms in ice have about the same activation energy.

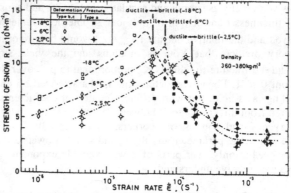

Fig.4 Temperature dependence of tensile strength of calmly deposited snow, and transition from brittle fracture to ductile fracture [25].

The theoretical connection between the curves for ice and the curves for snow are the subject of a separate publication [26], suffice it to say that the ice and snow curves tally within standard foam theory, confirming the structural evidence that snow behaves like an ice foam. Within the foam structure there are stress concentrations where the ice creeps faster, this does not change the strain rate sensitivity exponent n (which remains, about 3 or 4), nor the activation energy Q (which remains, about 0.6 eV), but causes a density dependence of the strain rate $d\varepsilon/dt$, which becomes proportional to $\rho^{[(3n+1)/2]}$ for intermediate densities. Notice the strong variation: for n = 4 the strain rate varies with the 6.5th power of the density of snow !

Although ice is a surprisingly strong material, given the fact that we come into contact with it usually at 99% of its melting point, snow is not. Moreover, because of its foamy nature, the elastic and plastic properties vary enormously with the density. This is to be expected in standard foam theory. Figs. 5a-b show Young's modulus E, and the yield strength σ_y as function of the density ρ. Of course, just as polymer, ceramic and biological foams (trabecular vs. cancellous bone) are to the left and below their solids, also snow is weaker and lighter than ice. Compared with the enormous variation with density, microstructural parameters have only a secondary influence, but can still cause scatter of almost an order of magnitude! If the structure is edgy as in Fig. 3a or has been rounded by metamorphosis to Fig. 3b is secondary. This is for the better because so far nobody has succeeded in characterizing the microstructure of tested snow samples, nor has anybody succeeded in artificially making natural snow.

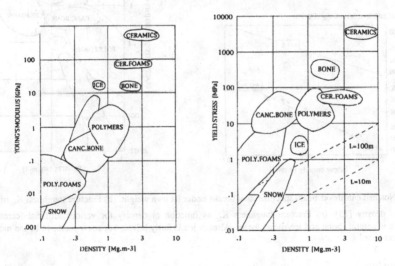

Fig.5: Mechanical properties of foams as function of density. (a) Young's modulus E. (b) Yield stress σ_y as function of density. The lines of slope one give the lengths L = σ_y/ρ of specimens of uniform cross section that break under their own weight when suspended.

Notice in Fig. 5b the lines of slope one, with a length L as parameter. This is the maximum length of free suspension for which material can support its own weight, it is kilometers for steel, around hundreds and tens of meters for ice and snow, respectively. Ice climbers rejoice about the former, the latter is the nemesis of skiers.

4. Fracture toughness K$_{Ic}$

Snow is so brittle that it easily fractures under its own weight. The most suitable testing geometry is the notched cantilever beam of Fig. 6a, loaded by body forces only. The beam of height b is harvested, shoved a distance D over its supporting edge, and cut of depth a is made from above. When the cut reaches a critical depth a$_{cr}$, the cantilever breaks off and is harvested in a bag so that the density can be determined. Using a formula K$_{Ic}$ (b, D, a$_{cr}$, ρ) the fracture toughness was measured as a function of density [15]. It is shown in Fig. 6b. In Fig. 6c it is compared with the toughness of other foams.

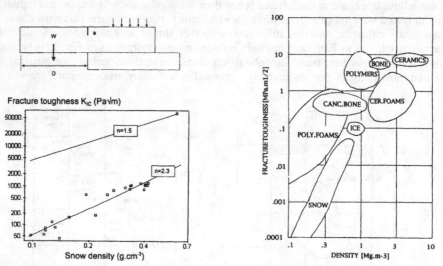

Fig. 6 (a) Notched cantilever beam specimen that breaks under its own weight. (b) Fracture toughness K$_{Ic}$ of snow as function of density [15]. (c) Fracture toughness K$_{Ic}$ as function of density for various artificial (ceramics and polymers) and natural (bone and ice/snow) foams. Already ice is fairly brittle, but snow seems to be the most brittle material ever tested.

5. Avalanches

For obvious reasons there has been considerable research on avalanches, from the meteorological aspect to the mechanical [27]. Institutes like the Schweizer Institut fuer Lawinenforschung in Davos and CEMAGREF in Grenoble enjoy a worldwide reputation. One distinguishes three types of avalanches:

5.1 Powder avalanches: these occur spontaneously within hours after a heavy snow fall. The individual flakes are not yet well connected and resemble a granular material of very low (<0.1) density. A fresh layer of perhaps several meters thickness sits on a weak layer. If this occurs on a steep slope, the whole arrangement simply does not hold and comes down. They can reach speeds of 300km/h, damage is done by the shock wave created, the appropriate theory is aerodynamics. Relatively few victims because the danger is obvious.

5.2 Ground avalanches: typically these occur once a year, in spring, always at the same spots. During the whole winter the snow gets more and more compact, densities can be as high as 0.7. Such avalanches travel at slow speed (around 10km/h) on shallow (as little as 10 degrees) slopes, but are of irresistible force. The appropriate theory is rheology. The dates of some of these avalanches have been reported in parish records for centuries, there is some evidence that they come down earlier and earlier in the year (global warming?). Construction should not be permitted in the area, there should be no victims, but there are. A number of 16.000 victims for one ground avalanche [28] is the exception.

5.3 Slab avalanches (avalanches de plaque, Schneebretter): these are triggered by tourists, and as skiing became more and more popular, the number of victims has increased to a few hundred per year, both in Europe and the US. To put this into perspective, a dozen climbers [29] die annually on the Matterhorn (Cervin). There is only a general understanding of these. Basic arguments, and experience, invoke the weight of the skier (and, less often, of the snow), a weak layer of bad adhesion below the slab, and the like. Material scientists have proposed the following theory:

6. Slab avalanches are brittle fracture of snow [30]

The key element for understanding what happens is the crown face left by the departing snow. There are no tear-outs, there is no plasticity, the surface left is faultless and testifies to brittle fracture. But if brittle fracture there is, the controlling parameter must be a stress intensity factor at a notch or crack. Suppose, for simplicity, that friction between the snow and the support (either the ground or another snow layer, marked by A in Fig. 7) is negligible, the weight of the slab is simply suspended at the top, where the stress is lower than the yield stress (otherwise the slab would have departed). The mechanical stress at the top of a slab of length L and thickness h inclined an angle a is

$$\sigma = (\rho g) \, L \sin\alpha \qquad (2)$$

Imagine a skier crossing a slab. As he has learned in skiing school, he puts all his weight on the lower ski, the sharpened edges of which he presses into the snow. The deformation around this indentation will be similar to the imprint of as punch in a metallic foam [31]. Close to the ski the deformation and strain rate is largest, so that the snow behaves in a brittle manner. Further (but only millimeters) away from the ski, where deformation and strain rate are less, the snow will behave in a plastic manner. The skier leaves a semielliptical imprint in the snow, the highest stress concentration is produced under his foot. Macroscopically it matters little if he leaves a notch or crack (with a stress intensity). There are formulae for calculating K_{Ic} as function of the length a of the crack (the length of the skis), the depth b of the cut, and the load. The weight of the skier (say 70kg) contributes only a small fraction to the stress, the important contribution is the weight of the snow (say 300kg per cubic meter). On the strength of materials side of the problem, there is a characteristic length.

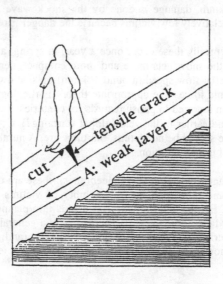

Fig.7 Skier triggering a slab avalanche. The layer A (possibly depth hoar) is supposed to provide no or little friction. The slab is pulled down by its own weight with the stress ρsinα. The skier makes a cut of depth b, which leads tu tension fracture.

$$D = (K_{Ic}/\sigma_y)^2 \qquad (3)$$

On the geometric side of the problem there is the depth b of the semielliptical cut made by the skier. Going through the mathematics, one arrives at the criterion that, up to a numerical factor of the order of unity, the semielliptical cut is unstable when

$$b > D. \qquad (4)$$

In this case the cut is unstable with respect to the formation of a tunnel crack of length 2a, between the top surface of the snow and the slab support. If the condition (3) is fulfilled and the tunnel crack is formed at all, it is, in turn, unstable with respect to propagation parallel to the slope, and thus propagates with the Raleigh wave velocity (a few hundred meters per second).

Qualitative conclusions are that it is better to cross low rather than high, to distribute the weight on both skies rather than one, and put the skies flat on the snow rather than in a sharp cut. Quantitative predictions are difficult. Only last year the fracture toughness of snow was measured [15], but what is required to judge safety in the field, is knowledge of the fracture toughness K_{Ic} and σ_y of the snow on site. Guided by Figs. 4 and 6 one can take values of 50 to 500 $Pa.m^{1/2}$ for densities between 0.1 and 0.3, and σ_y between 100 and 10 000 Pa, so that D varies between 250 and 2.5mm. The mechanical stress intensity is proportional to the load, which is proportional to the length of the slab and its density times the sine of the slope. With safety factors included. local variations of structure, inhomogeneities and other sources of trouble, one quickly gets a spread between meters and millimeters in the critical depth D of the trace not to be cut by the skier, information insufficient if he is supposed to risk his life. The materials science interpretation of the length D is the size of a processing zone, or small scale yielding zone, in which plasticity occurs. One could imagine little transportable devices for measuring that in the field. The theory presented overcomes the basic difficulty to explain slab avalanches: why are they triggered by a skier when his weight, say 100 kg, is negligible compared with the weight of the slab (300 kg/m^3)?

7. Other avalanche theories

In some models the shear strength of the weak interface is seen as the criteria [32. 33], the idea being that the weight of the skier, producing a line load of about $500kN.m^{-1}$, causes a stress exceeding the shear strength of some weak layer under him. Once interfacial support is lost, the subsequent tension fracture is placed above the skier. An alternative [34] is to invoke yielding of the snow because the weight of the slab. Intervention of the skier is not provided for. Since Fig. 5b shows that a slab of 50m length can be suspended vertically without ground support, this scenario seems to be unlikely. With reasonable values of L and τ_f, in eqn. (2), and Narita's [24, 25] yield stresses of Fig.4 one, comes to the conclusion that slabs of a few hundred meters length can be held by the yield stress at the top. Another alternative [34] is to invoke a (for simplicity circular) shear crack at a weak layer under a slab of thickness h. The stability criterion is

$$ah^2 = \left(4K_{IIc}/\pi\, \rho g\, \sin2\alpha\right)^2 \tag{5}$$

if the slab is not attached at the top at all. If K_{IIc} is somewhere close to K_{Ic} as measured [15], numerical values of ah^2 are about 1 $m^{3/2}$, clearly unrealistic. The absence of suspension at the top and adding the weight of the skier would make this a theory for the rutschblock test [35].

8. Conclusion

From the fact that, at high deformation rates, snow behaves in a brittle manner, it is possible to develop a theory of slab avalanches. If the skier cuts more deeply than a critical depth b into the snow, he creates under his feet a stress intensity higher than the snow can support. The snow slab fractures in tension under his feet, and the slab below him slides off, with or without him. This is called an avalanche.

56

Acknowledgements : Professors Gerard Michot, Nancy, France, and Takayoshi Suzuki, Tokyo, Japan, have, in turn, pointed out the possibility, impossibility, and possibility of measuring fracture toughness of snow, and developing a theory of avalanches. Without them this text would not have been written, as I acknowledge.

References

1. Hobbs, P.V., 1974, Physics of Ice. Oxford: Clarendon Press.
2. Schulson, E.M., 1999, The Structure and Mechanical Behaviour of Ice. JOM February 1999, 21-27.
3. Schulson, E.M., 2000. Ice: Mechanical Properties. Manuscript for the Encyclopaedia of Materials Science.
4. Nakaya, U., 1954, Snow Crystals. Cambridge: Harvard University Press.
5. Gubler, H., 1996, Physik von Schnee. Vorlesungsmanuskript ETH.
6. Mellor, M., 1974, A review of basic snow mechanics. IAHS Publ. no. 114, 251-291.
7. Bentley, W.A., and W.J. Humphreys, 1963, Snow Crystals, New York: Dover.
8. Dobrowolski, A.B., Historja Naturaina Lodu, Warsaw 1923.
9. The Encyclopaedia Britannica, eleventh edition, vol. 26, 1910-1911.
10. Colbeck, S.C., 1980, Thermodynamics old snow metamorphism due to variations in curvature. J. Glaciology 26, 291-301.
11. McClung, D., and Schaerer, P., 1998. The Avalanche Handbook, fourth printing. Seattle: The Mountaineers, p. 44.
12. Bader, H., et al., 1939, Der Schnee und seine Metamorphose. Davos, Switzerland.
13. Golubev, V.N., and Frolov, A.D., 1998, Modelling the change in structure and mechanical properties in dry-snow densification to ice. Annals of Glaciology, 26, 45-50.
14. Gibson, L.J., and Ashby, M.F., 1988, Cellular solids. Oxford: Pergamon Press.
15. Kirchner, H.O.K., Michot, G., and Suzuki, T., 2000, Fracture toughness of snow in tension. Phil. Mag. A 80, 1265-1272.
16. Brzoska, J.-B., Coleou, C., Lesaffre, B., Borel, S., Brissaud, O., Ludwig, W.. Boller, E., and Baruchel, J., 2000, 3D Visualization of Snow Samples by Microtomography at Low Temperature. ESRF Newsletter April 1999, 22-23.
17. Barnes, P., Tabor, D., and Walker, J.C.F., 1971, The friction and creep of polycristalline ice. Proc. Roy. Soc. Lond. A 324, 127-155.
18. Arakawa, M., and Maeono, N., 1997, Mechanical Strength of polycrystalline ice under uniaxial compression.Cold Regions Science and Technology 26, 215-229.
19. Weertman, J., 2000, Microstructural Mechanisms in Creep. In: Mechanics and Materials, edited by M.A. Meyers, R.W. Armstrong and H. Kirchner. New York, Wiley.
20. Ramseier, R.O., 1967 . J. Appl. Phys. 38, 2553 .
21.Yamamoto, Y., 1981, M.S. Thesis, Hokkaido University.
22. Hondoh, T., Iwamatsu, H., and Mae, S., 1990, Dislocation mobility for non-basal glide in ice measured by in situ X-ray topography. Phil. Mag. A 62, 89-102.
23. Weiss, J., and Schulson, E.M., 2000, Grain-boundary sliding and crack nucleation in ice. Phil. Mag. A 80, 279-300.

24. Narita, H., 1980, Mechanical Behaviour and structure of snow under uniaxial stress. J. Glaciology 26, 275-282.

25. Narita, H., 1983, An Experimental Study on the Tensile Fracture of Snow. Contribution no. 2625 of the Institute of Low Temperature Physics, Hokkaido University, Japan, pp. 1-37.

26. Kirchner, H.O.K., Michot, G., Narita, H., and Suzuki, T., 2000, Plasticity and fracture of snow: the brittle-ductile transition. Phil. Mag. A, in the press.

27. McClung, D., and Schaerer, P., 1998. The Avalanche Handbook, fourth printing. Seattle: The Mountaineers.

28. Morales, B., 1966, The Huascaran Avalanche in the Santa Valley, Peru. International Symposium on Scientific Aspects of Snow and Ice Avalanches, 5- 10 April 1965, Davos, Switzerland. Publication no. 69 de l'AIHS, Gentbrugge, Belgium.

29. Gsteiger, F., and Loppow, B., 1999, Das Gipfeltreffen. DIE ZEIT 18. 2. 99, 51-52.

30. Kirchner, H.O.K., and Michot, G., 2000, in preparation.

31. Olurin, O.B., Fleck, N.A., and Ashby, M.F., 2000, Indentation resistance of an aluminium foam. Scripta mat., to be published.

32. Foehn, P.M.B., 1987, The stability index and various triggering mechanisms. International Association of Hydrological Sciences Publication 162 (Symposium at Davos 1986 - Avalanche Formation, Movement and Effects), 195-214.

33. Jamieson, J.B., and Johnston, C.D., 1998, Refinements to the stability index for skier-triggered dry-slab avalanches. Annals of Glaciology 26, 296-302.

34. Louchet, F., 2000, A transition in dry snow slab avalanche triggering modes. Ann. Glaciology, in the press.

35. Ancey, C., 1996, Le Bloc Norvegien. La Montagne et Alpinism, 1/96, p. 64.

RANDOM FUSE NETWORKS: A REVIEW

ALEX HANSEN
NORDITA and Niels Bohr Institute
Blegdamsvej 17
DK-2100 Copenhagen, Denmark
and
Institutt for fysikk
Norges teknisk-naturvitenskapelige Universitet
N-7491 Trondheim, Norway

It is indeed a heterogeneous crowd of people who think fracture is fun. This only reflects the complexity and breadth of the problem itself. The wide range of approaches that this entails is, however, necessary to asymptotically reach full predictive understanding of fracture.

I will in this short review present an approach to slow brittle fracture that has its origin in the statistical physics community in the mid eighties. The background was a bunch of people that all had a training in the field of equillibrium *critical phenomena* which had reached its peak some years earlier and by the early eighties had begun to dwindle down into a well-established field where all the major discoveries were of the past [1].[1] Percolation theory, which was the favorite sub-field of many of these workers, experienced, however, a surge of interest and interesting discoveries in connection with dynamical properties near critical points. Perhaps the most striking of these discoveries from this time was the multifractal character of the current distribution of the random resistor network at the percolation threshold [2, 3, 4]. The distance in thought-space from the random resistor network to a model for the successive breakdown of a network of fuses under increasing load is not long, and, consequently, the first papers on the fuse model appeared [5, 6].

And, the physics community took an interest in fracture. Clearly the fuse model is a far cry from any conceivable model for real materials. However, in the best theoretical physics tradition, it was presented as a model for fracture stripped of all details that were thought to complicate the problem

---

[1] At least at the time when these words were written.

E. Bouchaud et al. (eds.), Physical Aspects of Fracture, 59–72.
© 2001 *Kluwer Academic Publishers. Printed in the Netherlands.*

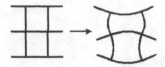

Figure 1. Possible deformation of a small beam lattice.

without bringing anything fundamental. As the understanding of the model would grow, so would the refinements of it.

Fuse models do indeed look suspect to the engineer, at least when presented in the way as done in the paragraph above. The engineer is faced with concrete situations where his or her calculations often is a matter of life or death. However, the philosophy behind numerical calculations in fracture mechanics is the following one: Even though materials at the smallest scale are not continuous, it is an excellent approximation to assume this at the scale where fracture and damage mechanics operate. However, in order to solve numerically the differential equations that describe the fracture process, it is necessary to discretize the continuum description of the problem. Thus, in reality one has the following chain of approximations: Original discrete material $\rightarrow$ continuum description $\rightarrow$ dicretization of continuum description. The favorite discretization scheme in fracture and damage mechanics is the *finite element method* (FEM) [7].

However, recently a direct approach which is *not* based on any continuum assumptions, but rather deal with the atoms themselves — *molecular dynamics* — is looking increasingly interesting [8]. Due to the need for extreme computer power, only very small systems (of the order of microns cubed) may still be studied, but the size increases with the increase of available computer power.

Returning to discretization schemes for the continuum description, *network models* constitute an alternate scheme to the FEM approach [9]. It is, however, not the fuse network we end up with. If the continuum description is based on the Lamé elastic equations, the discretized network may be the *central force* network [10], while the asymmetric Cosserat elastic equations corresponds to the *beam* network [11].

The central force lattice consists of springs that may rotate freely about the nodes. Thus, each node has two degrees of freedom, both translational. The beam network consists of bonds that are thin elastic beams which are rigidly connected at the nodes. In two dimensions, each node has three degrees of freedom, two translationally and one rotational, see Figure 1.

The fuse model does not ensue from discretizing any elastic problem. Rather, it may be viewed as a discretization of a continuum dielectrical material that suffers breakdown. The material we have in mind is an elec-

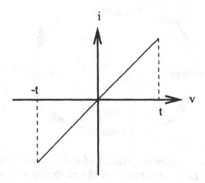

Figure 2. The voltage-current characteristics of a fuse with threshold t.

tric conductor. Each point in the material can sustain up to a maximum current density after which the material at this point becomes irreversibly an insulator.[2] We represent this material as a network of fuses as shown in Figure 4. Here each bond is a fuse with a voltage-current characteristics as shown in Figure 2. When the voltage difference v is less than a threshold t, the fuse acts as an ohmic resistor carrying a current $i = gv$, where g is its conductance. However, if $v \geq t$, the fuse turns irreversibly into an insulator.

However, the fuse model as studied so far in the literature has been so for its own right, and not as some discretization scheme. As stated in the introduction, the fuse model allows one to focus on one particular aspect of the fracture or breakdown process, while stripping away all others (such as elastic waves etc.). The feature that the fuse model focus on is this: there are two reasons that a fracture opens somewhere in a loaded material: (1) the material is weaker at that place or (2) the stress is higher at that place. Furthermore, (3) the fracture process will evolve the stress field and the stress field with induce further crack creation and growth, see Figure 3. The fuse model captures this interplay between the changing boundary conditions in the material and the changing stress field — or rather its electrical analog.

In most studies of the fuse network in the physics literature, the conductances of the fuses are all equal, while the thresholds t vary from fuse to fuse. They are typically chosen from some spatially uncorrelated distribution $p(t)$. In Figure 4, the fuses are placed at 45° with the bus bars. The reason for this is that when all fuses are intact, they will carry the

---

[2]Porous sand stones in oil reservoirs typically have clay particles deposited on the pore walls. If the flow rate exceeds the mobilization threshold for the clay particles, these may clog the pore throaths ([12]). This reduces the permeability of the sand stone and is a practical example of the kind of system that we have in mind with the electrical network model we here are developing.

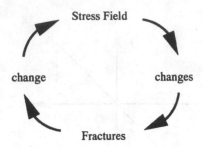

Stress Field

change   changes

Fractures

Figure 3. The stress field changes as a result of the creation and growth of cracks and the cracks are created and grow as a result of the changing stress field.

same current if all conductances are equal. The system is therefore easier to analyse.

The fuse network is initiated with all fuses intact. The voltage V across the network is then increased gradually and the fuse with the smallest threshold t burns out. The current redistributes itself instantaneously in the network. The voltage is then further gradually increased until a second fuse, and so on.

The redistribution of currents after a fuse blows gives rise to a competition between fuses with small thresholds and fuses that carry large currents as to which will blow next. The redistribution of currents leads to current enhancement at the crack tips. The model contains the essential characteristics of brittle fracture.

The process of "gradually increasing the voltage across the network" is *not* implemented literally on the computer. Rather, the breakdown process is implemented as follows. First, note that for the range of voltage differences V that does not lead to breakdown of fuses — i.e. the voltages that lie between two consecutive brekadowns — the voltage current relation is *linear.*

The network conductance si G, so that $I = GV$. The voltage difference between neighboring nodes i and j is v_{ij}, and the corresponding current is $i_{ij} = g_{ij}v_{ij}$, where g_{ij} is the conductance of the bond. The total power dissipated in the network is

$$VI = GV^2 = \frac{1}{2} \sum_{ij} g_{ij}v_{ij}^2 \,. \tag{1}$$

Thus, if we scale the voltage across the network $V \to \lambda V$, then the voltage across bond ij is scaled $v_{ij} \to \lambda v_{ij}$.

Set the voltage across the network equal to one. Identify the bond for which

$$\frac{1}{\lambda} = \max_{ij} \frac{v_{ij}}{t_{ij}} \,. \tag{2}$$

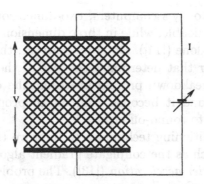

Figure 4. Each bond in this network is a fuse. The voltage V across the network, and the corresponding current I flowing through it, may be changed. As it is increased, the fuses will burn out one by one.

Thus, if we scale $V = 1 \to V = \lambda$, we find

$$\max_{ij} \frac{\lambda v_{ij}}{t_{ij}} = \lambda \cdot \frac{1}{\lambda} = 1, \tag{3}$$

which is the maximum voltage difference the bond which is next to break down can sustain. Therefore, $V = \lambda$ is the voltage difference across the network at which the fuse blows.

The breakdown algorithm is therefore

- 1. Calculate all v_{ij} given $V = 1$.

- 2. Identify $\max_{ij}(v_{ij}/t_{ij}) = 1/\lambda$.

- 3. Remove — break — the bond with the maximum ratio.

- 4. Test whether network still conducts. If yes: Goto 1, otherwise stop.

Once λ is known, we have the current the network carries when that particular fuse blows from Eq. (1), $I = G\lambda$. The conductance is also easily calculated from Eq. (1) when $V = 1$, $G = (1/2) \sum_{ij} g_{ij} v_{ij}^2$.

Note that there is no time in the breakdown process. In order to be able to compare the breakdown process in one network with that in a different network we need some kind of "clock." The most natural choice is to use the number of broken bonds, i.e. the damage. Another possibility is to use the decaying conductance of the networks.

The first step in the breakdown algorithm described above consists of solving the Kirchhoff equations to find all currents in the network. In our opinion, the most efficient algorithm to do this is the *conjugate gradient method.* ([13]). With today's computers, two-dimensional networks of size 1000×1000 are easily doable, while in three dimensions, $48 \times 48 \times 48$ is the largest that has been done ([14]). Do note, however, that it is not only the speed of the computer that determines the size of the networks that can be studied. As the breakdown process proceeds, the current distribution becomes very broad and it becomes difficult to properly determine the smallest currents due to round-off errors.

There are preconditioning techniques to improve the convergence rate of iterative solvers such as the conjugate gradient algorithm. One efficient preconditioner is *Fourier acceleration* ([13]). The problem is to find $\vec{V}$ from the matrix equation $D\vec{V} = \vec{B}$. In Fourier acceleration, one Fourier transforms D_{ij}, V_i and B_i, so that $\vec{r}_i$ — the spatial position of node $i \rightarrow \vec{k}$. We then define $\tilde{D} = |\vec{k}|\mathcal{F}D\mathcal{F}^{-1}|\vec{k}|$, $\tilde{B} = |\vec{k}|\mathcal{F}\vec{B}$ and $\vec{V} = \mathcal{F}^{-1}|\vec{k}|\tilde{V}$. Here, $\mathcal{F}$ is the Fourier transform. The original equation, $D\vec{V} = \vec{B}$ has now been transformed into the equation $\tilde{D}\tilde{V} = |\vec{k}|\mathcal{F}D\mathcal{F}^{-1}|\vec{k}|\hat{V} = \tilde{B}$. By storing the original matrix D rather than $\tilde{D}$, we preserve its sparseness. ($\tilde{D}$ is not sparse, having N^2 elements, where N is the number of nodes, while D has N elements.) The price one has to pay for this is the necessity to perform two fast Fourier transforms per iteration. However, even with the FFTs, the saving in time may be considerable. The reason why Fourier acceleration works is that $|\vec{k}|^2$ is approximatively the inverse of $\tilde{D}$. When no bonds have been removed, it is precisely the inverse. Thus, $|\vec{k}|\tilde{D}|\vec{k}|$ is not far from the unit matrix and hence, its coordination number (which is the ratio of the largest to the smallest eigenvalue) is close to unity. This ratio determines the convergence rate, the closer the number is to unity the faster the convergence.

We address in the rest of this article the question of how the fuse model behaves as a function of the distribution of breaking thresholds, $p(t)$ in the limit of large networks. Our discussion will follow closely that of [17]. We define $P(t) = \int_0^t dt' p(t')$ as the cumulative threshold distribution. We assume in the following a two-dimensional network of size $L \times L$.

It was shown numerically by de Arcangelis and Herrmann [18] that the current distribution right before the fuse network breaks apart is broad and multifractal when the threshold distribution is broad. "Broadness" means that the logarithmically binned histogram of the currents has the scaling form

$$N(i, L) \sim L^{f(\alpha)} , \tag{4}$$

where

$$i \sim L^{-\alpha} . \tag{5}$$

If $f(\alpha)$ is a more complicated function of α than of the form $a + b\alpha$, then the current distribution is multifractal [9]. Such a distribution indicates that the network at the final breaking stage is "critical", i. e. that there are no length scales in the problem apart from L. The broadness of the current distribution indicated by Eq. (4) does not develop suddenly in the breakdown process, but rather gradually. There is some correlation length, ξ, that marks the length scale at which the network crosses over from showing the scaling properties of a system with a narrow current distribution, and the scaling behavior of a system with a multifractal current distribution. Then, at a "time" in the breakdown process when the correlation length is less than L, the current distribution is

$$N(i, L, \xi) \sim L^{2+(f(\alpha)-2)\log\xi/\log L} , \tag{6}$$

where

$$i \sim L^{-1+(1-\alpha)\log\xi/\log L} . \tag{7}$$

This is a result that follows directly from finite-size scaling [15].

What we have acomplished through Eqs. (6) and (7), is to identify an intensive — in a thermodynamical sense — formulation of the development of the current distribution through the breakdown process. We may use the intensive variable

$$\tau = \frac{\log\xi}{\log L} \tag{8}$$

rather than n — the number of broken bonds — as a "time" parameter in the process. Likewise, an intensive current is

$$\alpha(\tau) = 1 - (\alpha - 1)\tau , \tag{9}$$

and an intensive histogram is

$$f(\alpha(\tau), \tau) = 2 - (f(\alpha) - 2)\tau . \tag{10}$$

These variables are those that describe the current distribution in the limit $L \to \infty$.

The breakdown process is governed by Eq. (2). Let us now write the threshold distribution in intensive variables rather than the extensive ones, t and $p(t)$. This we do in a way similar to that of the current distribution, Eqs. (4) and (5),

$$L^2 t\, p(t) = L^{f_t(\alpha_t)} \tag{11}$$

where

$$t \sim L^{-\alpha_t}. \tag{12}$$

The extra factor t in Eq. (11) is a result of binning the histogram logarithmically. Thus, the intensive thresholds and threshold distribution are

$$\alpha_t = \frac{\log t}{\log L} , \tag{13}$$

and

$$f_t(\alpha_t) = \frac{\log L^2 t\, p(t)}{\log L} , \tag{14}$$

in the limit of $L \to \infty$.

As the breakdown proceeds, τ grows from zero to one, and the current distribution evolves from a point $\alpha = -1$, $f = 2$ to a multifractal curve $f(\alpha)$. At the same time, the threshold distribution evolves. This, since the thresholds belonging to the bonds that burn out, cannot be picked anew. Thus, there is a τ dependence in both f_t and α_t also. However, while the distribution of currents become broader as the breakdown process evolves, the threshold distribution becomes narrower.

The rupture criterion, Eq. (2), becomes in these variables

$$\min \frac{i}{t} \to \min(\alpha_t(\tau) - \alpha(\tau)) . \tag{15}$$

We have no way of using this formulation to predict the detailed shape of $f(\alpha)$, since it is merely a rewriting of our starting point in different variables. However, it makes it possible to deduce some powerful statements on what kind of threshold distributions may allow for the evolvement of multifractality in the current distribution.

We have already introduced through Eqs. (13) and (14) the notion of a multifractal spectrum computed from a given distribution. Let us now show that when the distribution does not explicitly depend on the system size, its multifractal spectrum reduces to a very simple curve.

Suppose that the threshold distribution is bounded between $t_< \leq t \leq t_>$. Then, using extreme value statistics [16], we may estimate the smallest and largest thresholds, $t_<(L)$ and $t_>(L)$, that we expect to find among the L^2 bonds in the network,

$$\int_{t_<}^{t_<(L)} dt\, p(t) = \int_{t_>(L)}^{t_>} dt\, p(t) = \frac{1}{L^2} . \tag{16}$$

Quite generally, we have that $t_<(L) = t_< + \epsilon_<(L)$, and $t_>(L) = t_> - \epsilon_>(L)$, where both $\epsilon_<(L)$, and $\epsilon_>(L)$ tend to zero as $L \to \infty$. Thus,

$$\frac{\log t_<(L)}{\log L} \to 0 \quad \text{if } t_< \neq 0 , \tag{17}$$

and

$$\frac{\log t_>(L)}{\log L} \to 0 \quad \text{if } t_> \neq \infty . \tag{18}$$

This result shows that unless either $t = 0$, or $t = \infty$ is included in the threshold distribution, it is equivalent to *no disorder* in the limit $L \to \infty$; i. e. it is only a point $\alpha_t = 0$, $f_t = 2$.

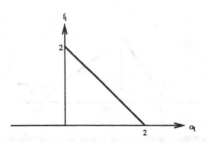

Figure 5. The uniform distribution between 0 and 1 is expressed in the intensive variables f_t and α_t.

Taking logarithmic binning into account, a flat distribution between zero and one, gives an $f_t - \alpha_t$ curve as shown in Figure 5. The Weibull distribution has the form

$$p(t) \sim t^{\mu-1}e^{-t^{\mu}}, \quad 0 \leq t \leq 1. \tag{19}$$

This probability distribution expressed in terms of $\log t$ rather than t, has the form

$$p(\log t) \sim t^{\mu}e^{-t^{\mu}}. \tag{20}$$

Now, using Eqs. (17) and (18), we find

$$f_t(\alpha_t) = \lim_{L \to \infty} \frac{\log L^2 p(\log t)}{\log L} = 2 - \mu\alpha_t - \lim_{L \to \infty} \frac{L^{\mu\alpha_t}}{\log L}. \tag{21}$$

The last term on the right-hand side of this equation approaches zero as $L \to \infty$, since $\alpha_t \leq 0$. Using Eq. (16), we find that the range of α_t is from zero to $2/\mu$.

Similarly, a power law on the interval $0 \leq t \leq 1$,

$$p(t) \sim t^{\phi-1}, \tag{22}$$

gives

$$f_t(\alpha_t) = 2 - \phi\alpha_t, \quad 0 \leq \alpha_t \leq \frac{2}{\phi}. \tag{23}$$

Comparing Eqs. (21) and (23), we see that the two distributions have exactly the same scale-invariant form, when identifying $\mu = \phi$.

In the general case, we can characterize the behavior of $p(t)$ close to zero and infinity by taking the limits

$$\lim_{t \to 0/\infty} \left(\frac{\log(tp(t))}{\log(t)} \right) = \phi_{0/\infty} \tag{24}$$

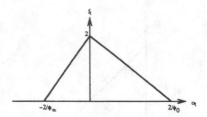

Figure 6. The general scale-invariant spectrum of a size-independent threshold distribution in the intensive variables α_t amd f_t.

These two numbers, ϕ_0 and ϕ_∞ are enough to construct the multifractal spectrum of $p(t)$. It consists in three points of coordinates $(-2/\phi_\infty, 0)$, $(0, 2)$, $(2/\phi_0, 0)$ joined by straight lines, see Figure 6. From this property, it is obvious that very different distributions can share the same spectrum. For instance an exponential distribution from zero to infinity, has the same spectrum that a uniform distribution between zero and one. Both of them are characterized by $\phi_0 = 1$, and $1/\phi_\infty = 0$ (Figure 5).

Not only the small threshold part of the distribution, but also the large threshold part plays a role in the breakdown pocess. If a very fragile bond can initiate the breakdown of the complete lattice, the role of very strong bonds is more subtle, since one very strong element can easily be avoided by the crack.

Let us give some arguments to understand how these strong bonds can play a determining role and obtain in addition a criterion for a non-trivial behavior. Let us consider here for the sake of simplicity a distribution of thresholds bounded by a non-zero lower value, and having a long tail up to infinity, say $p(t) = \beta t^{-1-\beta}$ for $t \in [1, \infty[$. Let us first assume that this tail is unimportant. We would expect thanks to the above presented argument, that, eventually after some transient stage, a single straight crack develops and finally disconnect the medium. This crack may start from a locally weak zone. Once the growth of this linear crack has started, then we know precisely where the crack is supposed to propagate. Let us call ℓ, the length of this crack, and let us suppose that the lattice size is infinite. We know that the current flowing at the tip of the crack, i_{tip}, scales as

$$i_{tip} \propto \sqrt{\ell}. \tag{25}$$

On the other hand, we can use the form of the distribution $p(t)$ to determine the value of the largest threshold, t_{max}, the crack will encounter

$$t_{max} \propto \ell^{1/\beta}. \tag{26}$$

Therefore, by forming the ratio of these two quantities, we can write the scaling of the macroscopic breaking current, $I(\ell)$, necessary to create a crack of length ℓ:

$$I(\ell) \propto \ell^{1/\beta - 1/2} . \tag{27}$$

From this result, we can distinguish two regimes according to the value of β. If $\beta > 2$, then $I(\ell)$ will decrease with increasing ℓ, otherwise it will increase. Let us note that we chose here a very particular scenario for the fracture process. Obviously, the current $I(\ell)$ is an upper bound for the breaking current.

This scenario should be compared with an alternative one, which is extremely simple, ie bonds break in increasing order of their threshold. This corresponds to a random dilution of the lattice. Since the distribution $p(t)$ is bounded below by a positive number (here 1), the breaking current is constant in the early stage of fracture, independent of the number of bonds broken.

When $\beta > 2$, the decrease of $I(\ell)$ with the number of bonds broken makes the straight crack scenario much more favorable than the random dilution. We indeed expect in this case that the fracture will only produce a single straight crack, ie it will be identical to what is expected in a disorderless material.

On the contrary, for $\beta < 2$, $I(\ell)$ will increase according to the first scenario, see Eq. (27). The physical meaning of this property is clear. Any linear crack will get arrested by a bond of very high strength, and thus the current has to be increased in order to follow the prescribed path. However, this increase is not realistic. We have seen indeed that the random dilution hypothesis does not need any increase of the external current. Thus, the first scenario cannot be followed. It is difficult to say whether the random dilution will be a good approximation of the process or not, but we can use the fact that a scenario gives an upper bound on the breaking current. Thus we expect to see in all cases that the breaking current will decrease or remain constant as the number of broken bonds increases. Simultaneously, we obtain that the number of bonds at the final stage of rupture varies faster than L where L is the system size. In the case of the random dilution, it goes as L^2, as we will show later.

Therefore, we see that the role of strong bonds may affect the development of the fracture, by arresting cracks, and forcing the nucleation of new ones, until finally, the crowding of microcracks will produce a local enhancement of the currents at the tip of some of them. Clearly this situation is much more difficult to analyze. However, we will show some data relative to numerical simulation of such distributions which suggest that the proportion of bonds broken when the fracture becomes unstable is finite, and approaches a constant when the systeme size increases, in contrast to what is seen for the opposite situation, ie a power-law distribution close to zero threshold.

Khang et al. [19] studied the case of a uniform distribution, $p(t) = 1/w$ for $t \in [1 - w/2, 1 + w/2]$. They observed that for all values of w less than 2, the uncontrolled fracture of the medium occured after a finite number of bonds were broken, independent of the system size. Khang et al. also noted that as w approached 2, a different behavior was found. Let us now rephrase this result in terms of scale invariant distribution. The spectrum of the distribution for $w < 2$ is reduced to a point, as noted previously since neither zero nor infinity are included in the distribution. We thus expect indeed to see the behavior of a disorderless material, as observed numerically and demonstrated theoretically in [19]. For $w = 2$, $p(t)$ is a uniform distribution between zero and two, and thus Eq. (24) shows that $\phi_0 = 1$. We see that the limit $w \to 2$ correspond to a change in the scale invariant part of the distribution as noted previously.

In Ref. [20], a similar problem was addressed with equivalent conclusions. A different distribution was also studied, ie a power-law distribution close to the origin. This is exactly the case of a variable ϕ_0 and a fixed $1/\phi_\infty = 0$. In this case, the conclusion was that for ϕ_0 less than 2, a disorderless behavior was observed, but for ϕ_0 greater than 2, a more complex scaling behavior could be expected. Again, we see that naturally the criterion for determining the scaling behavior of the model lies exclusively on ϕ_0 (since ϕ_∞ is fixed).

Duxbury et al. [6] studied the fracture of a randomly diluted medium. The distribution of breaking thresholds can be written as $p\delta(t - 1) + (1 - p)\delta(t)$, where δ is a Dirac distribution, and p is the fraction of present bonds. In this case, the scale–invariant part of the distribution cannot be obtained by the direct use of Eq. (24). However, since our interest is the effect of a change of scale, we can in this case apply a renormalization group argument to study the evolution of this distribution as the scale changes. It is a classical result from percolation theory, that if p is equal to the percolation threshold p_c, then the distribution $p(t)$ is invariant under rescaling. For values of p larger than p_c (ie the lattice is not yet broken in the initial state) upon rescaling, the disorder will disappear, and the effective value of p defined at a scale L will converge to 1 as L goes to infinity. Therefore, away from the percolation threshold, the spectrum of the distribution is equivalent to no disorder at all, ie in our language $1/\phi_0 = 1/\phi_\infty = 0$. Thus we expect that the scaling properties of the fracture of such systems is, at the limit of a large size lattice, that of a disorderless material — however still keeping the possibility open for logarithmic corrrections as suggested by [6], as these cannot be detected through dimensional arguments.

These were but some examples of how the multifractal analysis — which boils down to identifying the *intensive* variables in the problem — may be blended with extreme value statistics to produce extremely powerful results

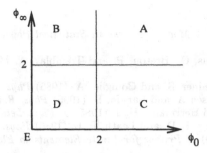

Figure 7. Different scaling regimes of brittle fracture in the fuse network may be summarized as follows: There are two relevant parameters: ϕ_0 and ϕ_∞. (A) Disorderless regime with one single crack developing. (B) Scaling regime with diffuse damage and localization. (C) Diffuse damage. (D) Strong disorder case where a finite fraction of the bonds in the lattice needs to be broken before breakdown. (E) Screened percolation.

on the the fracture process in the asymptotic limit of large systms.

The results of the discussion of behavior of the fuse model with different threshold distributions may be summarized in Figure 7 where we present a "phase diagram" for the model.

There are many aspects of the fuse model that we have not touched upon in this short review. One such aspect is that which concerns the *roughness* of the cracks that appear in the model [14, 21]. As with "real cracks" [22], the fracture surfaces are self affine with a roughness exponent that is independent of the underlying disorder (material). Even though the roughness exponent is different from that found in real materials (and there is no reason why they should be the same; the fuse model describes dielectrical breakdown, while real materials are elastic), the self organization of the fracture process, illustrated in Figure 3, give rise to not-understood mechanisms that in turn lead to the observed universality of the roughness. These processes should also be present in real materials.

We point the interested reader to Ref. [23] for a much extended review of the subject of this article.

There are many collaborators to thank collaborations past and present that have led to many of the results presented in this short review. Among them are Einar L. Hinrichsen, Per C. Hemmer, Hans J. Herrmann, and first and foremost Stéphane Roux, who has time and again thrown me into the deepest abyss of despair with his sharp comments, the latest one being that network models are *out*. This is, however, one opinion where he is mistaken.

I would also like to thank my present-day collaborators on these problems, george Batrouni, Olav Flornes, Jean Schmittbuhl and Bjørn Skjetne.

References

1. Reichl, L. E. (1988) *A Modern Course in Statistical Physics,* University of Texas Press, Austin.
2. Rammal, R., Tannous, C., Breton, P. and Tremblay, A. M. S. (1985) *Phys. Rev. Lett.* **54**, 1718.
3. de Arcangelis, L., Redner, S. and Coniglio, A. (1985) *Phys. Rev. B* **31**, 4725.
4. Batrouni,G. G., Hansen A. and Larson, B. (1996) *Phys. Rev. E* **53**, 2292.
5. de Arcangelis, L. and Herrmann, H. J. (1985) *J. Phys. Lett. (France)* **46**, L585.
6. Duxbury, P. M., Beale, P. D. and Leath, P. L. (1986) *Phys. Rev. Lett.* **57**, 1052.
7. Carroll, W. F. (1999) *A Primer for Finite Elements in Elastic Structures,* Wiley, New York.
8. Zhou, S. J., Lomdahl, P. S., Voter, A. F. and Holian, B. L. (1998) *Eng. Fract. Mech.* **61**, 173.
9. Herrmann, H. J. and Roux, S. *Statistical Models for the Fracture of Disordered Media* North-Holland, Amsterdam.
10. Feng, S. and Sen, P. N. (1984) *Phys. Rev. Lett.* **52**, 216.
11. Roux, S. and Guyon, E. (1986) *J. Physique Lett.* **46**, L999.
12. Wennberg, K. E., Batrouni, G. G., Hansen, A. and Horsrud, P. (1996) *Transport in Porous Media,* **25**, 247.
13. Batrouni, G. G. and Hansen, A. (1988) *J. Stat. Phys.* **52**, 747.
14. Batrouni G. G. and Hansen, A. (1998) *Phys. Rev. Lett.* **80**, 325.
15. Roux, S. and Hansen, A. (1989) *Europhys. Lett.* **8**, 729.
16. Gumbel, E. J. (1958) *Statistics of Extremes,* Columbia University Press, New York.
17. Hansen, A., Hinrichsen, E. L and Roux, S. (1991) *Phys. Rev. B* **43**, 665.
18. de Arcangelis, L. and Herrmann, H. J. (1989) *Phys. Rev. B* **39**, 2678.
19. Kahng, B., Batrouni, G. G., Redner, S., de Arcangelis, L. and Herrmann, H. J. (1988) *Phys. Rev. B* **37**, 7625.
20. Roux, S. and Hansen, A. (1990) *Europhys. Lett.* **11**, 37.
21. Hansen, A., Hinrichsen, E. L. and Roux, S. (1991) *Phys. Rev. Lett.* **66**, 2476.
22. Bouchaud, E., Lapasset, G. and Planès, J. (1990) *Europhys. Lett.* **13**, 73.
23. Krajcinovic, D. and van Mier, J. G. M. (2000) *Damage and Fracture of Disordered Materials,* Springer Verlag, Berlin.

ON MODELLING OF "WINGED" CRACKS FORMING UNDER COMPRESSION

Florian Lehner , *Koninklijke/Shell Exploratie en Productie Laboratorium P.O.Box 60, 2280 AB Rijswijk Z.H., The Netherlands*
e-mail: F.K.Lehner@siep.shell.com fax -. 31-70-311-2558

Mark Kachanov , *Department of Mechanical Engineering, Tufts University Medford, MA, 02155 USA*
e-mail: mkachano@tufts.edu - fax. 1-617-627-3058

Abstract. Understanding of inelastic behavior of brittle materials (like rocks) under compression requires studies of the underlying microscale deformation mechanisms. Microscopic observations indicate the existence of several distinct micromechanisms. Although their relative importance is unclear, a frictional sliding defect giving rise to tensile microcracking represents one major micromechanism in the brittle-elastic range. It has been studied quantitatively by a number of authors. We briefly discuss several of the proposed models.

1. Introduction

We suggest a simple model that is rigorously correct in both short and long wing limits. The model yields
- the stress intensity factor (SIF) at the wing tip; in conjunction with the condition $K_I = K_{IC}$, this serves to determine the wing length, and
- displacements associated with sliding and wing opening, needed for construction of the effective stress-strain relations for a solid with a number of winged cracks.

The configuration to be modelled is two-dimensional and is shown in fig. 1. From an initially present crack of length $2l_0$ and unit normal n, two tensile wing cracks (with a unit normal n') have grown to a length l, driven by a slip along the initial crack.

The remotely applied stress field is $\sigma = \sigma_{11}e_1e_1 + \sigma_{22}e_2e_2$ where e_1, e_2 are unit vectors. The largest principal stress σ_{11} is assumed to be compressive (negative); x_1 will be referred to as the axis of compression. The lateral principal stress σ_{22} may be either compressive or tensile; we denote the ratio $\lambda = \sigma_{22}/\sigma_{11}$.

In homogeneous and isotropic elastic solids, the wing crack kinks at an angle $\alpha_0 \approx 70°$ to the

73

E. Bouchaud et al. (eds.), Physical Aspects of Fracture, 73–75.

74

initial crack; as they grow longer, they gradually turn into the direction x_1 of maximum compression: $\alpha \to \pi/2 - \phi$ as $L \equiv l/l_0 \to \infty$

2. Traction-driven vs displacement-driven cracks.

As seen from the geometry of the configuration, K_I at the wing tip depends directly on the amount of sliding U along the original crack. (We shall operate with the *average* sliding

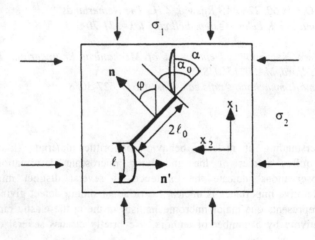

Fig. 1. The winged crack configuration

displacement, neglecting the impact of non-uniformities in U). As discussed by Kachanov (1982), the amount of slip is loading history-dependent and is not a unique function of applied stresses. However, in the loading regime of forward sliding, as well as in the regime of backsliding, U is a unique function of applied stresses. Then, the crack can equally be viewed as stress-driven and K_I - as a function of applied stresses. (In contrast, in those loading regimes when sliding is locked by friction, K_I is a unique function of U, but not of σ).

An important difference between the displacement-driven and stress-driven cracks is that the terms in K_I accounting for the stress σ_n, normal to the wings are different. This difference is best illustrated in the simplest approximation of long straight wings parallel to the axis of compression x_1.

For a *displacement-driven* crack, each wing reacts to $\sigma_n = \sigma_{22}$ as a separate crack of length l. Hence, the term accounting for the stress normal to the wings is $\sigma_{22}\sqrt{\pi.l/2}$ (in the

$$K_I = \frac{EU\cos\Phi}{2\sqrt{2\pi l}} + \sigma_{22}\sqrt{\frac{\pi l}{2}} \tag{1}$$

where $U\cos\phi$ is the width of the "wedge".

For a *stress-driven crack*, the term accounting for the stress σ_{22} is determined by the reaction of the *entire* crack configuration to σ_{22} hence, it is $\sigma_{22}\sqrt{\pi l}$ so that

$$K_I = \frac{(\tau + \mu\sigma_{nn})2l_0\cos\phi}{\sqrt{\pi l}} + \sigma_{22}\sqrt{\pi l} \tag{2}$$

where $(\tau + \mu\sigma_{nn})2l_0\cos\phi$ is the projection of the driving shear force on the sliding crack onto the direction normal to wings. Such a model was suggested by Zaitsev (1985) for the uniaxial loading $\sigma_{22}= 0$ and, in the form (2) for biaxial loading - by Kemeny and Cook (1984) (ZKC model). Equating K_I to K_{IC} yields the dimensionless wing length L as a function of the dimensionless stress $S = \sigma_{11}\sqrt{\pi l_0} / K_{IC}$ and the stress ratio $\lambda = \sigma_{22}/\sigma_{11}$

$$\frac{1}{S} = \frac{-\mu(1+\lambda)+(1-\lambda)(\sin 2\phi - \mu\cos 2\phi)}{\pi\sqrt{L}}\sin\phi - \lambda\sqrt{L} \tag{3}$$

for wings that grow from an initial crack of orientation ϕ. The stress driven model (1) shows an interplay of two factors: of the driving point forces $(\tau+\mu\sigma_n)2l_0\cos\phi$, whose influence is proportional to $l^{1/2}$ and of the stress σ_{22}, whose influence is proportional to $l^{1/2}$. Crack growth is always stable at small L. It remains stable as long as $\lambda\geq 0$, but becomes unstable at a certain critical length if $\lambda < 0$ (tensile σ_{22}).

This asymptotic result is independent of the wing length l. (An added correction term for moderately small wings, in the form $\sigma_n\cdot\sqrt{\pi l / 2}$ was proposed by Kachanov (1982).

Acknowledgement: The second author (MK) was supported by the Army Research Office through grant to Tufts University.

REFERENCES

Kachanov, M. (1982). Microcrack Model of Rock Inelasticity. Part I: Frictional Sliding on Pre-existing Microcracks. *Mechanics of Materials* 1, 19-41.

Rice, J.R. (1975). Continuum mechanics and thermodynamics of plasticity in relation to microscale deformation mechanisms. In: *Constitutive equations in plasticity* (Edited by A. Argon), MIT Press, 171-190.

Zimmerman, R. (1991). *Compressibility of sandstones*. Elsevier Science Publishers, Amsterdam, the Netherlands.

CONTINUUM DAMAGE AND SCALING OF FRACTURE

G. PIJAUDIER-CABOT[1], C. LE BELLEGO[2] AND B. GERARD[2]
1. – R&DO, Laboratoire de Génie Civil de Nantes – Saint Nazaire,
Ecole Centrale de Nantes, 1 rue de la Noe, BP 92101, F-44321 Nantes
cedex 3, France
2. – LMT-CACHAN and EDF R&D, ENS Cachan, 61 av. du pdt.
Wilson, 94235 Cachan cedex, France

ABSTRACT: This paper presents a simple application of continuum damage modelling to failure analysis. An isotropic damage model is described and its non local enhancement discussed. The calibration of the model parameters is addressed with the help of structural size effect data. Such a size effect is a feature of non local models and is also observed experimentally. We show that this scaling law of fracture can be a convenient tool for the purpose of model calibration, and in particular for the determination of the internal length which is introduced in non local models.

1. Introduction

Quasi-brittle materials such as concrete exhibit a non linear stress-strain response due to micro-cracking which may be modelled with continuum damage models. A consequence of such a progressive degradation of the material is strain softening. It is today well established that strain softening induces bifurcation and strain localisation. More importantly, strain softening produces failure without energy dissipation [1]. This is certainly not realistic and modifications of the constitutive relations are among the possible remedies. In this paper, we address these problems with the help of a simple isotropic damage model. After a brief description of the constitutive relations, the non local extension of the damage models is discussed. This modification of the constitutive relations is based on the introduction of an internal length in the material model. It induces a size effect which can serve as a way of calibrating the constitutive parameters.

2. Non local Continuum Damage

Let us consider, for the sake of simplicity, the case of tension dominated mechanical actions and restrict attention to the simplest possible damage model. In the reversible (elastic) domain, the stress-strain relation reads [5]:

$$\varepsilon_{ij} = \frac{1+v_0}{E_0(1-d)}\sigma_{ij} - \frac{v_0}{E_0(1-d)}[\sigma_{kk}\delta_{ij}], \tag{1}$$

77

E. Bouchaud et al. (eds.), Physical Aspects of Fracture, 77–84.

or
$$\sigma_{ij} = (1-d)C^0_{ijkl}\varepsilon_{kl}, \tag{2}$$

where σ_{ij} is the stress component, ε_{kl} is the strain component, E_0 and v_0 are the Young's modulus and Poisson's ratio of the undamaged isotropic material, and δ_{ij} is the Kronecker symbol. d is the damage variable which varies between 0 and 1. C^0_{ijkl} is the stiffness of the undamaged material. We shall examine non local (integral and gradient) solutions.

2.1 INTEGRAL CONSTITUTIVE RELATIONS

Consider for instance the case in which damage is a function of the positive strains (which means that it is mainly due to micro cracks opening in mode I). The stress strain relations are those given above and damage is function of the equivalent strain defined by Mazars [6]:

$$\tilde{\varepsilon} = \sqrt{\sum_{i=1}^{3}\left(\langle\varepsilon_i\rangle_+\right)^2}, \tag{3}$$

where ε_i are the principal strains. In classical damage models, the loading function f reads $f(\tilde{\varepsilon}, \chi) = \tilde{\varepsilon} - \chi$. The principle of non local continuum models with local strains is to replace $\tilde{\varepsilon}$ with its average [10]:

$$\bar{\varepsilon}(x) = \frac{1}{V_r(x)}\int_{\Omega}\psi(x-s)\tilde{\varepsilon}(s)ds \text{ with } V_r(x) = \int_{\Omega}\psi(x-s)ds, \tag{4}$$

where Ω is the volume of the structure, $V_r(x)$ is the representative volume at point x, and $\psi(x-s)$ is the weight function, for instance:

$$\psi(x-s) = \exp\left(-\frac{4\|x-s\|^2}{l_c^2}\right). \tag{5}$$

l_c is the internal length of the non local continuum. $\bar{\varepsilon}$ replaces the equivalent strain in the evolution of damage which is not detailed in this paper and can be found in Ref. [6]. The loading function becomes also $f(\bar{\varepsilon}, \chi) = \bar{\varepsilon} - \chi$, where χ is the hardening – softening variable.

It should be noticed that this model is easy to implement in the context of explicit, total strain models. Its implementation to plasticity and to implicit incremental relations is awkward. The local tangent stiffness operator relating incremental strains to incremental stresses becomes non symmetric, and its bandwidth can be very large due to non local interactions [9]. This is one of the reasons why gradient damage models, such as those described in the next section, have become popular over the past few years.

The internal length is an additional parameter, which is difficult to obtain directly by experiments. In fact, whenever the strains in specimen are homogeneous, the local damage model and the non local damage model are, by definition, strictly equivalent ($\bar{\varepsilon} = \tilde{\varepsilon}$). An approximation of the internal length was obtained by Bazant and Pijaudier-Cabot [2]. For standard concrete, the internal length lies between $3\,d_a$ and $5\,d_a$, where d_a is the maximum aggregate size of concrete

2.2 ISOTROPIC GRADIENT DAMAGE MODEL

A simple method to transform the above non local model to a gradient model is to expand the effective strain into Taylor series truncated for instance to the second order:

$$\tilde{\varepsilon}(x+s) = \tilde{\varepsilon}(x) + \frac{\partial \tilde{\varepsilon}(x)}{\partial x} s + \frac{\partial^2 \tilde{\varepsilon}(x)}{\partial x^2} \frac{s^2}{2!} + \dots \tag{6}$$

Substitution in Eq. (4) and integration with respect to variable s yields:

$$\bar{\varepsilon}(x) = \tilde{\varepsilon}(x) + c^2 \nabla^2 \tilde{\varepsilon}(x). \tag{7}$$

c is a parameter which depends on the type of weight function in Eq. (4). Its dimension is L^2 and it can be regarded as the square of an internal length. Substitution of the new expression of the non local effective strain in the non local damage model presented above yields a gradient damage model. Computationally, this model is still delicate to implement because it requires higher continuity in the interpolation of the displacement field. This difficulty can be solved with a simple calculation devised by Peerlings et al. [8]: let us take the Laplacian of the right and left hand-sides of Eq. (4). Because the Taylor expansion in Eq. (6) was truncated to the second order, higher derivatives can be assumed to be negligible. It follows that:

$$\nabla^2 \bar{\varepsilon}(x) = \nabla^2 \tilde{\varepsilon}(x) + \nabla^2 \left(c^2 \nabla^2 \tilde{\varepsilon}(x) \right) \approx \nabla^2 \tilde{\varepsilon}(x). \tag{8}$$

Therefore, Eq. (7) becomes $\bar{\varepsilon}(x) - c^2 \nabla^2 \bar{\varepsilon}(x) = \tilde{\varepsilon}(x)$. This implicit equation, which defines the non local effective strain as a function of the local effective strain, is the exact approximation of the integral relation in Eq. (4). The implementation of the gradient damage model becomes similar to the implementation of a thermo-mechanical (local) model in which the non local effective strain replaces the nodal temperatures.

3. Calibration of the mechanical model

The calibration of the model parameters in the integral or gradient damage model is facing the difficulty of getting at the same time the parameters involved in the evolution of damage and the internal length. Theoretically, it could be possible to calibrate the evolution of damage on tests in which the strain distribution remains homogeneous over the specimen. This is in fact impossible because whenever strain softening occurs, the strain and damage distribution happen to become strongly non homogeneous due to bifurcation and strain localisation during the failure process. It follows that without some simplifying assumptions such as the ones devised by Bazant and Pijaudier-Cabot [2], it is not possible to calibrate the model parameters without considering the computation of a complete boundary value problem up to failure. In the context of non local models, analytical solutions can hardly be derived, and it is the reason why a robust finite element implementation is a crucial stage in the exploitation of non local failure models.

Another characteristic of non local models is size effect. Because the constitutive relations contain a length scale, the response of geometrically similar specimens is not geometrically similar. At least the peak loads for different sizes do not scale with the geometrical similarity factor. In fact, the internal length scale controls the size of the fracture process zone (FPZ), in which micro-cracking occurs during failure. Hence, the size of the FPZ is a model characteristics, which is not dependent on the structure size. It follows that the ratio of the FPZ size to the structural size varies when geometrically similar specimens are tested. This variation produces a size effect, which may be modelled with simple equations proposed by Bazant [3]:

Let us consider the case of three-point bending of geometrically similar notched beams. We tested geometrically similar specimens of various height D = 80, 160, and 320 mm, of length L = 4D, and of thickness b = 40 mm kept constant for all the specimens. The length-to-height ratio is L/D = 4 and the span-to-height ratio is $l/D = 3$. These are mortar specimens, cast with the side of height in the vertical position, using a water-cement ratio of 0.4 and a cement-sand ratio 0.46 (all by weight). The maximal sand grain size is d_a = 3mm. The maximal strength is obtained with the formula:

$$\sigma = \frac{3}{2} \frac{Fl}{bD^2}, \qquad (9)$$

where b is the thickness of the beam, D is the height (mm), l the span (mm), and F the maximal load (N). F is obtained for each different size. A simple size effect formula for concrete and others materials is used:

$$\sigma = \frac{Bf_t'}{\sqrt{1 + D/D_0}}. \qquad (10)$$

σ is the maximal strength, D is the size of the beam (height in this case), f_t' is the tensile strength of the material, D_0 is a characteristic size which corresponds to a change of failure mechanism between strength of materials and fracture mechanics, and B is a geometry – related parameter. In a log – log plot, a strength of material criterion is represented by a horizontal curve and a LEFM (linear elastic fracture mechanics) criterion by a line with the slope $-1/2$. The two lines cut each other at the abscissa $D/D_0 = 1$. The larger the beam, the lower the relative strength, as observed on Figs. 3 and 5.

This size effect law provides also the fracture energy G_f as a result of the fit. A linear regression is used:

$$\frac{1}{\sigma^2} = aD + c \qquad (11)$$

and

$$Gf = \frac{k_0^2}{E} \frac{1}{a}.$$ (12)

E is the elastic modulus of the material, and k_0 is a geometry-related parameter.

Let us now look at finite element computations on the three sizes of specimens, and compare the load deflexion curves with the experimental data. Because there are experimental errors, which are inherent to the testing system, there are several sets of model parameters, including the internal length, which provide an acceptable fit of the tests data. Figures 1 and 2 show two of these fits. In the first set, the internal length is $l_c = 40mm$ and it is $l_c = 7mm$ in the second fit. Of course, the other parameters in the evolution law of damage are also quite different between the two fits (they can be found in [4]).

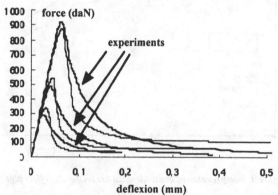

Figure 1. Fit of the mechanical model with the parameter set 1.

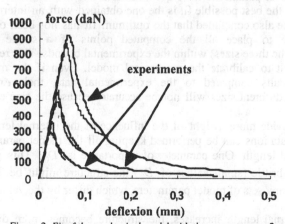

Figure 2. Fit of the mechanical model with the parameter set 2.

Globally these two fits are acceptable. If they were equally good, it would mean that the mechanical model cannot be calibrated objectively because there are several, quite

82

different sets of parameters which provide identical fits. The difference between these two fits ought to be magnified. For this purpose, the size effect method is quite adequate. Each of this fit can be analysed through a subsequent fit of the size effect formula in Eq. (10). Figure 3 shows the result of such interpretation for several quasi-identical fits. We have also reported on this plot upper and lower experimental bounds (borne sup – borne inf) between which the best fit should be. Note that on this figure, each set of computed data, corresponding to a given set of model parameters, has been fitted separately with the size effect formula. In particular, each set has a different value of D_0.

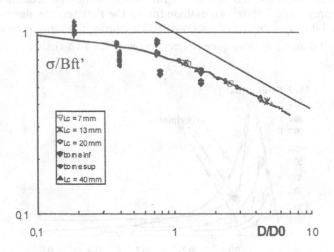

Figure 3. Interpretation of quasi-identical fits on the size effect plot.

The sets of model parameters have quite different internal lengths. It becomes clear on this figure that the best possible fit is the one obtained with an internal length equal to 40 mm. It can be also concluded that the optimum fit has not been obtained since it was not possible to place all the computed points of a single set (3 points corresponding to the three sizes) within the experimental bounds. This result shows that it is quite difficult to calibrate the mechanical model. Even if the model prediction provides good results compared to the experimental data, their *extrapolations* to structures of very different sizes will not be accurate because the size effect law is not well fitted.

In order to provide more insight of the influence of the internal length on the size effect plot, computations can be performed keeping all the model parameters constant, except the internal length. One parameter of importance is D_0. This parameter is an indication of how brittle (small values of D_0) the structure might be. Figure 4 shows its variation for many sets of model parameters, which differ by the value of the internal length only.

When the internal length increases, the structure becomes more ductile. The case where the internal length is very large compared to the structure size (e.g. larger than 40 mm on this plot), exhibits a secondary effect, which is very seldom. In fact, the fracture process zone is very large when the internal length is very large. It becomes larger that the beam thickness and the beam depth. It is much larger that the specimen notch and in this range, we believe that the simple size effect law implemented here is not applicable because notch sensitivity is entirely lost.

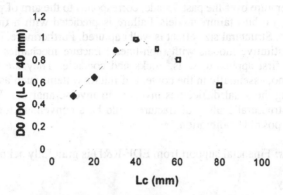

Figure 4 Influence of the internal length on D_0.

Finally, we could retain for future computations the best fit obtained according to the size effect plot, that is set 1. Figure 5 shows this fit compared with the experimental data. The relatively good agreement is now obtained because the parameters in the size effect formula are *those of the experiments*. In Fig. 3, the size effect formula was fitted with different values (D_0 especially) for each set of (experimental and computed) points. We obtained then a magnification of the discrepancies between the model predictions and the experiments which does not appear any more.

There is one point important to note: with a local model, it is not possible to predict numerically the size effect depicted in this paper. This kind of experiment provides also an indirect proof that an internal length ought to be introduced in the constitutive relations. It can be done with the help of a non local formulation or in the context of a cohesive crack model with the same success in the case of simple mode 1 crack propagation.

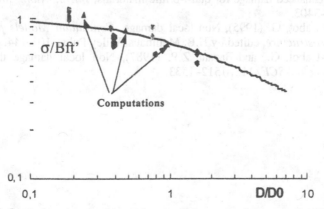

Figure 5. Comparisons between the experiments and the computations on the size effect plot.

4. Closure

The development of non local continuum models with local strains, which has been observed in the literature over the past decade, corresponds to the aim of providing sound and computationally robust failure models. Failure is predicted with a finite – non zero – energy dissipated. Structural size effect is well captured. Furthermore, it is possible to correlate such constitutive models with (non-linear) fracture mechanics [7]. Non local models have been first applied to soils, rocks and concrete. They are currently being applied to alloys too, essentially in the context of micro-systems such as MEMS. Their calibration is facing the usual difficulties involved in inverse analysis. We have shown in this paper that structural scaling of fracture could be a convenient tool, which helps at getting the best possible calibration.

Acknowledgements: Financial support from EDF-R&D is gratefully acknowledged.

5. Bibliography

[1] Bazant, Z.P. (1976), Intsability, ductility and size effect in strain softening concrete, *J. Engng. Mech. ASCE*, 102, 331-344.
[2] Bazant, Z.P., and Pijaudier-Cabot, G. (1989), Measurement of the characteristic length of non local continuum, *J. Engrg. Mech. ASCE*, 115, 755-767.
[3] Bazant, Z.P. and Planas, J. (1998), *Fracture and size effect in concrete and other quasibrittle Materials*, CRC Press.
[4] Le Bellego, C. (2001), Couplages chimie-mécanique dans les structures en béton attaquées par l'eau: Etude expérimentale et analyse numérique, Thèse de doctorat de l'ENS Cachan, France.
[5] Lemaitre J. (1992), *A Course on damage mechanics*, Springer Verlag.
[6] Mazars, J. (1984), Application de la mécanique de l'endommagement au comportement non linéaire et à la rupture du béton de structure, Thèse de Doctorat ès Sciences, Université Paris 6, France.
[7] Mazars, J. and Pijaudier-Cabot, G. (1996), From damage to fracture mechanics and conversely: a combined approach, *Int. J. Solids. Struct.*, **33**, 3327-3342,
[8] Peerlings, R.H., de Borst, R., Brekelmans, W.A.M., de Vree, J.H.P. (1996), Gradient enhanced damage for quasi-brittle materials, *Int. J. Num. Meth. Engrg.*, 39, 3391-3403.
[9] Pijaudier-Cabot, G. (1995), Non local damage, *Continuum Models for Materials with Microstructure*, edited by H.B. Muhlhaus, Wiley Pubs., 105-144.
[10] Pijaudier-Cabot, G., and Bazant, Z.P. (1987), Non local damage theory, *J. of Engrg. Mech. ASCE*, 113, 1512–1533.

DAMAGE OF CONCRETE: APPLICATION OF NETWORK SIMULATIONS

J.G.M. van MIER
Delft University of Technology Faculty of Civil Engineering and Geosciences Microlab,
P.O. Box 5048
2600 GA Delft, The Netherlands
e-mail: j.vanmier@ct.tudelft.nl

Abstract In the paper a combined experimental and numerical approach to understanding the mechanical behaviour of concrete is outlined. Concrete is a very complicated material with a distinct material structure at the micro-, meso- and macro-level. Heterogeneities are encountered at each of these levels. Quite obvious are both the structure of cement at the micro-level [10^{-8}-10^{-6} m] and the particle structure of concrete at the meso-level [10^{-3} m]. Less obvious is the heterogeneity at a more global level [10^{-1}-10^{-0} m], which is caused by wall effects, segregation, bleeding, non-uniform drying and compaction techniques during manufacturing of the material. The mechanical behaviour of concrete is the result of the interplay between stress-concentrations caused by the various heterogeneities at the three aforementioned levels, and the global loading applied to the structure under consideration. To describe the behaviour of concrete different types of models are needed; namely a model for hydration and structure development of cement, a model for moisture flow and a fracture model. The interaction between simple models and experiments is essential to come to a tool for the design of new types of concretes. In the paper attention focuses on the application of lattice type models for fracture of concrete under tensile loading. The model describes several observed physical mechanisms very well, and can be used to model size/scale effects. With regard to size/scale effects on fracture, the application of lattice type models is essential up to the transition length scale, after which the effects of heterogeneity at different levels of observation goes unnoticed.

1. Introduction

Concrete is the most widely used building material. Concrete is also a very cheap material. In addition, it is also probably one of the most complicated man-made materials. This combination of factors makes that concrete is very much "underresearched". Due to the low cost not much attention is given to optimization, and a phenomenological approach is favored above a more solid

E. Bouchaud et al. (eds.), Physical Aspects of Fracture, 85–97.
© 2001 *Kluwer Academic Publishers. Printed in the Netherlands.*

materials science approach. Recently, however, a change in attitude is observed. Due to the many additions to concrete, it is not possible anymore to speak about a material, but rather about a class of materials. This is where phenomenological approaches break down and where new methods that are common in physics and materials science must be introduced.

For fracture an approach based on mesoscopic and microscopic observations and models has been used over the past decade. Before that in the 1970's specific non-linear fracture models such as the Fictitious Crack Model (Hillerborg et al 1976), which is an extension of the plastic crack tip model by Dugdale and Barenblatt, were introduced for modelling concrete fracture at the macro-level. The experiments to determine the parameters for these models have been endless and still continue. Since infinite combinations of material compositions can be imagined, micro- and meso-level models for the assessment of fracture properties for this large range of concretes should feed this macroscopic approach.

Over the years attempts have been made to model fracture processes based on models where the structure of the material is incorporated directly in the discretization. Lattice type models, which were derived from statistical physics, have been applied, but also models based on classical continuum theory. Because, however, (quasi) brittle fracture is a localized phenomenon, it seems to make sense to use a discrete (numerical) fracture model. In the last case, two approaches seem feasible: one based on a combination of interface elements and continuum elements, the other based on network models where discrete cracks are introduced directly by removal of elements in the network at critical locations.

In the past decade, we have followed the network approach. The models that were developed have been applied for simulating fracture of concrete and sandstone under a variety of load paths. The current situation is that global tensile and shear loading can be simulated quite accurately. The comparison with experimental observations is good, and the correct fracture mechanisms are 'predicted'. In lattice type models the incorporation of frictional constraint still poses a problem. In spite of this, the models have been used as tool to design new fracture experiments, again with some success. Since the behaviour of concrete is prone to moisture flow, both during the hardening and hardened stage (hydration process, drying shrinkage and autogenous shrinkage), these matters cannot be neglected. Therefore, a suitable model for concrete includes a model for microstructure development of cement, moisture flow in all stages of the life-time of the material, coupling to shrinkage and a mechanical (fracture) model. Reality is still far from establishing this goal.

In this contribution an overview will be given of the various issues. Although relevant issues regarding structure development are mentioned, major part of the paper deals with fracture and the final consequence of fracture, size/scale effects. An essential ingredient of a realistic fracture model is that size/scale effects are incorporated. With such model, observations at a laboratory scale could be extrapolated to field scales. This requires insight in a number of factors that cause a deviation from linearity. In principle, models operating at the level of the material structure generate a size/scale effect. The price is however an enormous computational effort. Attempts should be made to define a suitable macroscopic model with a large range of validity.

2 Structure of cement and concrete

Concrete has a distinct structure both at the micro-level and the meso-level. At the micro-level (size/scale [10^{-6} m]) the structure of hardened cement is important. The structure of cement changes continuously in time, from the first hydration reactions till the moisture dependent properties at larger ages. The microstructure of cement is partially known, but much knowledge is still missing considering the complex three-dimensional structure of calcium silicate hydrates. Calcium hydroxide appears in different forms at different places, but most abundant in the interface zone with the aggregates. In addition, modern cements contain several admixtures such as fly ash or blast furnace slag. These admixtures are latently hydraulic, and will harden as well, be it at a slower pace than the Portland cement clinckers. Figure 1 shows an example of a hydrating cement grain. The platelet is hexagonal calcium hydroxide, the fibres are calcium silicate hydrates. At the meso-level (size/scale [10^{-3} m]) aggregates are part of the material structure, see Figure 2. Next to natural aggregates many different products are nowadays mixed into concrete, ranging from very small such as fillers (powders) and condensed silica fume (size/scale [10^{-8} m]) to artificial material aggregates in the same size range as sand and gravel.

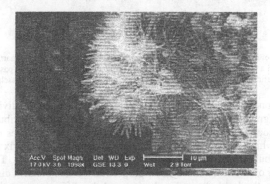

Figure 1: Microstructure of hydrating cement. ESEM image by Dr. M.R. de Rooij (microlab).

Several mechanisms must be considered to understand the complex behaviour of concrete. Since water forms an integral part of the cement structure, drying shrinkage and autogeneous shrinkage (also referred to as self-desiccation caused by hydration of cement) may lead to eigen-stresses in the material. Moreover, hydration proceeds differently near "walls" (the surfaces of aggregates can be interpreted as a wall), and an abundant amount of calcium hydroxide is deposited there. The grain distribution is also different near walls: fine grained particles are packed closer to the wall. This leads to heterogeneity of the cement structure, both before and after hydration. In the interface zone between cement and aggregates water transport is of importance as well. Before hydration, water will be absorbed at the surfaces of all particles, and will be present in pockets between aggregates. During hydration, the interface zones will therefore have a higher water/cement ratio, which will result in a more porous structure. It should be mentioned that during hydration quite some heat is produced, which may lead to thermal gradients in the material. These thermal gradients may cause initial damage as well.

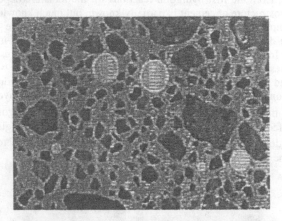

Figure 2: Particle structure of mortar. Sand grains and large pores are embedded in a cement matrix. The largest sand grains are 2 mm in diameter. In a two-dimensional image the sand grains seem to float in cement paste. In three dimensions however, the particles are stacked and all grains are in contact. Image by Mr. J. Bisschop (microlab).

At the meso-level concrete is regarded at a larger size/scale and individual aggregates are distinguished. The aggregates have a significant effect on the stress distribution in the material. Under mechanical load, stresses will be distributed according to differences in Young's moduli of the three phases of which the material is made: aggregate, cement matrix and interface zones. Larger pores and air bubbles are sometimes recognized as a fourth phase. The different material phases each have their own distinct mechanical properties and will break at distinct strength levels. The interplay between the local stress distribution and the different strength values of the three phases leads to a typical fracture process where crack nucleation, crack arrest and crack interaction all play a role. The heterogeneity of the material will lead to stable propagation in early stages of loading. However, at some stage the material heterogeneity is not capable for arresting the larger cracks. Fracturing will become unstable and will lead - in general - to a localized macro-crack.

At the macro-level a number of other factors further complicate the mechanical behaviour of concrete. Heterogeneity not only derives from the aggregates, but a large-scale heterogeneity is caused by segregation during hardening of the concrete. In general the upper surface is rich in water, whereas larger aggregates may settle preferentially near the bottom of the structure. The distribution has a significant effect on stiffness and strength distribution in the structure, which comes on top of the heterogeneity from the phase distribution. After casting concrete is compacted by means of all kinds of vibrators. The idea is to remove air as much as possible in order to obtain a dense material structure. The specific way of vibration will influence the material structure and thereby the strength and stiffness distribution as well.

Properties of concrete are usually determined at laboratory scale, which is in the order of 0.1 m till 10 m (maximum). Beyond that results from the laboratory must be extrapolated, which seems to make sense only when a physically correct theory is available. Above a certain transition length the heterogeneity from different sources at the micro-, meso- and macro-level will diminish and the material can be considered as a continuum. This is shown schematically in Figure 3

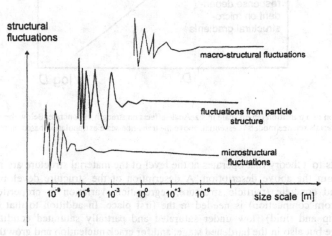

Figure 3: Continuum hypothesis at different size scales.

In Figure 4 the expected behaviour at different size scales is plotted. This sketch is based on the idea that above the transition length continuum fracture theories can be adopted to describe size/scale effects. Below the transition length, all the heterogeneities must be known in order to allow for extrapolation. Models that operate at the level of the material structure can be applied to assess the effect of the various heterogeneities. Most difficult to assess will be the effects of the large-scale (macroscopic) heterogeneity that was mentioned before. The dilemma is that the transition scale is most likely in the order of one to a few meters, which implies that a representative laboratory scale specimen should have that size. If the specimen size is smaller, consequently one has to revert to models in which the relevant heterogeneities have been incorporated.

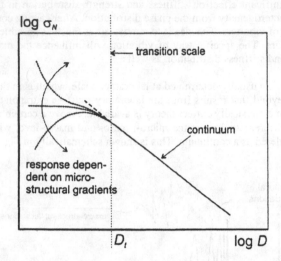

Figure 4: Definition of a transition length scale in size/scale effects on strength of concrete. Below the transition scale the application of materials science models is essential, above the transition scale extrapolation based on macroscopic theories seems feasible.

The ingredients to a theory that operates at the level of the material structure are multiple, as may be obvious from the above description. A description of the structure development in cement (hydration) and concrete (particle structure, segregation, variation in properties due to wave propagation from compaction) is needed in the first place. In addition to that also models for moisture (damp and fluid) flow under saturated and partially saturated conditions during the hardening stage but also in the hardened stage, and for crack nucleation and growth are essential.

3 Simple lattice model

Because the material 'concrete' is extremely complicated, it is sensible to look for simple (numerical) models that generate complex behaviour. Examples of suitable simulation models are cellular automata models or simple network models. For hydration and moisture flow cellular automata models seem appropriate (see for example Bentz et al. 1994 for a hydration model and Küntz et al. 2000 for a lattice gas model). For fracture simple lattice models based on developments in crystallography, linear elasticity and statistical physics seem quite appropriate. In this paper a lattice beam model will be presented only, and hydration and lattice gas models are left out. They are however important to develop a coherent overall model for the behaviour of concrete.

As early as 1941 Hrennikoff proposed a simple network model to obtain approximate solutions of problems that can not be solved by means of the theory of linear elasticity. In 1941 Hrennikoff's approach was not really applicable. Large systems could not be solved since sufficient computational power was lacking. This has changed significantly over the past decades. Towards the end of the 1980's statistical physicists started to work with lattice models again. Herrmann et al. (1989) proposed a beam model based on a square lattice. In the beam model crack growth is simulated by means of a simple Von Mises type fracture law.

Two essential changes were needed to render the model by Herrmann et al. suitable for concrete materials and structures. First of all, in order to generate correct values of the Poisson's ratio, a triangular lattice is needed. Next to that, since the structure of concrete at the meso-level is quite well understood, incorporating the distinct particle structure directly in the lattice has proven to yield more realistic results. In Figure 5 two examples where a regular triangular lattice with a beam length of 1 mm has been overlaid on two different concrete structures are shown. The difference between these two examples is in the aggregate density. Schlangen & Van Mier (1992) have incorporated the above modifications in a beam model. A standard finite element code was adopted for solving the set of equations. The model has since then be used as a tool for the design and analysis of fracture experiments on concrete and sandstone, see for example Schlangen (1993), Van Mier et al. (1995), Van Mier (1997), Vervuurt (1997) and Van Vliet (2000). In addition to the above changes, a method for parameter identification was developed. Next to the assessment of the elastic properties of the lattice elements, fracture strength must be assigned to beams falling in different phases. Most critical is the determination of the strength of the interface between aggregate and matrix, see Vervuurt (1997).

The way to operate the model is simple and straightforward. The test specimen is modelled including all loading platens, supports and so on. The finite element program that has been adopted is excellently suited for that purpose since it allows using continuum finite elements in areas where no cracks are expected to grow. The lattice part and continuum parts are connected by means of "tyings", which describe displacement continuity along the edges of the lattice and the continuum parts. Because the procedures for the determination of the properties of the lattice elements were all described in the aforementioned publications, the interested reader is referred to those sources.

4 Material structure effects on fracture

A single example is shown in these pages. The main reason for using the lattice is to come to a better understanding of fracture processes in cement, concrete and rock. Over the years the model has proven to give realistic results for the fracture of concrete under tension and shear.

Compressive failure still poses a problem. The Von Mises criterion does not include a criterion for compressive failure. Experimenting with a Mohr-Coulomb criterion has so-far not resulted in better results. For example, simulations of a Brazilian splitting test on a concrete disc have shown that the splitting mechanisms can be very well described by means of the Von Mises criterion. However, the development of a small cone enclosed by shear cracks near the support, which is an essential element in the break down process of the Brazilian test, is missed. Curiously, the cone is found in simulations with a Mohr-Coulomb criterion, but then the tensile splitting crack does not appear, see Lilliu et al. (1999). The computation fails because the cone is completely detached from the rest of the disc.

An example where no experimental information has been obtained yet is shown in Figure 5. A material model containing different amounts of aggregate particles is constructed. Around the aggregates interface zones with a constant thickness are included. In fact, the interfaces are 1 mm thick, which is an over-estimation of reality (from microscopic observations interface thickness between 5 and 50 μm is reported). However, this is of little concern for the stiffness, whereas the strength limit will remain the same, and the relative effect on the composite behaviour will be computed correctly. Depending on the specific particle content, continuous paths of bond zones can form in the structure. This will happen when the particle content exceeds the percolation threshold, which lies somewhere between an aggregate area fraction of 0.34-0.48, see Van Vliet (2000). These values hold for the case when aggregates are distributed according a continuous (Fuller) distribution. When the percolation threshold for the bond beams is exceeded, the composite behaviour is completely governed by the properties of the bond beams. This implies a low composite strength, see Figure 5 (lower row).

The contrasting case is when percolation of bond beams does not occur. This occurs when a relatively small amount of aggregates are included as in the example of Figure 5 (upper row). In this case, initially several bond beams will fracture, but extension of the bond cracks in the matrix is prohibited due to the higher strength of the matrix. A situation develops where many bond beams fail before the first matrix beam reaches its maximum strength. As soon as this occurs, macroscopic failure of the composite system cannot be avoided. The load-elongation diagram shows in this case a larger pre-peak non-linearity. This regime is completely determined by the number of bond beams that must be excluded, and depends directly on the number of aggregates in the model. Intermediate particle contents will lead to a mechanical behaviour in between of these extremes.

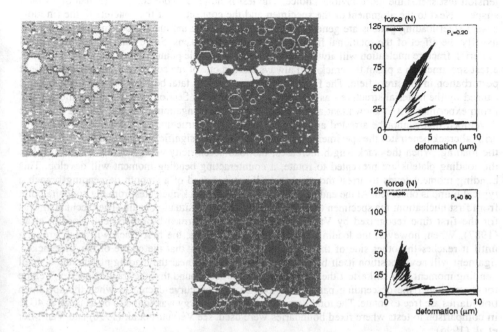

Figure 5: Effect of particle content on the fracture process under uniaxial tension, after Van Vliet (2000). In the upper row crack growth at the peak of the load-displacement diagram is shown (left), the fracture pattern at the end of the simulation (middle), and the load-elongation diagram for the case with low particle content (right). In the second row similar diagrams are shown for the case with high particle content. Tensile loading is applied in vertical direction. The P_k values refer to the theoretical aggregate area before interface zones are assigned and a lattice overlay with a regular triangular lattice with 1-mm beam length has been made. Thus, the real aggregate area is smaller, see Van Vliet (2000) for full detail.

The results of Figure 5 have not been verified/falsified by experiment. As a matter of fact, this would be an impossible task, unless a composite is made with cylindrical aggregates. This points to one of the largest restrictions of the current lattice models, which are mostly 2-dimensional plane stress models. In principle, extension to three dimensions is possible, but at the cost of larger computational effort, see for example Schlangen & Van Mier (1994).

5 Boundary condition effects in tensile testing and size/scale effects

The above example shows the effect of material structure (limited to the case of particle content) only. The loading is tensile. For the determination of fracture properties of concrete the uniaxial tension test seems the most obvious choice. The test is however more complicate than one would suspect. Next to the alignment of the specimen and the connection of the concrete to the (in most cases) steel loading platens are generally acknowledged to be the main problems. Not noticed is usually the effect of the rotational freedom of the loading platens. Since concrete is a disordered material, fracture nucleation will always take place from a single point around the circumference of a test specimen. In a prism the crack will always start in one of the comers as a result of the slight perturbation in the stress field. The first crack is generally not fatal because stress-re-distributions caused by the material structure - as described in the previous Section - may stabilize the growth. From experiments it is known that at the peak of the load-elongation diagram a critical crack has developed that cannot be arrested anymore by the material structure, see Van Mier (1997). This critical crack penetrates the specimen from one side and has a significant effect on the centricity of the loading. When the crack length increases, the load eccentricity will gradually increase. When the loading platens are prevented to rotate, a counteracting bending moment will develop. This bending moment tends to arrest the crack growth, and instead of a smooth descending branch a typical bump is observed. At the end of the bump, a second crack nucleates from the opposite point from first nucleation: the specimen tries to create a symmetric situation again. The mechanism was for the first time recognized by Van Mier (1986), and confirmed by others, e.g. Hordijk et al. (1987). When, however, the loading platens are free to rotate, the first crack will keep growing until it reaches the other side of the specimen. Note however that the centre of the 'un-cracked' ligament will never position itself between the rotation points near the loading platen, and a small bending moment will still affect the experiment. It has been found that the fracture energy, i.e. the total area under the descending part of the load-elongation curve decreases when the specimen boundaries are free to rotate. The total decrease of fracture energy was found to be as large as 40 % in comparison to tests where fixed boundaries were used, see Vervuurt et al. (1996) and Van Mier et al. (1996).

The crux of the story is that the external loading affects the fracturing of the tensile specimen. The same mechanism is found in compressive failure of concrete, where in addition frictional effects play a role after the cracks have developed. The mechanical behaviour of concrete is therefore dependent on the local stresses generated by the structure of the material, and the more global stress distribution that emerges from the specific test set-up. The test set-up then includes the global loading, the specimen shape and boundary conditions.

The lattice model also generates a scale effect. When the mesh is kept constant, but the global structure size increases, a decrease of global tensile strength is observed, in agreement with experimental observations, see Van Vliet (2000). In the experiments Van Vliet used freely rotating boundaries, such that a lower bound of the fracture energy was measured. The specimens were loaded in uniaxial tension with a small eccentricity. Dog-bone shaped specimens were used to avoid notches, and an electronic system was used to allow for stable test control. The results from the experiments showed an initial increase of tensile strength followed by a subsequent decrease for larger sizes. The size range was 1:32, the largest specimen having a length of 2.40 m. The initial strength increase could be explained from wall effects from the grain distribution near the specimen edges, but also from non-uniform drying and eigen-stresses in the material at loading, Van Vliet & Van Mier (2000). The idea is summarized in Figure 4. It must be verified in future tests, not only in full 3D but for other than tensile loading conditions as well.

6 Concluding remarks

The approach that was followed over the years was one where a close observation of fracture in laboratory scale specimens and network type modelling at the meso-level was used to come to a better understanding of fracture processes in concrete. The combination of experiments and numerical models is considered essential for progress in the field. Matters that have been elucidated are the following. A clear explanation for the shape of the softening curve in the load-elongation diagram of concrete has been obtained. After a stage of isolated microcracking, a fatal macrocrack starts to propagate. In the wake of the macrocrack, isolated crack face bridges are still capable of stress transfer, see Van Mier (1991). The crack face bridges are visible in the simulations of Figure 5. The steep part of the softening curve is associated with the growth of the macrocrack through the cross-section of the structure under consideration, whereas the tail is associated to bridging phenomena. Under compression friction plays a more dominant role in addition to crack nucleation and propagation. The way specific concrete will fracture is further determined between the relative strength and stiffness of the various material phases. In conventional gravel concrete, the interface between cement matrix and aggregates is extremely weak, and grossly determines the observed behaviour. Size/scale effects are better understood, and the existence of a transition length scale has been proposed, see Figure 4. The transition length scale is defined as the length scale where heterogeneity in the material at micro- (cement), meso(aggregate structure), and macro-level (large-scale heterogeneity from bleeding and segregation, wall effects and eigen-stresses from drying) have become small relative to the size of the structure.

The future attention in concrete technology should focus on a more appropriate description of the interfacial transition zone. Fracturing in the interface zone depends on moisture transport. Also affected by the presence and flow of water is the formation of the cement structure in the interface zone. Research programs are currently underway to get a better insight in these matters. Since moisture flow is important, any theory about concrete is incomplete when a flow model is missing. Again, efforts are underway in different institutes to better grasp the physics of moisture flow in relation to structure development of cement. As was mentioned earlier (Section 4) it is important to develop three-dimensional models. Limits on computational resources often restrain such developments. It means that ways must be found to accelerate the numerical solution techniques.

The final goal is to develop a universal theory of concrete behaviour that can be applied for the design of new types of concretes.

Acknowledgement

The work that is - very globally - described in this contribution is the result of collaboration in the past decade with my colleagues, former and present students, M.R.A. van Vliet, E. Schlangen, A. Vervuurt, B. Chiaia, A. Arslan, M.B. Nooru-Mohamed, J.H.M. Visser, R.A. Vonk, H.S. Rutten, H.J. Fijneman, H. Sadouki, J. Bisschop, G. Lilliu, M.R. de Rooij, Ch. Shi and D. Jankovic. Much of the experimental work was carried out in collaboration with A.S. Elgersma and G. Timmers.

References

Bentz, D.P., Coveney, P.V., Garboczi, E.J., Kleyn, M.F. and Stutzman, P.E. (1994), Cellular automaton simulations of cement hydration and microstructure development, Modelling Simul. Mater. Sci. Eng., 2, 783-808.

Herrmann, H.J., Hansen, H. and Roux, S. (1989), Fracture of disordered elastic lattices in two dimensions, Phys. Rev. B, 39, 637-648.

Hordijk, D.A., Reinhardt, H.W. and Cornelissen, H.A.W. (1987), Fracture mechanics parameters of concrete from uniaxial tensile tests as influenced by specimen length, in Proc. SEMIRILEM Int'l. Conf. 'Fracture of Concrete and Rock', S.P. Shah & S.E. Swartz (eds), SEM, Bethel (CT), USA, 138-149.

Hrennikoff, A. (1941), Solutions of problems of elasticity by the framework method, J. Appl. Mech., 12, A169-A175.

Küntz, M., Van Mier, J.G.M. and Lavall6e (2000), A lattice gas automaton simulation of the non-linear diffusion equation: a model for moisture flow in unsaturated porous media, Transp Porous Media, in press.

Lilliu G., Van Mier J.G.M. and Van Vliet M.R.A. (1999), Analysis of crack growth of the Brazilian test: Experiments and lattice analysis, In Proc. ICM8 Progress in Mechanical Behaviour of Materials, Vol. I 'Fatigue and Fracture', Ellyin J. and Provan J.W. (eds.), 273-278.

Schlangen, E. (1993), Experimental and Numerical Analysis of Fracture Processes in Concrete, PhD thesis, Delft University of Technology,.

Schlangen E. and Van Mier J.G.M. (1992), Experimental and numerical analysis of the micromechanisms, of fracture of cement-based composites, Cem. Conc. Comp. 14, 105-18.

Schlangen, E. and Van Mier, J.G.M. (1994), Fracture simulations in concrete and rock using a random lattice, In Computer Methods and Advances in Geomechanics, Siriwardane, H. & Zaman, M.M. (eds.), Balkerna, Rotterdam, 1641-1646

Van Mier, J.G.M. (1986), Fracture of concrete under complex stress, HERON, 31, 1-90.

Van Mier, J.G.M. (1991), Mode I fracture of concrete: Discontinuous crack growth and crack interface grain bridging, Cem. Conc. Res., 21(1), 1-15.

Van Mier, J.G.M. (1997), Fracture Processes of Concrete, CRC Press, Boca Raton (FL), USA, 448 p.

Van Mier, J.G.M., Schlangen, E. and Vervuurt, A., (1995), Lattice type fracture models for concrete, Chapter 10 in Continuum Models for Materials with Microstructure, H.B. Mühlhaus (ed.), Wiley, Chichester, 341-377.

Van Mier, J.G.M., Schlangen E. and Vervuurt, A. (1996), Tensile cracking in concrete and sandstone, Part 2: Effect of boundary rotations, Mater. Struct. (RILEM), 29, 87-96.

Van Vliet, M.R.A. (2000), Size Effect in Tensile Fracture of Concrete and Rock, PhD thesis, Delft University of Technology.

Van Vliet, M.R.A. and Van Mier, J.G.M. (2000), Effect of strain gradients on the size effect of concrete in uniaxial tension, Int. J. Fract. 95, 195-219.

Vervuurt, A. (1997), Interface Fracture in Concrete, PhD thesis, Delft University of Technology.

Vervuurt, A., Schlangen, E. and Van Mier, J.G.M. (1996), Tensile cracking in concrete and sandstone, Part 1: Basic instruments, Mater. Struct. (RILEM), 29, 9-18.

DEGRADATION IN BRITTLE MATERIALS UNDER STATIC LOADINGS

Y. BERTHAUD
Laboratoire de Mécanique et Technologie, ENS de Cachan,
CNRS, Université Paris 6
61 avenue du Président Wilson, 94235 Cachan Cedex

Abstract

The behavior of rock like materials under quasi-static loading is of major importance (i) to understand the degradation process (ii) to determine the appropriate model in view of finite element predictions. The localization that occurs prior to the peak of the load displacement Curve imposes to examine the constitutive equations in order to match both the homogeneous response and the localized one.

1. Introduction

Brittle materials like concrete rocks are widely used for civil engineering constructions Their behavior under static loadings has been Widely Studied [1-5]. It is well known that micro porosity and microcracking exists in these materials due to the fabrication process. The microcraking can be observed directly using replica technique [6] or its consequences on the propagation of ultrasonic waves for example [7]. These existing microcracks can propagate around the aggregates or inside the matrix depending on the interface properties. The evolution of the density of microcrack or micro pores is responsible for modifications of the elastic stiffness and for the apparition of permanent strains (visible after each unloading path). Other phenomena are also observed: non-linear unloading paths or stiffness recovery. Depending on the size of the specimen and on the applied boundary conditions, homogeneous Or localized failure occurs [5,8]. In the latter case it is necessary to investigate this process using optical techniques such as stereophotogrametry or speckle laser to detect the inception of cracking. The onset of localization is always found to occur prior to the peak of the load displacement curve for uniaxial compression with classical cylinder specimens. The mode of localization (shear band or not) has been carefully analyzed as the constitutive equations are very sensitive to this aspect. We propose different procedures to identify mechanical models. A first one developed in [9] considers a given Drüker Prager model. The different characteristics of the localization process are input into the model to deduce its main parameters i.e. the friction angle and the cohesion. Another approach deals with a more general model, including scalar damage and non-associated flow rules that has been compared to the experimental results.

2. Initial state for rock like materials
The initial microcracking (see Fig. 1) existing in brittle materials can be seen using replica, which are observed under SEM and analyzed using classical stereological parameters [5].

E. Bouchaud et al. (eds.), Physical Aspects of Fracture, 99–109.

Figure 2 shows that the initial microcracking present at the surface of the concrete specimen is isotropic. Within the elastic domain the applied stress induces an elastic closure of the microcracks that perturbs the propagation of ultrasonic waves and leads to a reversible anisotropy linked to the stress state [7, 10, 11].

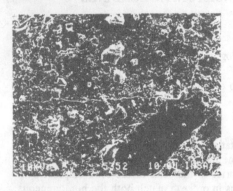

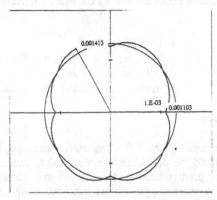

Figure 1. Microcracking in concrete

Figure 2. Stereological analysis of the cracking density versus the orientation (unit.-μ m^{-1})

3. Degradation for uniaxial tension compression tests

As the load increases the microcracks evolve. The cracking pattern at different stages of a uniaxial compression test can be revealed by X-Ray technique [12] or imagined through acoustic emission or ultrasonic wave propagation. The classical axial cracking can produce (under particular conditions) an axial splitting of the specimen that is commonly adopted as the homogeneous response of such materials (Fig. 3) [13].

Figure 3. Axial splitting [13]

The stress strain curve of brittle materials is classically determined using slender cylindrical specimens (height to transverse size ratio equal to 2). The figure 4 gives an example of the stress strain for the longitudinal, transverse and volumetric strains. We can first observe the stiffening zone at the beginning of the curve due to crack closure followed by a linear regime. Then a non-linear hardening zone exists in which cracks develop. The softening branch corresponds to an intense dilatation of the specimen. One can also observe the variation of the longitudinal and lateral stiffness of the material for each unloading-reloading process.

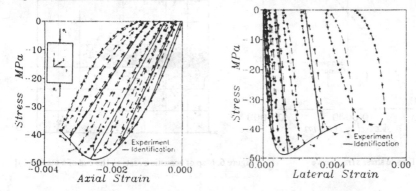

Figure 4. Stress strain curve for compression [171

The macroscopic measure of the stiffness within the linear part of the reloading process does not account for the microscopical phenomena that exist in the material (for more details see [14, 15]). Another interesting point is the apparition of permanent strains that can be measured after each unloading path. They are anisotropic and increase with the degradation. This can be partially explained at the beginning by the release of internal stresses due to the hydration process (in the case of concrete) and then by the fact that created cracks cannot close.

The behavior of brittle materials is difficult to handle under tensile loadings due to early and unstable microcracking. The localized failure is may be delayed adopting special experimental set up such as proposed in [16, 171. In this system the aluminum alloy bars stuck on the concrete specimen remain elastic while the concrete damages. Assumptions of stress homogeneity in the central part of the specimen, measurements of the global load and of the local strain lead to the computation of the stress strain curve in tension with the beginning of the reloading curve in compression. Figure 5 shows that the stiffness decreases and that permanent strains are created. The recovery of the initial stiffness is due to the closure of the previously created cracks. Additional tests have been made using acoustic emission to locate the acoustic events (see Fig. 6) [181]. The maps of acoustic events for different stages of the load displacement curve have been recorded. They show clearly a non-homogeneous damage starting at a point and propagating from another.

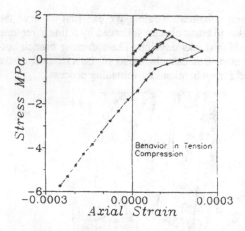

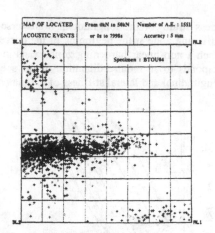

Figure 5. Tensile stress slrain curve   Figure 6. Map of acoustic events al lhe end of a tensile [18]

4. Localization in uniaxial compression tests for concrete or sand stone specimens

4.1. OPTICAL OBSERVATIONS

This case has been widely studied because it is possible to examine the evolution of the strain field at the surface of the specimen during the test. For this, two optical techniques have been used: the stereo photogrammetry and the speckle laser technique. In the first case successive photos of the specimen are recorded on photographic plates for different mechanical stages. Then they are compared on a stereo comparator. A local mismatch between two identical zones of two plates produces a perception of relief. The local displacement is given by the value of the relative movement of one plate in order to make this relief disappear. The accuracy reaches 2-3 micrometers (for more details see [19, 20]). The second technique called speckle laser is also very simple. An extended laser beam illuminates the rough surface of the specimen. The reemitted light, which interferes in the whole space, is recorded on a holographic plate. As the specimen is loaded, the speckle pattern moves in the same way as the surface roughness. A second photo is recorded on the same plate. In a second step a focused laser beam is used to illuminate a local zone of the plate. Due to the small displacement recorded on the film, Young's fringes appear on a screen, the spacing and the orientation of which being related to the local displacement on the surface of the specimen. The accuracy is nearly I micrometer (directly related to the wavelength of the He Ne laser). In the last step the strains are computed using classical interpolation methods.

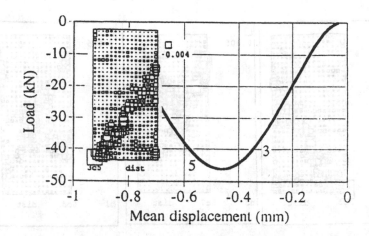

Figure 7. Load displacement curve and typical incremental strain field

Among a lot of tests we consider here a compression test that has been made using the two optical techniques applied on two opposite sides of the same concrete specimen. The load displacement curve is given in figure 7. The incremental strains fields deduced from the incremental displacement maps are comparable and match the mean strain given by classical strain gauges. Distortion and surfacic strains (difference and sum of the two eigen strains respectively) show that localization occurs always around the peak of the load displacement curve. A more refined analysis confirms that the inception of localization (on a macroscopic sense) always starts before the peak.

The softening behavior appears to be a post localized failure behavior. Different questions remain (i) what is the influence of the boundary conditions? (ii) is the localized zone shearing or not? (iii) what is the role of the microstructure?

(i) E.H. Benaija [20] tested four possibilities: high or reduced friction between the specimen and the platen and free rotation or not of the upper platen. In all cases a localized zone appeared before or around the peak load.

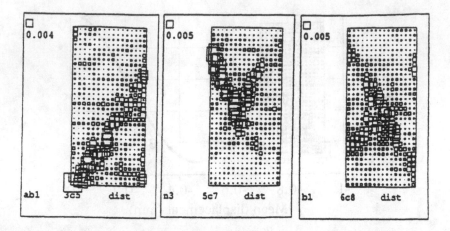

Figure 8. Incremental distorsion fields for various boundary conditions on concrete specimen. The increments are
located around or just after the peak

ab l: no rotation, reduced friction; n: free rotation, high friction;
b: no rotation high friction

(ii) The process of localization is not so clear: the observation of the strain fields indicates a first stage in which a high distortion takes place before an intense dilatation occurs. This can be understood as a similar process to that observed in sand specimens. Due to the cohesion of concrete or rocks, cracking is the only possibility for the material to accommodate the imposed displacement. The beginning of the process looks like that observed in sand specimens while in a second stage the particular behavior of brittle materials is responsible for the intense cracking and as a consequence for the intense and localized crack opening. We have decided to retain this latter phenomenon as the characteristic of localization in these brittle materials. The ratio between the normal strain jump and its tangential counterpart is about 2.5.

(iii) The width of the localized band is directly proportional to the size of the typical heterogeneity of the material [5].

4.2. REPLICA OBSERVATIONS

We have performed various tests using both the replica technique and stereophotogrammetry in order to follow the microcracking process inside and outside the localization band. The specimen has been covered with replicas on one face. After the test is finished, the optical technique gives us the location of the band on the surface of the specimen and also the stage at which it appeared. Doing so it has been possible to analyze replicas concerned or not by the localization process.

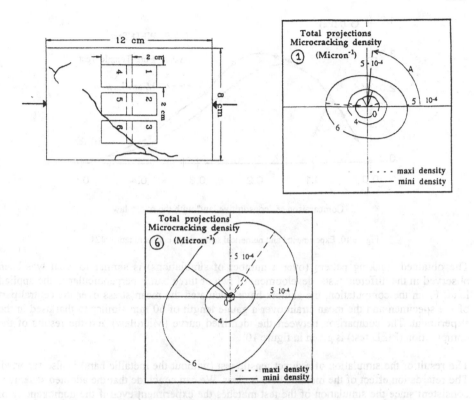

Figure 9. Location of replica and microcracking density

Figure 9 shows the location of the replicas on the specimen, the cracking pattern at localization and the microcracking density. Map 1 is related to replica 1 and map 6 to replica 6. Number 0, 4 and 6 are respectively related to different states: initial, just before the peak and after the peak. It is clear that (i) the cracking density increases in both cases (in and out the band), (ii) the mean orientation of the cracks is different. The mean orientation of created microcracking in the band matches that of the band while a classical splitting is observed outside the band. It must be remarked that the cracks concerned by this technique are microcracks and not macrocracks. The visible cracks at the end of the test (even within the band) are generally axial cracks related to a post-localized propagation.

5. Toward the identification of constitutive equations

5.1 TENSILE TEST

Considering that a constitutive equation exists and may be identified from the experimental stress strain curve (Fig. 5) D. Breysse and N. Schmidt [21] made numerical simulations of this tensile test including some statistics in the damage model (damage threshold in particular).

106

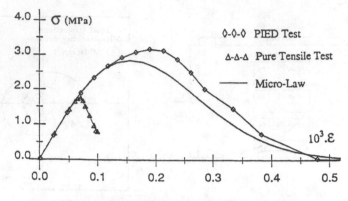

Comparaison of "constitutive law" with the micro law

Figure 10. Experimental and numerical stress strain curves in tension[21]

The obtained cracking pattern (over a number of simulations) is similar to what was been observed in the different tests: development of two or three bands perpendicular to the applied load. From the computation, the authors have calculated the mean stress over the central part of the specimen and the mean strain over a gauge length of 60 mm similar to that used in the experiment. The comparison between the identified curve (Microlaw) and the results of the computation (PIED test) is given in figure 10.

The result of the simulation of a direct tensile test (without the metallic bars) is also reported. The retardation effect of the metallic bars is clear. We can conclude that the adopted scheme is consistent since the simulation of the test matches the experiment even if the homogeneity of the damage field is not achieved.

5.2. COMPRESSION TEST

Another way supposes that a classical Mohr Coulomb model applies for rock material in compression. A classical Rice's localization analysis gives the theoretical relation between the hardening modulus, the characteristics of the model and the experimental data (tangent moduli at localization, angle of the band, initial Young's modulus and Poisson ratio). Similar relations exist for the theoretical angle and the experimental one. It is so possible to plot the evolution of the hardening parameter along the test, considering the Mohr Coulomb model and the experimental data, and to determine the friction angle for example (Fig. 11)[9]. The 35-degrees value is realistic for sandstone.

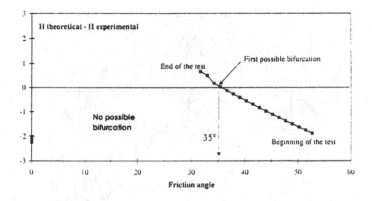

Figure 11. Determination of material Parameters from a localized test

6. Relation between modeling and experiments

The goal of this part of the work, done in cooperation with C. Comi and R. Billardon, was to propose a simple model including damage variable and able to describe the main features of the homogeneous response of brittle materials (evolution of the stiffness and apparition of inelastic stains) and also to capture the localization process i.e. inception, angle and mode of localization [22]. A scalar variable was input in the expression of the free energy, an elliptic load surface (in the first versus the second stress invariant space) was written allowing the description of the hardening-softening effect. The expression of the dissipation potential allows three possibilities for the permanent strains. In the first case there is no permanent strains (classical elasticity coupled to damage); the second case deals with spherical permanent strains and the last one considers more general strains. The identification of the different versions of the model was made using a curve obtained in compression on a prismatic specimen, considering both the hardening and the softening branch, even if this latter is not representative. Then the bifurcation analysis gives the theoretical angle of the band, the inception reported on the curve (black arrow). It is compared to the experimental data (mean value of 50' for the orientation of the band and inception of localization before the peak). It can be seen in figure 12 that it was impossible to match both the angle and the inception of localization, even with such a general modeling.

108

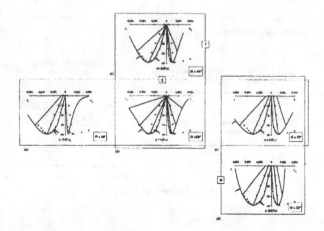

Figure 12. Comparison between experimental and theoretical results for localization in compression [22].

The hypothesis of isotropic damage is possibly a limitation. The interested reader may refer to recent publications to understand the effect of anisotropic damage coupled to elasticity including friction and unilateral conditions on the localization parameters [23]. It seems to be a possible way to reproduce the experimental data.

7. Conclusions

We have shown in this paper (i) that the behaviour of rock like materials is experimentally very difficult to determine because of the interactions between the specimen and the testing system (i.e. stiffness of the testing machine, boundary conditions), (ii) that load displacement curves may be used to deduce stress strain relations in a limited domain before the peak of the load displacement curve due to strain localization, (iii) that special devices may give a good estimation of the existing constitutive response, (iv) the localization characteristics are of interest either to identify classical models or to give information on the structure of the thermodynamical model.

Acknowledgements: The results presented in this paper have been obtained within a GRECO collaboration together with J.M. Torrenti, E.H. Benaija (LCPC Paris), A. Bascoul, M. Massat (LMDC Toulouse) and S. Rainlani, D. Fokwa, C. Fond (LMT Cachan). Cooperation with C. Coini (Milano), R. Billardon, N. Schmidt and D. Breysse (LMT have to be mentioned.

References

1.   Torrenti, J.M. (1987) Comportement multiaxial du béton, aspects expérimentaux et modélisation, Thèse de doctorat, École Nationale des Ponts et Chaussées (France).
2.   Ramtani, S. Berthaud, Y. and Mazars, J. (1992) Orthotropic Behavior of Concrete with Directionnal Aspects: Modeling and Experiments, Nuclear Eng. And Design, 133, 1, 97-111.

3. Horii, H. and Nemat-Nasser, S. (1985) Elastic fields of interacting inhomogeneities, International Journal of Solids and Structures, 21, 7, 731-745.
4. Vonk, R. (1992) Softening, of concrete in compression, Technishe Universiteit Eindhoven.
5. Benaija, E.H., (1992) Application de la stéréophotogrammétrie au béton : cas de la compression simple, Thèse de Doctorat, École Nationale des Ponts et Chaussées (France).
6. Bascoul, A. Benaija, E.H. Berthaud, Y. Torrenti, J.M. and Zizi, Z. (1993) Analysis of localization in concrete through stereophotogrammetry, speckle laser and replica, Cement and Concrete Research, 23, 6, 1340-1350.
7. Berthaud, Y., (1991) Damage in Concrete Via an Ultrasonic Technique - Part I Experiment, Cement and Concrete Research, 21, 1, 73-82.
8. Read, H.E. and Hegemier G.A. 1984) Strain softening of rock, soil and concrete - A review article, Mech. Mater., 3, 271-294.
9. Berthaud, Y. (1991) Mesures et modélisations de l'endommagement des matériaux, Mémoire d'habilitation à diriger des recherches, Université Paris 6 (France).
10. Sayers, C. (1988) Stress induced ultrasonic wave velocity anisotropy in fractured rock, Ultrasonics, 26, 6.
11. Marigo, J.J. (1980) Propagation des ondes ultrasonores et microfissuration du béton, Thèse de 3e cycle, Université Paris 6 (France).
12. Robinson, S.R. (1965) Methods of detecting the formation and propagation of microcracks in concrete, Proc. of the Int. Conf. on the Structure of Concrete, London.
13. Van Mier, J. G. M. (1984). Strain softening under multiaxial loading conditions, Technische Wetenschappen aan de Technische Hogeschool Eindhoven.
14. Andrieux, S. (1983). Un modèle de matériau microfissuré, application aux roches et aux bétons. Thèse de l'École Nationale des Ponts et Chaussées (France).
15. Boudon-Lussac, D. Hild, F. and Pijaudier-Cabot, G. (1999) Tensile damage in concrete: analysis of experimental technique, 125, 8, 906-913.
16. Bazant, Z.P. and Pijaudier-Cabot, G. (1989) Measurement of characteristic length of non local continuum, J. of Engng Mechanic, ASCE, 115, 4, 755-767.
17. Ramtani, S. (1990) Description du comportement du béton endommagé par un modèle d'endommagement anisotrope avec description du caractère unilatéral, Thèse de doctorat, Université Paris 6 (France).
18. Berthaud, Y. Ringot, E. and Fokwa D. (1991) A Test for Delaying Localization in Tension, Experimental Investigation, Cement and Concrete Res., 21, 5, 928940.
19. Desrues, J. (1984) Localisation de la déformation dans les matériaux granulaires, Thèse d'État, Université de Grenoble (France).
20. Torrenti, J.M. and Benaija, E.H. (1990) Stereophotogrametry : a new way to study strain localization in concrete under compression, 9th International Conference on Experimental Mechanics, Ed. by Copenhague (Denmark), Technical University of Denmark Pub., 4, 1346-1354.
21. Breysse, D. and Schmitt, N. (1991). A Test for Delaying Localization in Tension, Numerical interpretation through a probabilistic approach, Cement and Concr. Res., 21, 6, 963-974.
22. Comi, C. Berthaud, Y. and Billardon, R., (1995) On localization in ductile-brittle materials under compressive loadings, European J. Mech. /Solids, 14, 1, 19-43.
23. Dragon, A. and Haim, D. (1995) Anisotropic damage by (micro)crack growth and associated 3D localized failure mechanisms, Proc 10th ASCE Engng. Mechanics Conf. (Boulder, Colorado), vol. 2, Stein Sture Ed., New York, 702-705.

STUDY OF THE BRITTLE-TO-DUCTILE TRANSITION IN CERAMICS AND CERMETS BY MECHANICAL SPECTROSCOPY

R. SCHALLER[1] and G. FANTOZZI[2]

[1] Ecole Polytechnique Fédérale de Lausanne, Institut de Génie Atomique, CH-1015 Lausanne, Switzerland

[2] Institut National des Sciences Appliquées de Lyon, GEMPPM-UMR CNRS 5510, Bât. 502, F-69621 Villeurbanne Cedex, France.

1. Introduction

Ceramics and cermets are refractory materials particularly well suited for applications at high temperature. High temperature plastic deformation of fine-grained ceramics, such as Al_2O_3, ZrO_2 or Si_3N_4, has been interpreted as mainly due to grain boundary (GB) sliding [1 - 5]. In this process, the amorphous intergranular phase, which results from the sintering aids, plays an important lubricating role. The appearance of plasticity due to GB sliding above a certain temperature defines the brittle-to-ductile transition [6].

Anelastic deformation precedes plastic deformation and consequently may be the first manifestation of the brittle-to-ductile transition in ceramics. Mechanical spectroscopy, which refers to anelasticity [7], consists in measuring the internal friction of materials. It is very sensitive to microstructure evolutions, and then well suited for studying a brittle-to-ductile transition, in which energy dissipation is involved.

2. Mechanical spectroscopy

Anelasticity [7] appears in the following simple experiment (Fig.1a). A stress σ, of low intensity, is applied abruptly to a specimen at time, $t = 0$, and held constant while the strain, ε, is recorded as a function of time. One observes the instantaneous elastic strain, $\varepsilon_e = J_u \sigma$, where J_u is the unrelaxed compliance, and the anelastic strain, ε_a, which increases with time from zero to an equilibrium value ε_a^∞. When equilibrium is reached, $\varepsilon = \varepsilon_e + \varepsilon_a^\infty = J_r \sigma$, where J_r is the relaxed compliance. This evolution from one equilibrium state to a new one, under an applied stress, is called anelastic relaxation, and may be defined by two parameters, the relaxation time τ, and the relaxation strength Δ:

$$\Delta = \frac{\varepsilon_a^\infty}{\varepsilon_e} = \frac{J_r - J_u}{J_u} \qquad (1)$$

If, the stress is removed after a certain delay, one can observe the instantaneous recovery of the elastic strain and with the relaxation time τ, the recovery of the anelastic strain. In such an experiment, strain is completely recoverable.

From a microscopic point of view, anelastic strain can be interpreted as being due to the movements of structural defects (elastic dipoles, dislocations, interfaces) from one equilibrium position defined at $\sigma = 0$

E. Bouchaud et al. (eds.), Physical Aspects of Fracture, 111–121.

to another one defined at σ . The relaxation strength, •, is then proportional to the concentration of defects, which are relaxing; the relaxation time τ accounting for their mobility.

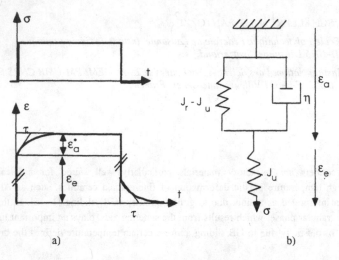

Figure 1. a) Elastic ε_e and anelastic ε_a strain of a solid submitted to stress σ.
b) Rheological model corresponding to the anelastic behavior of figure 1a.

From a rheological point-of-view, the solid can be represented by the model shown in Fig. 1b. The spring "J_r - J_u" in parallel with the dashpot of viscosity η is essential to the recoverable nature of the anelastic strain. If this spring does not operate, the anelastic strain does not reach an equilibrium value and is then no longer completely recoverable.

The anelastic behavior in figure 1 is representative of the standard anelastic solid [7], the equation of which is:

$$\varepsilon + \tau \dot{\varepsilon} = J_r \sigma + \tau J_u \dot{\sigma}$$

(2)

Measurements performed in the way described in figure 1 are very delicate, and it is more convenient to use in practice dynamic methods for measuring the relaxation parameters. If an alternative stress, of circular frequency ω, $\sigma = \sigma_0 \exp(i\omega t)$ is applied to the system, the linearity of the stress-strain relationships assures us that strain ε is periodic with the same frequency: $\varepsilon = \varepsilon_0 \exp[i(\omega t - \phi)]$, where ϕ is the phase lag of strain behind stress (due to anelasticity). Introducing these expressions of σ and ε in equation (2) leads to the following relationship:

$$\varepsilon = \left[\left(J_u + \frac{J_r - J_u}{1 + \omega^2 \tau^2} \right) - i \cdot \left(\frac{(J_r - J_u) \cdot \omega \tau}{1 + \omega^2 \tau^2} \right) \right] \cdot \sigma = \left[J_1(\omega) - i J_2(\omega) \right] \cdot \sigma = J*(\omega) \cdot \sigma \quad (3)$$

where J_1 and J_2 are, respectively, the real and the imaginary part of the complex compliance $J*$.
In metals and ceramics $J_r - J_u \ll J_u$, and the mechanical loss angle ϕ is given by:

$$\tan \phi = \frac{J_2}{J_1} = \Delta \frac{\omega \tau}{1 + \omega^2 \tau^2}$$

(4)

and the variation of the dynamical compliance $\delta J/_J$ due to anelasticity by:

$$\frac{\delta J}{J} = \frac{J_1(\omega) - J_u}{J_u} = \Delta \frac{1}{1 + \omega^2 \tau^2} \tag{5}$$

Mechanical loss, tanϕ, presents a maximum as a function of $\omega\tau$, centered at $\omega\tau = 1$. This maximum or peak gives us the relaxation strength, •, (height of the peak) and the relaxation time, τ, (position of the peak on the $\omega\tau$ axis). If several relaxation mechanisms are activated, the material exhibits a *mechanical loss spectrum* composed of damping peaks, which inform about microstructure dynamics. As most of the relaxation mechanisms are thermally activated,

$$\tau = \tau_0 \cdot \exp\left(\frac{H}{kT}\right) \tag{6}$$

H being the activation enthalpy. As a consequence, tanϕ can be also measured as a function of temperature holding ω constant and the curve tanϕ = tanϕ (T) is the mechanical loss spectrum described as a function of temperature. For an example, see the mechanical loss spectrum of a zirconia specimen in figure 2.

The internal friction IF (or internal damping) is defined as:

$$IF = \frac{1}{2\pi} \cdot \frac{W_{diss}}{W_{el}} \tag{7}$$

where W_{diss} is the energy dissipated in a volume unit during one cycle of vibration and W_{el} is the maximum stored elastic energy per unit volume.

$$W_{diss} = \oint \sigma \cdot d\varepsilon = \pi \cdot \sigma_0 \cdot \varepsilon_0 \cdot \sin\phi = \pi \cdot J_2 \sigma_0^2 \tag{8}$$

$$W_{el} = \int_0^{\sigma_0} \sigma \cdot d\varepsilon = \frac{1}{2} J_1 \sigma_0^2 \tag{9}$$

It follows immediately that: $\qquad\qquad IF = \tan\phi \qquad\qquad$ (10)

3. The mechanical loss spectrum of zirconia

Figure 2 shows a typical mechanical loss spectrum (tanϕ and elastic shear modulus G=1/J) measured as a function of temperature in fine-grained zirconia (2Y-TZP: fully tetragonal zirconia stabilized by 2 mol % of yttria) [8, 9]. The measurements were performed at a frequency of 1 Hz by means of a forced torsion pendulum described elsewhere [10]. The mechanical loss spectrum of TZP zirconia is composed of a damping peak at about 350 K associated with a modulus defect, and of an exponential increase in damping associated with a steep decrease in the shear modulus at high temperature.

The 350 K peak is a relaxation peak, which obeys the equations (4) and (5). Values of H = 0.93 eV and $\tau_0 = 0.7 \, 10^{14}$ have been obtained by an Arrhenius plot (eq. (6)). They are characteristic for point defect relaxation, and then the peak was interpreted as due to the reorientation of "oxygen vacancy - yttrium cation" elastic dipoles under the influence of the applied stress [11].

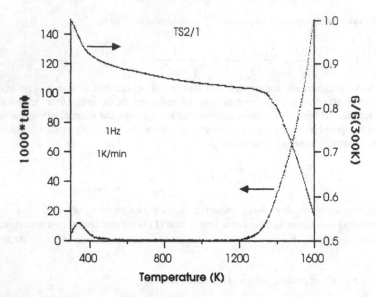

Figure 2. Mechanical loss (tanΦ) and elastic shear modulus (G) as functions of temperature in 2Y-TZP zirconia (measurements performed in a forced torsion pendulum at 1 Hz) [8].

The exponential increase in damping associated with the steep decrease in the shear modulus at high temperature account for the appearance of ductility in the ceramic material: "brittle-to-ductile" transition. In fact an exponential increase in tanφ means an increase in the anelastic strain, which does not reach an equilibrium value: onset of creep. Creep rate can be described by the classical power law relation:

$$\acute{\varepsilon}(\sigma, T) = A \frac{\sigma^n}{d^p} \exp\left(-\frac{H_{act}}{kT}\right) \qquad (11)$$

where n is the stress exponent, σ the applied stress, p the grain size parameter, d the grain size, H_{act} the apparent activation enthalpy, k the Boltzmann's constant and A a material constant. Equation (11) with equations (7) to (10) can be used to calculate the internal friction IF or the mechanical loss tanφ,

$$\tan \phi = A' \frac{G \cdot \sigma^{n-1}}{\omega \cdot d^p} \exp\left(-\frac{H_{act}}{kT}\right) \qquad (12)$$

where G is the shear modulus, ω the circular frequency and A' a material constant.

Equation (12) has an exponential form, which fits well to the experimental results obtained in the high temperature domain where creep is enhanced. Plotting ln(tanφ) as a function of 1/T leads to a straight line, the slope of which would give the activation enthalpy. However the value of H_{act} obtained by this way is an apparent value, which is too small. The correct value of 586 kJ/mol [9] for the activation enthalpy was obtained from the "temperature-frequency" shifts of the mechanical loss spectra. This value corresponds well to the ones obtained by creep tests in the low stress regime, i.e. 560-580 kJ/mol [12, 14].

In order to determine the origin of damping, it is possible to compare the mechanical loss spectrum of the ceramic with the spectra of the components: grains and GBs.

For zirconia the comparison in figure 3 of the high temperature spectra of poly- and single-crystals shows clearly that the high temperature background is due to GBs [15].

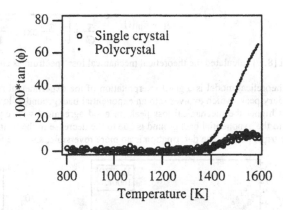

Figure 3. Mechanical loss as a function of the temperature in a poly- (dots) and in a single- crystalline zirconia (markers) [15]

Figure 4. Model of hexagonal grains separated by an intergranular layer of viscosity η

A theoretical model was developed by Lakki [8, 9] to account for this mechanical loss as due to the relative sliding of grains of hexagonal shape (Figure 4). If u is the relative displacement of two grains, τ the applied shear stress, η the interface viscosity, δ the intergranular layer thickness and $K \cdot u$ the restoring force due to the limiting grains at triple junctions, the motion equation is:

$$\tau = \frac{\eta}{\delta} \dot{u} + K u \tag{13}$$

The mechanical loss $\tan\phi$ is then directly obtained from the phase lag between stress τ and strain ε:

$$\varepsilon = \frac{\tau}{G} + \frac{u}{d} = \varepsilon_0 \cos\left(\omega t - \phi\right) \tag{14}$$

And:

$$\tan\phi = \frac{G}{d} \frac{\omega \frac{\eta}{\delta}}{\left(\frac{KG}{d} + K^2\right) + \omega^2 \frac{\eta^2}{\delta^2}} \tag{15}$$

$\tan\phi$ has the form of a peak with a maximum at

$$\omega = \frac{\delta}{\eta} \sqrt{\frac{KG}{d} + K^2} \tag{16}$$

However, the high temperature mechanical loss spectrum in zirconia does not exhibit a peak, but an exponential increase with temperature (Fig. 2). In fact, isothermal measurements performed at 1600 K show that the spectrum is composed of a peak and an exponential form at lower frequencies (Figure 5a).

It is possible to account for this spectrum by considering that the restoring force "Ku" is reduced at high temperature, where creep takes place. Assuming that K is proportional to the shear modulus $G(t) = \tau/\varepsilon(t)$,

$$K(t) = c_1 \frac{\tau}{\varepsilon_{el} + \varepsilon_{pl}(t)} = c_1 \frac{\tau}{\frac{\tau}{G} + \left[A \frac{\tau^2}{d} \exp\left(-\frac{E_{act}}{kT}\right)\right] \cdot t} \tag{17}$$

Lakki [8, 9] calculated the theoretical mechanical loss spectrum as shown in figure 5b.

The theoretical model is a good interpretation of the experimental results as shown in figure 5a: a grain boundary peak, which evolves into an exponential background at lower frequency. In figure 5a, finer the grains higher the mechanical loss peak, in good agreement with equation (15). The transition from the peak to the exponential background is due to the decrease in the restoring force "K" due to the weakening of the triple points. The deformation is no more reversible, and creep occurs.

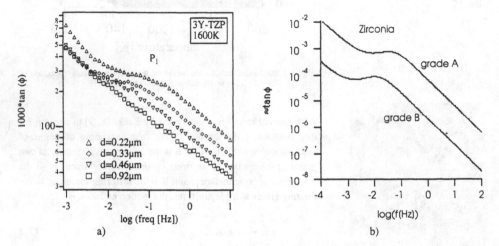

Figure 5. Mechanical loss spectra of 3Y-TZP zirconia measured at 1600 K for different grain sizes d (a), and (b) theoretical spectra calculated from equations (15) and (17).

Al_2O_3 with a grain structure similar to TZP zirconia, exhibits a similar mechanical loss spectrum at high temperature. However the level of damping is lower than in 2Y-TZP zirconia because most of the GBs in Al_2O_3 are devoid of any amorphous phase. Damping due to GB sliding in Al_2O_3 has been interpreted by a GB dislocation model [8, 16].

In Si_3N_4 the sintering aids generate an important volume fraction of amorphous phase (of YSiAlON type glass). This phase is not only located as thin films between the grains but also at the triple points where large glassy pockets have been observed. These pockets are responsible for a well defined mechanical loss peak at about 1250 K [17]. Comparing the peak characteristics, with the α-relaxation in a YSiAlON glass, Donzel [18] was able to conclude that the peak is due to the α-relaxation in the glassy pockets. Near the glass transition the pockets soften, but the ceramic skeleton remains elastic. As a consequence the exponential background and the creep rate are lower than in 2Y-TZP or Al_2O_3.

Ceramic-metal composites, like for instance WC-Co cemented carbides, show mechanical loss spectra similar to those of ceramics [19 - 21]. In figure 6a are reported the mechanical loss spectra of WC-11wt% Co and of the same material after extraction of the Co binder phase by chemical etching. A mechanical loss peak is observed in WC-11wt% Co superimposed on an exponential high temperature background (curve 1). These two components of the spectrum have disappeared with Co (curve 2), and the conclusion is that the dissipation mechanisms take place in the cobalt phase. From in situ transmission electron microscopy observations it was possible to interpret the damping mechanisms as due to dislocations in Co [22]. When the dislocation movements are limited by a restoring force, they give rise to the damping peak. At higher temperature, these dislocations do not experience a restoring force and the damping increases exponentially.

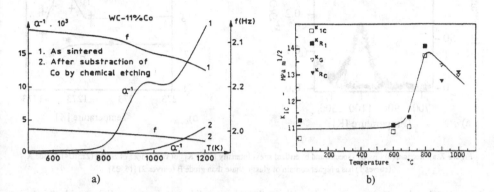

a)   b)

Figure 6. a)Mechanical loss (Q-1) and vibration frequency (f) of WC-11wt%Co, as sintered (curves 1) and after removal of the Co binder phase by chemical etching (curves 2) [21, 22].

b) Critical stress intensity factor K_{Ic} (MPa $m^{1/2}$) as a function of temperature in WC-6%Co [23].

4. Mechanical loss and toughness

An exponential increase in mechanical loss at high temperature means a bad creep resistance of the material, because the anelastic or microplastic strain has no limit. On the other hand, damping mechanisms are positive factors for improving toughness in hard and brittle materials. As a mechanical loss peak is associated with a restricted motion of structural defects, it cannot be associated with the onset of creep. It accounts for the ability of the material to dissipate locally a part of the vibration energy and hence to improve toughness by crack propagation blunting.

As a matter of fact, the mechanical loss peak in WC-Co (Fig. 6a) appears in the same temperature range where an increase in toughness has been observed (Fig. 6b) [23, 24].

Figure 7a shows the mechanical loss spectra of two grades of zircon (ZrSiO4) [15, 25] differing by the content of the intergranular glassy phase. Figure 7b shows that a K_{Ic} maximum is observed only in the grade, which exhibits a mechanical loss peak [25].

118

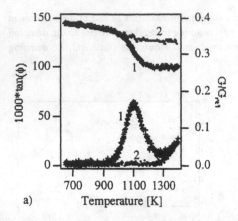
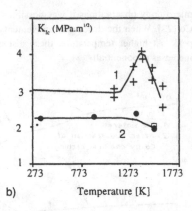

a)   Temperature [K]

b)   Temperature [K]

Figure 7. a) Mechanical loss Q^{-1} spectra and b) critical stress intensity factor K_{Ic} of two grades of zircon (ZrSiO4). Grade A (curves 1) has a higher contain of glassy phase than grade B (curves 2) [15, 25].

As mechanical loss accounts for energy dissipation in the solid, one can try to link it with the strain energy release rate G:

$$G_{Ic} = -\frac{dU}{h \cdot da} \tag{18}$$

where dU represents the potential energy release when the crack grows of "da", the specimen thickness being of "h". In brittle materials, an increase in the potential energy release ΔG_{Ic} can be due to an increase in the energy dissipation ΔW_{diss} in a volume V in front of the crack tip. When ΔW_{diss} is due to internal friction mechanisms, the equation (7) can be used for calculating ΔW_{diss}. Then:

$$\Delta G_{Ic} = -\frac{\Delta(dU)}{h \cdot da} = \frac{V \cdot \Delta W_{diss}}{h \cdot da} = \frac{2\pi \cdot V \cdot W_{el}}{h \cdot da} \cdot \Delta \tan \phi \tag{19}$$

where W_{el} is the maximum stored elastic energy and $\Delta \tan \phi$ the increase in damping, respectively. Considering that the volume V is a cylinder of radius r and height h, which is translated of da when crack propagates, the increase ΔG_{Ic} of the strain energy release rate G is:

$$\Delta G_{Ic} = 4\pi \cdot r \cdot W_{el} \cdot \Delta \tan \phi \tag{20}$$

Taking into account the height of the mechanical loss peak $\Delta \tan \phi$ (Fig. 7a), and considering that crack propagation in zircon is enhanced under an applied stress of 200 MPa, one obtains $\Delta G_{Ic} = \sim 7$ to 70 Jm^{-2} depending on the value of the zone radius $r = \sim 0.1$ to 1 mm respectively. On the other hand, in the case of elastic materials, it is possible to derive ΔG_{Ic} from the increase in K_{Ic} associated with the toughness peak. One obtains $\Delta G_{Ic} = 32$ Jm^{-2}, value of the same order of magnitude as the one predicted from mechanical loss peak. In fact, it is not evident to get a quantitative value of ΔG_{Ic} from the mechanical spectroscopy (equation (20)) because of the estimation of the radius r of the dissipative zone, which has to be larger than the plastic zone considered in the models of fracture mechanics. However, from a qualitative point of view, it appears that a mechanical loss peak is beneficial for toughness. For instance, it has been shown that the ability of Si3N4 to dissipate energy in the glassy phase decelerates strain accumulation under cyclic loading and consequently improves fatigue resistance [26, 27].

Besides the simple consideration that a mechanical loss peak may be associated with a K_{Ic} peak in hard materials, a question arises as it concerns the peak temperatures. K_{Ic} peak appears at a temperature higher (by more than 100 K) than the mechanical loss peak measured at a frequency of 1 Hz. Crack propagation in hard and brittle materials is associated with the emission of acoustic waves of much higher frequency (1kHz to 20 kHz) than the frequency of a torsion pendulum. In brittle polymers it has been shown that the emission spectrum associated with crack propagation covers a range between 100 Hz and 10 kHz, and peaks at about 5 kHz [28]. Taking into account the activation enthalpy of the mechanical loss peak (Fig. 7a), it is possible to calculate the frequency for shifting this peak to the K_{Ic} peak temperature (Fig. 7b). One obtains a frequency of 6 kHz, which is in good agreement with the crack propagation frequency spectrum of brittle materials.

5. Conclusions

In ceramics and cermets, the high temperature brittle-to-ductile transition gives rise to a mechanical loss peak, which evolves, into an exponential background at higher temperature [9, 22]. From the viewpoint of mechanical spectroscopy the brittle-to-ductile transition can be decomposed into two stages: a brittle-to-"tough" transition associated with the relaxation peak and a "tough"-to-ductile transition associated with the exponential background. Structural ceramics and cermets have to be tough and creep resistant. In other words they have to exhibit a mechanical loss peak with a low level of the high temperature background.

In the case of WC-Co, it was shown that the mechanical loss was due to dislocation motion in the Co phase [22]. By alloying the cobalt binder phase it was possible to stabilize the dislocation microstructure and improve the creep resistance without loosing toughness. In the case of Si_3N_4, it was shown that the mechanical loss peak was due to the α-relaxation in the pockets of YSiAlON type glass [17, 18]. Crystallization of these glassy pockets by thermal treatments led to an improved creep resistance [29]. Fine-grained Al_2O_3 and 2Y-TZP present a mechanical loss peak, which is not well resolved with respect to the exponential background. In these ceramics the regular shape of the grains does not provide a strong restoring force in the GB sliding process and then extended creep can be observed. In 2Y-TZP, GB sliding is lubricated by amorphous intergranular films and anelastic relaxation is affected by the glass transition in these films [9]. As these films are very stable, one way to limit GB sliding is to introduce hard particles or whiskers, for instance of SiC type, at the triple points [30]. In Al_2O_3, the GBs are devoid of any amorphous phase, and then GB peak was interpreted by a model of GB dislocations [16]. It was found that doping alumina with yttria modifies the GB dislocation mobility. The high temperature mechanical loss decreases with yttrium addition [31] and so does the creep rate. It was also observed that additions of SiC whiskers on the GBs suppress extensive GB sliding and reduce the creep rate [32].

To conclude, the brittle-to-ductile transition in hard materials like ceramics or ceramic composites can be studied by mechanical spectroscopy. The correct research strategy is to compare carefully the damping spectrum with the mechanical behavior determined by other techniques (creep tests, toughness tests). Moreover transmission electron microscopy observations are of particular interest in modeling the damping mechanisms. Damping is important for toughness and for fatigue resistance of ceramics [33].

Acknowledgements
This work has been partially supported by the Swiss National Science Foundation.

120

References

1   Wakai,F., Sakaguchi, S., and Matsuno, Y. (1086) Superplasticity of Yttria-Stabilized Tetragonal ZrO2 Polycrystals, *Advanced Ceramic Materials* **1**, 259-263.

2   Carry, C. and Mocellin, A., (1987) Structural Superplasticity in Single Phase Crystalline Ceramics, *Ceramics International* **13**, 89-98.

3   Duclos, R., Crampon, J., and Amana, B., (1989) Structural and Topological Study of Superplasticity in Zirconia Polycrystals, *Acta Metall.* **37**, 877-883.

4   Boutz, M.M.R., Winnubst, A.J.A., Burggraaf, A.J., Nauer, M., and Carry, C., (1994) Low Temperature Superplastic Flow of Yttria Stabilized Tetragonal Zirconia Polycrystals, *J. Europ. Ceram. Soc.* **13**, 103-111.

5   Gasdaska, Ch. J. (1994) Tensile Creep in an in Situ Reinforced Silicon Nitride, *J. Am. Ceram. Soc.* **77**, 2408-2418.

6   Rouxel, T. and Wakai, F., (1993) The Brittle to Ductile Transition in a Si,N,/SiC Composite with a Glassy Grain Boundary Phase, *Acta Metall. Mater.*, **41**, 3203-3213.

7   Nowick, A.S. and Berry, B.S. (1972) *Anelastic Relaxation in Crystalline Solids*, Academic Press, New York.

8   Lakki, A. (1994) *Mechanical spectroscopy of fine-grained zirconia, alumina and silicon nitride*, PhD thesis, EPF - Lausanne, No 1266.

9   Lakki, A., Schaller, R., Nauer, M.,and Carry, C. (1993) High temperature superplastic creep and internal friction of yttria doped zirconia polycrystals, *Acta Metall. Mater.* **41** 2845-2853.

10 Gadaud, P., Guisolan, B., Kulik, K., and Schaller, R. (1990) Apparatus for high-temperature internal friction differential measurements, *Rev. Sci. Instrum.* **61**, 2671-2675.

11 Weller, M. (1994) Mechanical loss measurements on yttria- and calcia-stabilized zirconia, *Journal of Alloys and Compounds* **211/212**, 66-70.

12 Bravo-Leòn, A., Jiménez-Melendo, M., and Dominguez-Rodriguez, A. (1992) Mechanical and microstructural aspects of the high temperature plastic deformation of yttria-stabilized zirconia polycrystals, *Acta Metall. Mater.* **40**. 27172726.

13 G. Schoeck, E. Bisogni and J. Shyne, *Acta Metall.*, *12* (1964) 1466.

14 Nauer, M. and Carry, C. (1990) Creep parameters of yttria doped zirconia materials and superplastic deformation mechanisms, *Scripta Metall. Mater.* **24**, 1459-1463.

15 Donzel, L. (1998) *Intra- and Intergranular High Temperature Mechanical Loss in Zirconia and Silicon Nitride*, PhD thesis, EPF, Lausanne, No 1851.

16 Lakki, A., Schaller, R., Carry, C. and Benoit, W. (1998) High Temperature Anelastic and Viscoplastic Deformation of Fine-Grained MgO-doped Al₂O₃, *Acta Mater.*, **46**, 689-700.

17 Lakki, A., Schaller, R., Bernard-Granger, G., and Duclos, R. (1995) High temperature anelastic behaviour of silicon nitride studied by mechanical spectroscopy, *Acta Metall. Mater.* **43**, 419-426.

18 Donzel, L., Lakki, A., and Schaller, R. (1997) Glass transition and α-relaxation in YSiAlON glasses and in Si₃N₄ ceramics studied by mechanical spectroscopy, *Phil. Mag. A*, **76** , 933-944.

19 Ammann, J.J. (1990) *Etude des propriétés mécaniques du matériau composite WC-Co par frottement intérieur*, PhD thesis, EPF, Lausanne, No 861.

20 Gardon, M., (1993) *Etude par spectrométrie mécanique de l'évolution structurale à haute température du composite WC-Co*, PhD thesis, INSA, Lyon, No 93 ISAL 0013.

21 Schaller, R., Ammann, J.J., and Bonjour, C. (1988) Internal Friction in WC-Co hardmetals, *Mat. Sci. Eng. A*, **105/106**, 313-321.

22 Schaller, R., Mari, D., Maamouri, M, and Ammann, J. J. (1992) Mechanical Behaviour of WC-11wt%Co Studied by Bending Tests, Internal Friction and Electron Microscopy, *J. of Hard Materials* **3**, 351-362.

23 Fantozzi, G., Si Mohand, H., and Orange, G. (1986) High temperature mechanical behaviour of WC - 6 wt% Co cemented carbide, in E.A. Almond, C.A. Brookes and R. Warren (eds), *Science of Hard Materials*, Inst. Phys. Conf. Ser., **75**, pp. 699-712.

24 Johanesson, B. and Warren, R. (1986) Fracture of harmetals up to 1000 °C, in E.A. Almond, C.A. Brookes and R. Warren (eds), *Science of Hard Materials*, Inst. Phys. Conf. Ser., **75**, pp. 713-723.

25 Carbonneau, X. (1997) *Etude des propriétés thermomécaniques de mullite zircone et de zircon*, PhD thesis, INSA, Lyon, No 97 ISAL 0105.

26 Roebben, G. (1999) *Viscous energy dissipation in silicon nitride at high temperature*, PhD thesis, KUL, Leuven.

27 Roebben, G., Donzel, L., Stemmer, S., Steen, M., Schaller, R., and Van der Biest, O. (1998) Viscous energy dissipation at high temperatures in silicon nitride, *Acta Mater.* **46**, 4711-4723.

28 Woo, L., Westphal, S., and Ling, M.T.K. (1994) Dynamical Mechanical Analysis and Its Relationship to Impact Transitions, *Polymer Eng. and Sci.* **34**, 420-423.

29 Besson, J.L., Rouxel, T., and Goursat, P. (1998) Ductility and creep resistance of a silicon nitride ceramic, *Scripta Mater.* **39**, 1339-1343.

30 Descamps, P., O'Sullivan, D., Poorteman, M., Descamps, J.C., Leriche,.A., and Cambier, F., Creep Behaviour of Al_2O, - SiC Nanocomposites, *J. Eur. Ceram. Soc.* **19**, 2475-2485.

31 Lakki, A., Schaller, R., Carry, C., and Benoit, W. (1999) High-Temperature Anelastic and Viscoplastic Deformation of Fine-Grained Magnesia- and Magnesia/Yttria-Doped Alumina, *J. Am. Ceram. Soc.* **82**, 2181-2187.

32 Xia, K. and Langdon, T. G. (1995) High temperature deformation of an alumina composite reinforced with silicon carbide whiskers, *Acta Metall. Mater.* **43**, 1421-1427.

33 Roebben, G., Steen, M., Bressers, J., and Van der Biest, O. (1996) Mechanical fatigue in monolithic non-transforming ceramics, *Progress in Mater. Sci.* **40**, 265-331.

DUCTILE FRACTURE

FRACTURE OF METALS
Part II : Ductile Fracture

Dominique FRANCOIS
Laboratoire Matériaux – École Centrale de Paris
UMR CNRS 8579
Grande Voie des Vignes
92295 Châtenay Malabry Cedex (France)

André PINEAU
Centre des Matériaux – École des Mines
UMR CNRS 7633
BP 87 – 91003 Evry Cedex (France)

Abstract

Ductile fracture results from the nucleation, growth and coalescence of cavities. Nucleation starts from inclusions. Mechanics of inclusions is used to write the nucleation criterion. The growth rate can be derived from the deformation of a cavity in a plastic or viscous material. Plastic potentials of a porous plastic solid yields the rate of increase of the porosity. Recent studies introduce anisotropy of plastic behaviour or of morphology in the analysis. Coalescence of cavities occurs by flat dimple mode or by void sheet instability. The criteria used are either a critical growth rate of porosity or a condition of plastic instability. Finite elements calculations together with the preceding criteria are used to calculate the growth of a crack.

Keywords

ductile fracture / cavities/ voids / dimples / nucleation / growth / coalescence / local approach / fracture mechanics / porous materials

1.   Introduction

Ductile fracture is essentially connected to glide mechanisms. This is exemplified in the extreme case of a deeply notched tensile specimen made of a non-hardening material (Fig. 1a).

A single glide band develops inclined at 45° to the tensile axis. In the case of a hardening material the plastic deformation takes place on several glide systems (Fig. 1b) and is more diffuse than in the preceding case.

E. Bouchaud et al. (eds.), Physical Aspects of Fracture, 125–146.

126

a)   b)

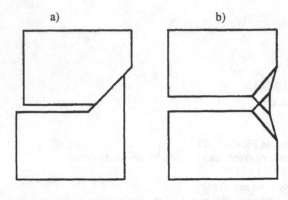

a

Figure 1. a) Development of a single shear band at the tip of a deeply notched tensile specimen made of a non hardening material. b) More diffuse plastic shear deformation in a hardening material.

In both cases the separation of the specimen in two pieces is due to the full extent of these shearing mechanisms. Considering a smooth tensile bar, at the maximum load a local necking appears, the strain hardening of the material is not matching anymore the decrease of the load bearing area of the specimen owing to plastic deformation. This plastic instability leads to the separation of the specimen in two pieces, in some cases by complete extent of the neck down to a point. In most cases however the fracture starts sooner, by the development of a crack in the mid-section of the neck, ending again by a shearing mechanism. The characteristic cup and cone fracture results (Fig. 2a).

An examination at a lower scale, by sectioning along the axis of the broken specimen and the use of a microscope, reveals the presence of holes elongated in the tensile direction and a shattered crack surface (Fig. 2b).

a)   b)

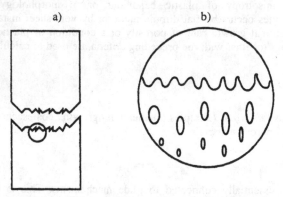

Figure 2. a) Section through the tensile axis of a broken specimen showing the typical cup and cone fracture. b) View at a lower scale of the longitudinal midsection through the fracture surface showing cavities.

Looking down at the crack surface through a scanning electron microscope shows characteristic dimples (Fig. 3). Thus the fracture in the mid-section of the test piece results from the growth of microscopic cavities which join together.

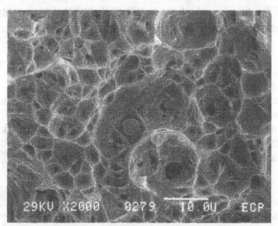

Figure 3. Typical dimples seen on the fracture surface through a scanning electron microscope. Austenitic stainless steel. A carbide particle is clearly seen in the center of the picture.

The study of ductile fracture deals with the nucleation, the growth and the coalescence of these cavities. These three aspects will now be described in more details. It must be emphasized that at this scale the mechanisms are essentially gliding on crystallographic planes owing to the displacement of dislocations. However they take place at a time such that the plastic deformation is rather homogeneous and can be treated by continuum mechanics approaches.

2. Nucleation of Cavities

We have seen in the general introduction that the creation of a cavity nucleus requires a very high stress. In pure ductile materials the deformation is not heterogeneous enough for large stress concentrations to reach the atomic bonds strength. However inclusions provide areas where this is the case.

Hard inclusions are obstacles to dislocations. Pile-ups can produce large stress concentrations in the inclusion itself and at its boundary, large enough to cleave it or for interface decohesion. It can be shown that a slip line of some ten microns length can trigger cleavage in a brittle inclusion. However such an event assumes quite an heterogeneous deformation, such that it is likely that the cleavage initiated in the inclusion will spread out in the outside matrix. The situation for cavity nucleation corresponds to a more homogeneous deformation, and, unless the interface is very weak, as it is the case in spheroidal graphite cast iron [1], it will take place when the plastic deformation of the matrix has raised enough the stress in the inclusion.

128

The problem can be treated in the framework of continuum solid mechanics. Eshelby [2] showed that the stress $\sigma_{ij}{}^I$ in an ellipsoidal inclusion, in linear elasticity, is homogeneous and related to the applied remote stress Σ_{ij} by a stress localization tensor which is a function of the elastic constants of the matrix and of the tensor of Eshelby, S_{ijkl}, depending itself on the Poisson's ratio of the matrix and on the shape of the inclusion. For a spheroidal inclusion, this simplifies to:

$$\sigma_{ij} = \frac{5\mu^I}{3\mu + 2\mu^I} \Sigma_{ij} \tag{1}$$

if the Poisson's ratio takes the value 1/2, which will be the case in plasticity. μ^I and μ are the shear moduli of the inclusion and of the matrix, respectively.

This was extended to plasticity [3] by considering that the behaviour was that of an elastic medium with Poisson's ratio equal to 1/2 and Young's modulus to the linear hardening slope E_p. In such case:

$$\sigma_{ij} = \Sigma_{ij} + E_p \left(S_{ijkl}^{-1} - \delta_{ijkl}\right) E_{kl} \tag{2}$$

for a rigid inclusion. E_{ij} is the remote strain. Other cases and the derivation of corresponding formulae are detailed in a comprehensive fashion in the thesis of Bourgeois [4].

This expression yields in the case of axial symmetry :

$$\sigma = \Sigma_m + \frac{2}{3} \Sigma_{eq} + k E_p E_{eq}^p \tag{3}$$

where Σ_m is the hydrostatic stress $(\Sigma_z + 2\Sigma_r)/3$, Σ_{eq} the equivalent Von Mises stress $(\Sigma_z - \Sigma_r)$ and $E_{eq}{}^p$ the equivalent plastic strain. k is a shape parameter depending on the excentricity $s = a/b$ of the inclusion, defined as the ratio of its axial to its radial dimensions (Fig. 4). Table 1 gives the values of k for different cases.

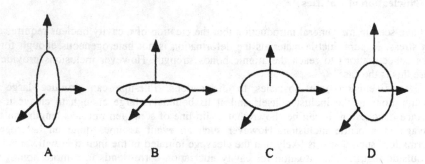

Figure 4. Expression of the shape parameter k for the stress concentration in rigid ellipsoidal inclusions for various values of their excentricity expressed as the ratio s of the axial to the radial dimension. Cases A and B, s>1 ; cases C and D, s<1. The loading is axisymmetric.

case	k	k
prolate inclusion (s>1)		s = 7
direction : longitudinal (A)	$\dfrac{2}{3}\left[\dfrac{1}{3}\dfrac{1+2s^2}{2\log(2s-1)-1}-1\right]$	4,66
transverse (B)	$\dfrac{1}{2}\left[\dfrac{1}{9}\dfrac{1+2s^2}{2\log(2s-1)-1}+1\right]$	1,83
spherical inclusion (s=1)	1	
oblate inclusion (s<1)		s = 1/7
direction : face (C)	$\dfrac{2}{3}\left(\dfrac{4}{3\pi s}-1\right)$	1,31
side (D)	$\dfrac{2}{3}\left(\dfrac{10}{3\pi s}-1\right)$	4,28

Table 1. Values of the shape factor k for different excentricities s of inclusions (A,B,C and D refer to Fig. 4).

As $E_p E_{eq}{}^P = \Sigma_{eq} - R_p$, R_p being the yield strength, the fracture criterion can be written, as proposed by Beremin [5]:

$$\Sigma_R + k\left(\Sigma_{eq} - R_p\right) = \sigma_d \tag{4}$$

where σ_d is the fracture stress of the inclusion or of the interface. Σ_R is the remote applied principal stress which initiates the fracture of inclusions.

As shown in the table the worst case is that of the longitudinal direction of elongated inclusions. Indeed manganese sulfides in steels, which were elongated by plastic deformation during hot working, fracture by cleavage when the plate is loaded in the longitudinal direction. Recent numerical calculations [6] showed that the stress concentration in the inclusion is larger than the stress concentration at the interface, especially for elongated particles. This would favour fracture of inclusions with respect to decohesion. The relative value of the respective critical stresses plays of course a role. Impurities migrating to the interface can embrittle it (Hydrogen or Phosphorus at the interface of cementite as an example [7]).

It should be checked that the strain energy released when an inclusion breaks is at least equal to the fracture energy. As the first is proportional to the volume of the inclusion whereas the second is proportional to its surface, the criterion is size dependent and requires an applied stress the larger the smaller the size of the inclusion [8]. Thus it supersedes the stress criterion for small inclusions. However this happens for smaller sizes than the usual ones.

A size independent critical stress criterion (Eq. 4) being relevant, it would appear that the dimensions of inclusions has no influence. In that case it would be expected that nucleation would be a discontinuous process. However studies in which the proportion of broken inclusions was measured as a function of their size [9] showed that, at a given level of the applied load, this fraction is an increasing function of the size of the inclusions (Fig. 5) (see also references [10] and [11]). This can be explained by the stochastic nature of the fracture of brittle inclusions. The fracture stress distribution was found to be consistent with a Weibull law. The same was found in aluminium-silicon alloys (Fig. 6) [12].

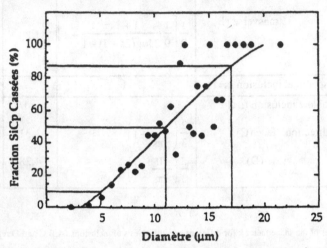

Figure 5. Fraction of broken SiC particulates as a function of their size in an Al based composite [9].

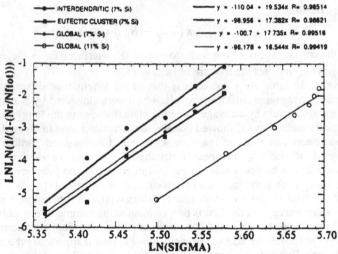

Figure 6. Application of the Weibull statistics to different eutectic morphologies in AlSi alloys (Al7Si0.3Mg and Al11Si) [12].

It is usually observed that the nucleation of cavities is a continuous process [13]. Thus Walsh, Jata and Starke [14] showed that in 2134 type aluminium alloys the number of voids per unit area increased linearly with the strain. On the other hand, in spheroidized steel Kwon and Asaro [15] concluded that the continuous nucleation in that case was stress controlled rather than strain controlled.

A particular case is the cleavage nucleation of ductile voids which takes place in embrittled duplex stainless steels. Once nucleated the cleavage cracks initiated in the ferrite phase which are blocked at interfaces with austenite grow by plastic deformation of this later phase. This is again a case of continuous strain controlled nucleation [13, 16].

3. Growth of Cavities

3.1. GROWTH OF A SINGLE CAVITY

Once nucleated the cavities grow owing to the plastic deformation of the matrix. The first studies are due to McClintock [17] who considered a single cylindrical hole, either in a perfectly plastic or in a viscous material [18]. He showed that the growth rate was proportional to the increment of the equivalent plastic strain and to an exponential function of the stress triaxiality ratio Σ_m/Σ_{eq}. Rice and Tracey [19] found the same for a spherical cavity of radius R in a remote uniaxial stress field with superimposed hydrostatic stress :

$$\frac{\dot{R}}{R} = 0.283 \exp\left(1.5\frac{\Sigma_m}{\sigma_0}\right)\dot{E}^p_{eq} \qquad (5)$$

σ_0 being the flow stress. The preexponential factor was later corrected by Huang [20] who proposed 0.427 instead of 0.283, for large stress triaxiality ratio.

The work of McClintock had shown that an increase of the Norton exponent n for a viscous medium decreased the cavity growth rate. This was confirmed by Budiansky, Hutchinson and Slutsky [21].

3.2. POROUS PLASTIC MATERIALS

3.2.1. *The model of Gurson*

To take account of the interaction between cavities it is convenient to study the plastic behaviour of a porous material [22]. Gurson [23, 24] derived the yield surface of a perfectly plastic porous material. He showed that the plastic potential could be written, for spherical cavities :

$$\Psi(\Sigma) = \frac{\Sigma_{eq}^2}{\sigma_0^2} + 2fch\left(\frac{3}{2}\frac{\Sigma_m}{\sigma_0}\right) - \left(1 + f^2\right) \qquad (6)$$

where σ_0 is the yield strength and f the volume fraction of cavities. This reduces to the Von Mises potential for a non-porous material. Figure 7 shows the Gurson's yield surface. As expected it shrinks when the porosity increases and a porous material yields under hydrostatic tension.

132

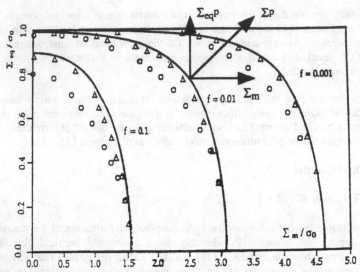

Figure 7. Gurson's yield surface for various levels of porosity in the equivalent stress versus hydrostatic stress plane. Exact criteria for a hollow sphere under homogeneous strain rate (triangles) or homogeneous stress (circles) are shown for comparison. The strain increment vector is normal to the yield surface.

The normality rule allows to derive the strain increment which is the combination of a distorsion at constant volume and of a volume increment. As the volume of the plastic material (excluding the cavities) remains constant :

$$\frac{\dot{V}}{V} = \frac{\dot{f}}{1-f} \qquad (7)$$

and the rate of growth of cavities, equal to $(\dot{V}/V)/3f$, is given by :

$$\frac{\dot{R}}{R} = \frac{\dot{E}_{eq}^p}{3f} \frac{\partial \Psi/\partial \Sigma_m}{\partial \Psi/\partial \Sigma_{eq}} = \frac{2\sigma_0}{\Sigma_{eq}} sh\left(\frac{3}{2} \frac{\Sigma_m}{\sigma_0}\right) \dot{E}_{eq}^p \qquad (8)$$

which is quite similar to the law derived for a single cavity. It is independent of the porosity.

Leblond [25] demonstrated rigorously that the potential of Gurson encompasses the true yield potential of a hollow sphere under homogeneous deformation on its surface. Under hydrostatic tension the potential of Gurson reduces to the exact solution. When the hydrostatic stress is zero it reduces to :

$$\Sigma_{eq} = (1 - f)\sigma_0 \qquad (9)$$

which is demonstrated to be larger than the exact solution.

3.2.2. *Other approaches*

Variational approaches construct bounds for the yield potential [26, 27]. Continuum damage mechanics, based on the thermodynamics of irreversible processes, provide models for porous materials [28, 29].

Lemaitre as well as Rousselier considered potentials such that there was no coupling between damage, that is porosity, and plasticity, damage affecting the elastic part of the potential only. Under those conditions, Lemaitre proposes :

$$\Psi = \frac{1}{2E(1-D)} \, \Sigma_{eq} \left[\frac{2}{3}(1+v) + 3(1-2v)\left(\frac{\Sigma_m}{\Sigma_{eq}}\right)^2 \right] \tag{10}$$

Where v is the Poisson's ratio and D the damage parameter which can be considered equal to the porosity.

The model of Rousselier resembles more the model of Gurson :

$$\Psi(\Sigma) = \frac{\Sigma_{eq}}{(1-f)\sigma_0} + \frac{\sigma_1}{\sigma_0} fD \exp\left[\frac{\Sigma_m}{(1-f)\sigma_1} \right] \tag{11}$$

where D and σ_1 are parameters depending on the material.

Such models are helpful to imagine improvements to the potential of porous materials.

3.2.3. *Improvements to the model of Gurson*

Tvergaard [30] modified the model of Gurson by the introduction of parameters which account for the interaction between cavities and allow to better fit experiments :

$$\Psi(\Sigma) = \frac{\Sigma_{eq}^2}{\sigma_0^2} + 2f^* q_1 ch\left(\frac{3}{2} q_2 \frac{\Sigma_m}{\sigma_0} \right) - \left[1 + \left(q_1 f^* \right)^2 \right] \tag{12}$$

f^* remains equal to f below a certain critical value f_c above which its dependence on f is accelerated to take account of the effect of void coalescence.

Garajeu [31] brought the model of Gurson closer to the exact solution by using the elastic field for a hollow sphere under deviatoric loading. This multiplies the first term in the potential by some function of the porosity.

The yield potential for ellipsoidal cavities was derived [32-34]. Its expression includes parameters which depend on the shape of the cavity characterized by $S = \log(s)$, (a and b analogous to the ones in Fig. 4). For the most frequent case of elongated cavities (s>1) the criterion takes the form :

$$\frac{\Sigma_{eq}^2}{\sigma_0^2} + 2q_w fch\left(\frac{\kappa \Sigma_h}{\sigma_0} \right) - q_w^2 f^2 = 0 \tag{13}$$

where q_w is a parameter given by :

134

$$q_w = 1 + 2(q_1 - 1)\frac{s}{1 + s^2}$$ (14)

q_1 being the parameter introduced by Tvergaard. According to Garajeu q_1 is of the order of 1.6. In any case q_w decreases when s increases from 1 for a spherical cavity to infinity for a cylindrical one. The stress Σ_h is expressed as:

$$\Sigma_h = \alpha(\Sigma_{xx} + \Sigma_{yy}) + (1 - 2\alpha)\Sigma_{zz}$$ (15)

α being a function of the porosity and of the shape parameter S. For spherical cavities $\alpha = 1/3$, so that $\Sigma_h = \Sigma_m$, and for cylindrical cavities $\alpha = 1/2$ so that the coefficient κ increases from 1.5 for spherical cavities to 3 for cylindrical ones.

Benzerga [35] in collaboration with Pardoen checked the Gologanu-Leblond model by numerical calculations of unit cells. The main conclusions are the following :

    - in uniaxial stress the porosity reaches a limit when the plastic deformation increases,

    - for a stress triaxiality ratio of 1, the porosity keeps increasing, the more so the smaller S (Fig.8).

A better agreement of the Gologanu-Leblond model with the cell calculations was obtained with $q_1 = 1.3$ (instead of 1.6) for small initial porosities (f of the order of 0.01%).

    - for high triaxialities (ratio equal to 3), the increase of the porosity, which is quite rapid, becomes independent of the shape of the cavities (Fig.9). For S>1 the cavities flatten.

Benzerga [35] studied the effect of plastic anisotropy. He showed that in the model of Gurson the equivalent stress must be replaced in that case by the equivalent stress of Hill :

$$\Sigma_{eq} = \frac{3}{2}S_{ij}H_{ijkl}S_{kl}$$

where S_{ij} is the stress deviator and H_{ijkl} the tensor of anisotropy.

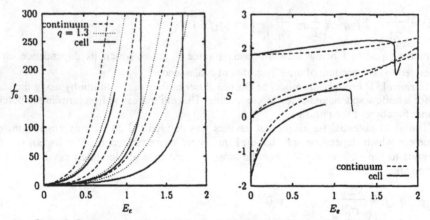

Figure 8. Comparison between the Gologanu-Leblond model of the increase of porosity and of change of shape of cavities with the results of cell calculations for a stress triaxiality ratio of 1 and small porosity $f_0 = 0.01$ [35].

On each figure the 3 sets of curves correspond to 3 different values of the initial shape factor, S.

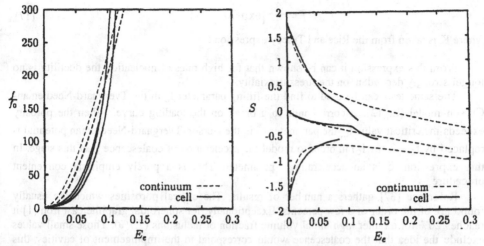

Figure 9. Same as figure 8 for a stress triaxiality ratio of 3 [35].

Often two populations of dimples can be observed on the fracture surface. This corresponds to the existence of inclusions of different sizes nucleating two populations of cavities. The interaction between them was studied, particularly by Perrin [36]. For that purpose he investigated the growth of a cavity within a sphere made of a porous material of Gurson.

4. Coalescence of cavities

4.1. CRITICAL GROWTH PARAMETER

Now that the law for the growth of cavities is established, a sensible assumption for the fracture of the material to take place is that these cavities coalesce for some critical value of the void growth $(R/R_0)_c$. As the initial porosity f_0 is equal to the volume fraction of inclusions, there exists a relation between the critical void growth and the critical porosity f_c :

$$\frac{f_c}{f_0} = \left(\frac{R}{R_0}\right)_c^3 \tag{16}$$

The model of Rice and Tracey for instance (Eq. 5), allows us to determine the critical void growth by measuring the fracture elongation and by integration of equation 5. This is usually achieved using axisymmetric notched specimens for which the integration is carried out from finite elements calculations. The acuity of the notch controls the stress triaxiality ratio. In the case of continuous strain-controlled nucleation the rate of nucleation A_n should be incorporated [13] leading to an evolution of the volume fraction of voids given by :

136

$$f = \frac{A_n}{K} \left[\exp(K\varepsilon - 1) \right] \tag{17}$$

where K is taken from the Rice and Tracey expression :

From this expression it can be shown that for high rates of nucleation the ductility is no longer strongly dependent on the stress triaxiality.

The same tests can be used to find the fitting parameter f_c in the Tvergaard-Needleman-Gurson model (Eq.12). It corresponds to a break on the loading curve. When the porosity exceeds the critical value f_c, the parameter f^* in the Gurson-Tvergaard-Needleman potential is replaced by $f_c + \delta (f - f_c)$ in order to model the acceleration of coalescence past this stage. In this expression, δ is an accelerating parameter. This is a purely empirical convenient procedure.

Reference [37] gathers a number of results. The critical porosities which are usually found are of the order of a few percents. Exceptionally, for spheroidal graphite cast iron [1] it reaches 20% for a rather high initial volume fraction of inclusions (7.7%). Those small values exclude the idea that the coalescence would correspond to the impingement of cavities; this would mean 68% for densely packed cavities, 52.4% for a distribution on a cubic lattice [38]. The examination of fracture surfaces confirms that the mechanism is different and corresponds in fact to plastic instability of the ligaments between voids.

Now the distribution of inclusions, and as a consequence the distribution of cavities, are not uniform. There are local zones where there are clusters. The higher porosity there promotes coalescence, so that the local critical value is higher than the one calculated as if the distribution was homogeneous (see the paper of Achon, Ehrström and Pineau in reference [37]) (Fig. 10).

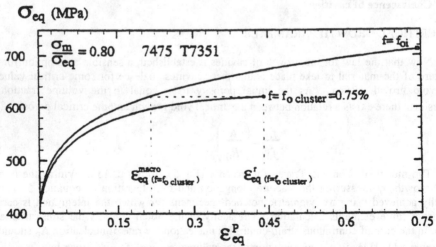

Figure 10. Sketch illustrating the effect associated with the growth of cavities and their uneven distribution (Achon, Ehrström and Pineau in [37]).

4.2. PLASTIC INSTABILITY

4.2.1. *Mechanisms*

The experimental determination of the critical parameters, as just explained, should allow in principle to predict the fracture elongation of any structure provided the evolution of the stresses and strains are determined during loading. However these critical values cannot be predicted from the knowledge of the microstructure. Furthermore it is often found that the critical growth rate, for instance, depends on the hydrostatic stress, or on the stress history [38]. What is needed for a better knowledge of the critical conditions for the coalescence of voids is an analysis of the plastic instability. This has not been studied as much as the growth stage. A recent thesis [35, 40] gives an excellent review of what has been achieved and proposes a model which incorporates a number of aspects, such as anisotropy of shape as well as of plastic behaviour.

Before giving more details about the models, a sketch of the mechanisms of coalescence which are observed is useful. The essential ones are a flat dimple mode and a void-sheet mode, as described by Faleskog and Fong Shih in reference [37] (Fig. 11). The first one leads to the idea that the cavities have a tendency to align on voided layers perpendicular to the highest extension. The second one results from plastic shear bands which either trigger cavitation or is triggered by it. In either cases the process is self-cataclysmic.

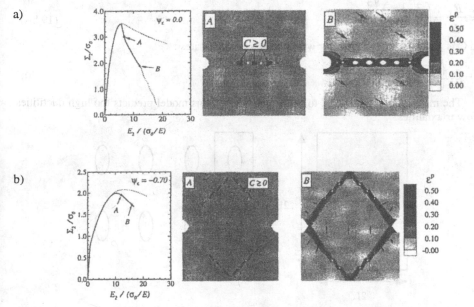

Figure 11. The two modes of coalescence : a) flat dimples mode b) void sheet mode (Jonas, Faleskog and Fong Shih in [37]).

138

4.2.2. *Model based on localization* [35]

Leblond and Perrin [41] applied the theory of localization of a dilatant material [42] to the Gurson material. They assumed axisymmetric loading conditions, perfect plasticity, progressive concentration of voids in parallel layers separated by compact regions. The voids spacing ratio is $\lambda = H/L$ (Fig.12), the void aspect ratio is s (Table 1).

The coalescence criterion is then, neglecting a term in which the Young's modulus is in the denominator :

$$\frac{\Sigma_m}{\sigma_0} + \frac{2}{3}\frac{\Sigma_{eq}}{\sigma_0} = \frac{\sigma_{xx}}{\sigma_0} + q\lambda f s h \left(\frac{1}{2}\frac{\Sigma_m}{\sigma_0} + \frac{1}{3}\frac{\Sigma_{eq}}{\sigma_0} + \frac{\sigma_{xx}}{\sigma_0} \right) \tag{18}$$

where σ_{xx} is the local radial stress on the layer.

4.2.3. *Models based on plastic limit loads* [40]

The basic model was proposed by Thomason [43]. It consists in calculating the plastic limit load of the layer of cavities. This yields :

$$\frac{\Sigma_m}{\sigma_0} + \frac{2}{3}\frac{\Sigma_{eq}}{\sigma_0} = \left(1 - r^2\right)C \tag{19}$$

r is the ratio of the radial dimension of the void to the distance between voids in the layer.

$$r = \frac{R_x}{L} = \left(\frac{3}{2}\frac{\lambda}{s} f \right)^{1/3} \tag{19 bis}$$

and C is a plastic constraint factor which Benzerga [35] writes :

$$C = 0.1 \left(\frac{1-r}{sr} \right)^2 + 1.2\sqrt{\frac{1}{r}} \tag{20}$$

The material being assumed to be incompressible, this model predicts too high ductilities at low triaxialities.

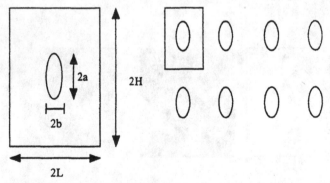

Figure 12. Definition of the particles spacing and aspect ratios : $\lambda = H/L$: s = a/b.

This led Zhang and Neimi [44] to use the model of Gurson to calculate both the evolutions of the porosity and of the equivalent stress, as a function of the load history characterized by the hydrostatic stress. This remained imperfect as it neglected the elongation of the voids.

The last step was to take care of that shape evolution by using the model of Gologanu-Leblond instead of the model of Gurson [40].

Figure 13 shows results obtained with the preceding models for various triaxiality ratios [40]. It highlights the deficiency of the model of Thomason which predicts too high ductilities at low triaxialities and too low ones at high triaxialities. The model of Zhang and Neimi predicts the possibility of coalescence in uniaxial stress, whereas this does not take place in reality ; necking is needed. This is what predicts the model which takes account of the elongation of the cavities during deformation ; this elongation hampers coalescence. A critical porosity independent of the stress triaxiality would provide a straight line with a slope of -3/2 on that diagram. It can be seen that it represents a passable approximation.

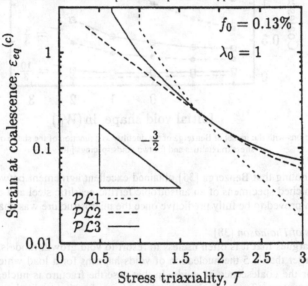

Figure 13. Calculations of the ductility as a function of the stress triaxiality ratio, using various models of coalescence [40] : model of Thomason PL1 [43] : model of Zhang and Neimi PL2 [44] ; model of Benzerga PL3[40].

Figure 14 shows the interesting fact that at high triaxialities the shape of the voids has no influence on the ductility.

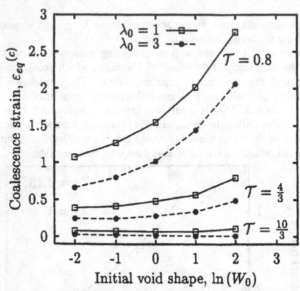

Figure 14. Calculations with the model of Benzerga of the ductility as a function of the shape of voids for various stress triaxialities and for two voids spacings [40].

It is worth noting that Benzerga [35] obtained excellent agreement between experiments on cylindrical notched specimens of an anisotropic ferritic-pearlitic steel and calculations with his model, which proved to be fully predictive once the microstructure was characterized.

4.2.4. *Objections of Thomason* [38]

Thomason argues that it is often useless to resort to void growth models. He objects that at triaxialities larger than 1.5 the nucleation of voids happens for a load which is higher than the limit load for the coalescence of voids. In that case the fracture is nucleation controlled. However from equation (3) giving the critical condition for the nucleation of cavities, and equations (18), (19) and (20) giving the critical condition for their coalescence, it is easy to draw the diagram shown on figure 15. Applied to spherical inclusions ($k = s = 1$), if the ratio of the fracture stress to the yield stress is of the order of 3, a ratio H/L of the order of 1, it is found that a volume fraction of at least 0.83% of inclusions is needed for the hypothesis of Thomason to be verified at stress triaxialities equal to 3. The corresponding volume fraction of prolate inclusions ($s = 7$, $k = 4.66$) is found to be 0.86% ; it is 1.53% for oblate inclusions ($s = 1/7$, $k = 4.28$).

Furthermore, according to figure 13, the condition of plastic instability of Thomason yields too low a ductility for a stress triaxiality ratio of 3. It means that the coalescence criterion should be shifted up on the diagram (Fig. 15). It thus appears that the previous volume fractions of inclusions are minimums. It is clear that if the ratio of the fracture stress of inclusions to the yield strength is less than 3, still larger volume fractions of inclusions would be required. In most practical cases nucleation should precede coalescence and the growth of cavities must then be evaluated.

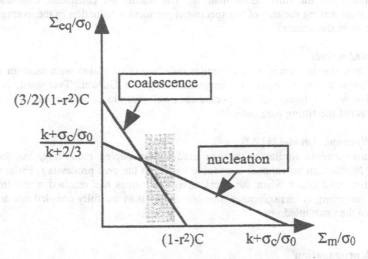

Figure 15. Diagram comparing the nucleation criterion (Eq. 4) and the coalescence criterion by plastic instability (Eqs. 18, 19 and 20).

4.2.5. *Voids sheet instability*

Voids sheet instability is due to shear localization and can be treated following the formalism of Rudnicki and Rice [42]. A number of studies were concerned with plane strain conditions. [35, 40, 45-49]. They showed that voids sheet instability could not be predicted with the model of Gurson with isotropic hardening. The introduction of either kinematic hardening or the nucleation of new cavities was needed. However observations on plane strain tensile specimen did not reveal cavities in shear bands at the onset of instability [35, 50]. The model of Rousselier[29] predicts shear localization without the nucleation of new cavities. So does the model of Benzerga [35].

4.3. MODELLING OF DUCTILITY

4.3.1. *Statistical aspects*

The void nucleating inclusions are far from being evenly distributed and this has a strong influence on the three steps of ductile fracture. In many instances clusters of inclusions are found in which the nucleation rate is quite high, where there is a strong interaction between voids as they grow and where coalescence is easy. This leads to size effects in the same manner as for brittle cleavage fracture[51, 52].

4.3.2. *Uncoupled models* [53]

A large number of grains are generated with a random nucleation rate according to the distribution function determined experimentally. The evolution of the volume fraction of voids is then integrated using a post-processor routine applied on the stress-strain evolution at each point computed for the fully dense material. The results are consistent with observations showing that increasing the size of the specimens produces a reduction in the average ductility and a decrease in the scatter.

4.3.3. *Coupled models*

This time, starting again with a similar distribution of grains with random nucleation rates, the stress-strain evolution is computed using the Gurson, Tvergaard, Needleman, potential. The work softening of the specimens is well reproduced in this way. This procedure avoids the use of the fitting parameter f^*.

4.3.4 *Locally coupled model* [51,52]

The procedure is similar to the one used for uncoupled modelling, but the Gurson, Tvergaard, Needleman potential is introduced locally in the post-processing. Failure initiation is assumed to take place when the local equivalent stress has reached a maximum. This method of modelling is less computer time consuming than the fully coupled one and is more realistic than the uncoupled one.

5. Crack propagation

A common method for the prediction of crack propagation, when the mechanism is ductile tearing, is the use of the J-Δa curve, J being the characterizing parameter of Rice. However it is found that the comparison of results obtained on specimens of different shapes, as well as the transposition of the results obtained on laboratory specimens to the prediction of the behaviour of a crack in a structure, are faced with plastic constraint effects. To the HRR field at the tip of the crack, rigorous in small scale yielding, must be added a Q stress to take account of these effects as already indicated in Part I. The local approach to fracture mechanics should solve these problems. Crack propagation can be modelled by finite elements calculations with either a post-processor fully uncoupled using of a critical local parameter, $(R/R_0)_c$ or f_c, or, in a fully coupled manner, the incorporation of a plastic potential of porous material in the formulation itself, Gurson's or Rousselier's. Since the determination of the Q stress requires a refined finite elements calculation anyway, the local approach is not more costly and has the advantage of emphasizing the influence of the microstructure of the material. It can be hoped that predictions of crack propagation will be possible without the introduction of any adjustable parameter, if a good description of the microstructure is available [13].

A number of applications of the local approach to fracture mechanics are described in reference [37] (Bauvineau, L., Burlet, H., Eripret, C., and Pineau, A., for a carbon-manganese steel ; Dong, M.J., Berdin,C., Béranger, A.S., and Prioul, C., for a spheroidal graphite cast iron; Eisele, U., Seidenfuss, M., and Pitard-Bouet, J.M., for a 10 MnMoNi55 steel and shallow crack; Zhang, Z.L., and Skallerud, B., for embedded cracks ; Achon, P., Ehrstrom, J.C., and

Pineau, A., for an aluminium alloy ; Giovanola, J.H., Kirkpatrick, S.W., and Crocker, J.E., for a high strength HY130 steel and scale effects). Roudier and François investigated the local approach to the dynamic fracture of TA6V [54]. The main difficulty in these calculations is to adopt an accurate mesh size of the finite elements at the tip of the crack. A common feature is to adjust this size so as to match the experimental results on a particular specimen, and then to stick to that mesh size whatever the specimen or the structure studied. However, in the local approach to fracture, the mesh size is considered as a parameter characteristic of the material, such as the mean distance between inclusions or clusters of inclusions.

When the fracture stress of inclusions or of their interface is so small that the material can be considered as porous from the beginning, the porosity shields the crack tip so that the stresses are reduced. On the other hand, the fracture toughness is also lowered. The two effects compensate and thus the net effect is negligible [55]. However when the crack propagates, the damage wake produces a strong R curve effect as was observed on spheroidal graphite cast iron [56]. Bui and Ehrlacher [57] predicted such a behaviour from a theoretical analysis of the situation at the crack tip for a perfectly plastic damaging material. Li and Pan extended the HRR analysis to a pressure sensitive material for which the yield stress is a linear function of the hydrostatic stress [58].

Recent studies [52, 59, 60] concentrate on the use of the local approach to fracture mechanics to model the Charpy test. At the lower shelf of the brittle-ductile transition of steels fracture can be predicted as explained in part I of the present paper. In the transition region, propagation of a crack by ductile tearing precedes fracture by cleavage. In that case it is needed to model this propagation using the model of Gurson and thus to follow the evolution of the stresses at the tip of the crack to determine the onset of cleavage fracture.

6.   Conclusion

Complete models for the nucleation, the growth and the coalescence of cavities exist. The thesis of Benzerga as an example showed that a full modelling could be achieved, even in complicated situations of anisotropic behaviour, with the sole introduction of a description of the microstructure and of the constitutive equations of the material. There are materials for which this represents particular difficulties; duplex stainless steels are an example as the microstructure needs to be studied at different scales, that of the Widmanstäten arrangement of the two phases, that of the austenite colonies, that of the ferritic grains. Even in such a case the local approach to fracture mechanics is applicable, provided a close junction is achieved between mechanics of solids and metallurgy.

References
1.   Dong, M.J., Prioul, C., and François, D. (1997) Damage Effect on the Fracture Toughness of Nodular Cast Iron: Part I. Damage Characterization and Plastic Flow Stress Modelling, *Metall. and Mat. Trans. A*, **28A**, 2245-2254.
2.   Eshelby, J.D., (1957)   The Determination of the Elastic Field of an Ellipsoidal Inclusion and Related Problems, *Proc. Royal Soc. London.* **A241**, 376-396.
3.   Berveiller, M., and Zaoui, A., (1979) An Extension of the Self-Consistent Scheme to Plastically Flowing Polycrystals, *J. Mech. Phys. Solids*, **26, 325 344.**

144

4.   Bourgeois, N.,(1994) Caractérisation et modélisation micromécanique du comportement et de l'endommagement d'un composite à matrice métallique : Al/Si$_p$, *Thèse*, École Centrale de Paris, Châtenay-Malabry.

5.   Beremin, F.M., (1981) Experimental and Numerical Study of the Different Stages in Ductile Rupture : Application to Crack Initiation and Stable Crack Growth, in *Three Dimensional Constitutve Relations of Damage and fracture (Nemat-nasser Ed.)*, Pergamon, 157-172.

6.   Lee, B.J., and Mear, M.E., (1999) Stress Concentration Induced by an Elastic Spheroidal Particle in a Plastically Deforming Solid, *J. Mech. Phys. Solids*, 47, 1301-1336.

7.   Smith, C.L., and Low, J.R., (1974), *Met. Trans.*, 5, 279.

8.   Tanaka, K., Mori, T., and Nakamura, T., (1970) Cavity Formation at the Interface of a Spherical Inclusion in a Plastically Deforming Matrix, *Phil. Mag.*, 21, 267-279.

9.   Llorca, J., Martin, A., Ruiz, J., and Elices, M., (1993) Particulate Fracture during Deformation of a Spray Formed Metal-Matrix-Composite, *Metal. Trans. A*, 24A, 1575-1588.

10.   Cox, T.B., and Low, J.R., (1974) An Investigation on the Plastic Fracture of AISI 4340 and 18 Nickel-200 Grade Maraging Steel, *Met. Trans.*, 5, 1457-1470.

11.   Antretter, T., and Fischer F.D., (1996) Critical Shapes and Arrangements of Carbides in High-Speed Tool Steel, *Mat. Sci. Eng.*

12.   Kamenova, T., Doglione, R., Douziech, J.L., Berdin, C., and François, D., (1997) Evolution of Silicon Particles Breaking in Hypo- and Pseudo-Eutectic Al-Si Alloys, in *Recent Advances in Fracture (Rao, K. Madihihara, Andrew B. Geltmacher, Peter Matic and Kuntimaddi Sadananda Eds.)*, TMS Warrendale, 351-359.

13.   Pineau, A., (1992) Global and Local Approaches of Fracture- Transferability of Laboratory Test Results to Components, *Topics in Fracture and Fatigue (Argon , A.S. Ed.)*, Springer-Verlag, 6, 197-234.

14.   Walsh, J.A., Jata, K.V. and Starke Jr., E.A., (1989) The Influence of Mn Dispersoid Content and Stress State on the Ductile Fracture of 2134 Type Al Alloy, *Acta Metall.*, 37 N°11, 2861-2871.

15.   Kwon, D. and Asaro, R.J.,(1990) *Met. Trans.*, 21A, 117-134.

16.   Joly, P., Cozar, R. and Pineau, A., (1990) Effect of Crystallographic Orientation of Austenite on the Formation of Cleavage Cracks in Ferrite in Aged Duplex Stainless Steel, *Scripta Metall..Mater.*, 24, 2235-2240.

17.   McClintock, F.A., and Argon, A., (1966) *Mechanical Behavior of Materials*, Addison-Wesley Pub., Reading.

18.   McClintock, F.A., (1965) Effects of Root Radius , Stress, Crack Growth rate on the Fracture Instability, *Proc. Royal Soc.*, A285, 58-72.

19.   Rice J.R., and Tracey, D.M., (1969) On the enlargement of Voids in Triaxial Stress Fields, *J. Mech. Phys. Solids*, 17, 201-217.

20.   Huang, Y., (1991), *J. Applied Mech.*, 58, 1084-1086.

21.   Budiansky, B., Hutchinson, J.W., and Slutsky, S., (1982) Void Growth and Collapse in *Viscous Solids, in Mechanics àof Solids, the Rodney Hill 60th Anniversary Volume (Hopkins, H.G., and Sowell, M.J., Eds.)*, Pergamon Press, 13-45.

22.   Green, R.J., (1975) A Plasticity Theory of Porous Solids, *Int. J. Mech. Sci.*, 14, 215-224.

23.   Gurson, A.L., (1975) Plastic Flow and Fracture Behavior of Ductile Materials Incorporating Void Nucleation, Growth and Interaction, *PhD. Thesis*, Brown University, Providence.

24.   Gurson, A.L., (1977) Continuum Theory of Ductile Rupture by Void Nucleation and Growth. Part I/ Yield Criteria and Flow Rules for Porous Ductile Media, *J. Eng. Mat. Tech.*, 99, 2-15.

25.   Leblond, J.-B., and Perrin, G., (1977), *Cours École Polytechnique*, Palaiseau.

26.   Ponte-Castaneda, P., and Zaidman, M., (1994) Constitutive Models for Porous Materials with Evolving Microstructure, *J. Mech. Phys. Solids*, 42, 1459-1495.

27.   Da Silva, M.G., and Ramesh, K.T., (1997) The Rate Dependent Deformations of a Porous Pure Iron, *Int. J. Plasticity*, 13 N°6-7, 587-610.

28.   Lemaitre, J., (1985), A Continuous Damage Mechanics Model for Ductile Fracture, *J. Eng. Mat. Tech..*, 107, 83-89.

29.   Rousselier, G., (1981) Finite Deformation Constitutive Relations Including Fracture Damage, in *Three Dimentional Constitutive Equations of Damage and Fracture (Nemat-Nasser Ed.)*, Pergamon Press, 331-355.

30. Tvergaard, V., (1981) Influence of Voids on Shear Band Instabilities under Plane Strain Conditions, *Int. J. Frac.*, **17**, 389-407.

31. Garajeu, M., (1995) Contribution à l'étude du comportement non linéaire de milieux poreux avec ou sans renforts., *Thèse*, Université de Marseille.

32. Gologanu, M., (1997) Étude de quelques problèmes de rupture des métaux, *Thèse*, Université Paris 6.

33. Gologanu, M., Leblond, J.-B., and Devaux J., (1993) Approximate Models for Ductile Metals Containing Non-Spherical Voids- Case of Axisymmetric Prolate Ellipsoidal Cavities, *J. Mech. Phys. Solids*, **41**, 1723-1754.

34. Gologanu, M., Leblond, J.-B., and Devaux J., (1994) Approximate Models for Ductile Metals Containing Non-Spherical Voids - Case of Axisymmetric Oblate Ellipsoidal Voids, *J. Eng. Mat. Tech.*, **116**, 290-297.

35. Benzerga, A.A., (2000) Rupture ductile de tôles anisotropes. Simulation de la propagation longitudinale de fissures dans un tube pressurisé, *Thèse*, École Nationale Supérieure des Mines de Paris.

36. Perrin, G, (1992) Contribution à l'étude théorique et numérique de la rupture ductile des métaux, *Thèse*, École Polytechnique.

37. *Local Approach to Fracture 86-96*. (1996), Pineau, A., and Rousselier, G., Euromech-Mecamat 96, 1st European Mechanics of Materials Conference, Fontainebleau,.

38. Thomason, P.F. (1998) A View on Ductile-Fracture Modelling, *Fatigue Fract. Eng. Mat. Struc.*, **21**, 1105-1122.

39. Lautridou, J.-C, and Pineau, A., (1981) Crack Initiation and Stable Crack Growth Resistance in 508 Steels in Relation to Inclusion Distribution, *Eng. Fract. Mech.*. **15**, 55-71.

40. Benzerga, A.A., Besson, J. and Pineau, A. (1999) Coalescence-Controlled Anisotropic Ductile Fracture, *J. Eng. Mat. Struct.*, **121**, 221-229.

41. Leblond, J.-B., and Perrin, G., (1991) Analytical Study of the Coalescence of Cavities in Ductile Fracture of Metals, in *Plasticity. 3rd Symposium on Anisotropy and Localization of Plastic deformation*, 233-236.

42. Rudnicki, J.W., and Rice, J.R., (1975) Conditions for the Localization of Deformation in Pressure-Sensitive Dilatant Materials, *J. Mech. Phys. Solids*. **23**, 371-394.

43. Thomason, P.F.. (1968) A Theory for Ductile Fracture by Internal Necking of Cavities, *J. Inst.Metals*, **96**, 360-365

44. Zhang, Z.L., and Neimi, E., (1994) Analysing Ductile Fracture using Dilational Constitutive Equations, *Fatigue Fract. Eng. Mat. Struct.*, **17**, 697-707.

45. Tvergaard, V., Needleman, A., and Lo, K.K., (1981) Flow Localization in the Plane Strain Tensile Test, *J. Mech. Phys. Solids*. **29**, 115-142.

46. Tvergaard, V , (1982) Influence of Void Nucleation on Ductile Shear Fracture at a Free surface, *J. Mech. Phys. Solids*, **30**, 399-325.

47. Tvergaard, V., (1987) Effect of Yield Surface Curvature and Void Nucleation on Plastic Flow Localization, *J. Mech. Phys. Solids*,. **35** N°1, 43-60.

48. Sage, M., Pan. J., and Needleman, A., (1982) Void Nucleation Effects on Shear Localization in Porous Plastic Solids, *Int. J. Fracture*, **19**, 163-182.

49. Becker, R., Smelser, R.E., and Richmond, O., (1989) The Effect of Void Shape on the Development of Damage and Fracture in Plane Strain Tension, *J. Mech. Phys. Solids*, **37** N°1, 11-129.

50. Rivalin, F., (1998) Rupture ductile dynamique des aciers pour gazoducs, *Thèse*, École des Mines de Paris.

51. Decamp, K., Bauvineau, L., Besson, J. and Pineau, A., (1997) Size and Geometry Effects on Ductile Rupture of Notched Bars in a C-Mn Steel : Experiments and Modelling, *Int. J; of Fracture*, 88, 1-18.

52. Devillers-Guerville, L., Besson, J. and Pineau, A., (1997) Notch Fracture Toughness of a Cast Duplex Stainless Styeel : Modelling of Experimental Scatter and Size Effect, *Nuclear Engng. and Design*, 108, 211-225.

53. Pineau, A., (1995), Effect of Inhomogeneities in the Modelling of Mechanical Behaviour and Damage of Metallic Materials, in *Mechanical Behaviour of Materials*, *Bakker, A. Ed.Seventh International Conference on Mechanical Behaviour of Materials - ICM7*, 1-22.

54. Roudier, Ph., and François, D., (1996) Dynamic Fracture Toughness Measurements and Local Approach Modelling of Titanium Alloys, *Fatigue Fract. Eng. Mat. Struct.* **19** N°11, 1317-1327.

55. Evans, A.G., and Faber, K.T., (1983), *J. Am. Ceram. Soc.*, **67** (4), 255.

56. Dong, M.J., Prioul, C., and François, D., (1997), Effect of Fracture Toughness of Nodular Cast Iron: Part-Damage Zone Characterization Ahead of a Crack Tip, *Metall. Mat. Trans. A*, **28A**, 2255-2262.

57. Bui, H.D., and Ehrlacher, A., (1981) Propagation of Damage in Elastic and Plastic Solids, in *Advances in Fracture Research, François, D., Ed.,* Pergamon, 533-551.

146

58. Li, F.Z., and Pan, J., (1990) Plane-Strain Crack-Tip Fields for Pressure Sensitive Dilatant Materials, *Trans. ASME, J. Appl. Mech.*, **57**, 40-49.
59. Rossol, A., (1998) Détermination de la ténacité d'un acier faiblement allié à partir de l'essai Charpy instrumenté, *Thèse*, École Centrale de Paris.
60. Tahar, M. (1998) Application de l'approche locale de la rupture fragile à l'acier 16MND5 : Corrélation Résilience-Ténacité et Probabilité de rupture bimodale. Thèse, École des Mines de Paris

FRACTURE MECHANICS OF METALS: SOME FEATURES OF CRACK INITIATION AND CRACK PROPAGATION

A.J. KRASOWSKY
Professor and Department Head
Institute for Problems of Strength
National Academy of Sciences of Ukraine,
Timiryazevskaya st., 2
Kiev 01014 Ukraine

Abstract

The main features are considered concerning the crack initiation and crack propagation events during the fracture toughness measurements at static and dynamic loading with respect to perfectly brittle, quasi-brittle and ductile behavior of material. The relationships are described between seven different stages of loading process and corresponding stages of the crack tip plastic zone evolution as well as the appearances of fracture surface of specimen. Quantitative stereofractographic analysis of the stretched zone is performed on the fracture surface of different steels. The effect of temperature, loading rate, stress state mode and specimen size on the stretched zone formation as well as on the crack initiation is studied. In particular, if plane-strain conditions are satisfied the stretched zone parameters are independent of the specimen size variation. A correlation is obtained of the stretched zone height and width with fracture toughness of materials. The stretched zone height being in better correlation with fracture toughness of the material than its width. In contrast to other methods of metals fracture toughness measurement, the stretched zone as a subject for investigation has a number of unique properties. The method of stereoscopic measurement of the stretched zone geometry by superimposing the mating profiles of fracture surfaces reveals the scattering of results comparable or even lower than a scattering at conventional fracture toughness measurements. The direction of initial crack propagation differs essentially from the main direction of crack growth, during testing of standard specimens for the fracture toughness measurement. This peculiarity of crack initiation is explained as initial crack propagation into one of two symmetrical plastic zone fans along the trajectories of maximal damage of material by the plastic deformations. One of the main conclusions was drawn: the fracture toughness dependence on temperature and loading rate can be quantitatively described within the framework of the interaction between thermally activated plastic flow inside of plastic zone and the athermal brittle fracture process involved the micro-crack nucleation at some characteristic distance ahead the crack tip. A plane strain transition temperature for a given specimen thickness is defined for ferritic steels as the temperature along the

E. Bouchaud et al. (eds.), Physical Aspects of Fracture, 147–166.

K_{Ic} transition curve where the ASTM validity limit is reached. This temperature is linearly related to the specimen thickness logarithm under both static and dynamic loading. For a given steel, transition temperatures deduced from other tests were found to correspond to various characteristic values of the non-dimensional crack tip plastic zone size. The different relationships are described to deduce transition temperatures of various tests from the knowledge of either the K_{Ic} transition curve or from two other transition temperatures. It could provide a method to decide when the stress relief is needed after welding plates of a given thickness. It should also find its use in studying irradiation effects

1. Introduction

Fracture toughness is the most important mechanical property for many materials including metals, especially in case of the brittle fracture danger. The experimental procedure for evaluation of this property is regulated by modern standards and codes. These documents prescribe usually the testing of specimens with an artificial crack of normal opening (i.e. with a crack of mode I). The diagram load, P, versus crack mouth opening displacement, v, is usually recorded during experiment. Such kind of diagram is shown schematically on the Fig.1 with compact tension specimen as an example. Let us consider the principal stages of the fracture process and their relation to the P-v diagram, Fig.1.

The whole P-v diagram up to the point 7 is typical for ductile materials which failure after entire plastic flow of the specimen ligament. The fracture of the more brittle material is observed on the more earlier stage of the P-v diagram. Point 1 corresponds to the perfectly brittle fracture of material. Practically, almost whole set of fracture appearances pointed out in Fig. 1 by the numbers 1...7 can be observed on the same material (e.g. carbon steel) at different temperatures and loading rates. The following stages can be observed for the plastic zone evolution at the crack tip (Fig.2) as well as for their appearance on the fracture surface (Fig.3). The numbers of stages correspond approximately each other on the Figures 1 and 2 and can be characterized as follows.

1. Perfectly brittle fracture. Plastic zone at the crack tip is absent. The stretched zone and the sub-critical crack growth zone are absent on the fracture surface, too. The criteria of linear fracture mechanics predict well the crack initiation.

2. Quasi-brittle fracture. Plastic zone at the crack tip is localized corresponding to the small scale yielding condition. The stretched zone is slightly visible on the fracture surface, usually only in the direction normal to the fracture surface. Zone of the stable sub-critical crack growth is absent on the fracture surface. The criteria of linear fracture mechanics are usually acceptable in order to predict the crack initiation.

3. Point of intersection of the beam OB with the P-v diagram (beam OB corresponds to the condition tan$<v$OB = 0.95tan$<v$OA). This point suggested by standards to be corresponded to the crack initiation moment. For the P-v diagram extending to the right of point 3, there are many evidences that load P_Q corresponds to the crack tip plastic

zone evolution but not to the very crack initiation. In case when the fracture instant corresponds to the point 3 of *P-v* diagram, Fig.1, the crack tip plastic zone is localized, Fig.2. Stretched zone is pronounced on fracture surface whereas zone of the stabile sub-critical crack growth is absent, Fig.3. The application possibility of the linear fracture mechanics criteria have to be evidenced.

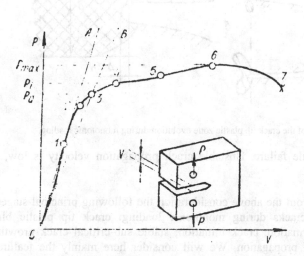

Figure 1. Schematic representation of the of the *P-v* diagram: OA-linear elastic deformations, OB-the beam which slope is 5 % lower than the slope of beam OA.

4. Point corresponding to the actual crack initiation as well as to the final stage of the stretched zone formation. The crack tip plastic zone is developed significantly. The criteria of non-linear fracture mechanics are required in order to predict the crack initiation

5. The stretched zone formation is completed, the stable sub-critical crack growth is followed by the quasi-brittle fracture appearance on fracture surface, Fig.3. The crack tip plastic zone is well developed, the plastic blunting of crack tip is significant. Unstable catastrophic crack propagation has quasi-brittle character with the high enough crack velocity, when fracture appears between points 4...6 on the *P-v* diagram.

6. Ductile failure preceded by the large geometry changes at the crack tip (plastic blunting). Plastic zone occupy all specimen's ligament. All typical zones pointed out in the Fig.3 are presented at the fracture surface. The last zone of unstable crack propagation is cowered by the dimples of ductile failure. Unstable crack propagation velocity is dependent on the elastic energy of loading system but generally is not high.

150

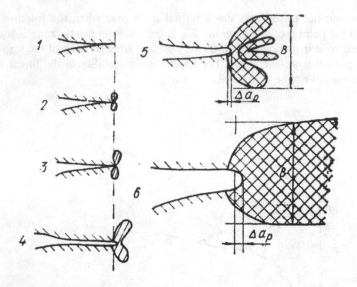

Figure 2. The stages of the crack tip plastic zone evolution during monotonic loading.

7. Perfectly ductile failure. Unstable crack propagation velocity is low, typical for ductile failure.

One can see from the above consideration the following principal stages of failure emanating from cracks during monotonic loading: crack tip plastic blunting and stretched zone formation; crack initiation; stable sub-critical crack growth; unstable, catastrophic crack propagation. We will consider here mainly the features of these stages, particularly stages relating to the crack initiation which event is responsible for the material fracture toughness property. The effects of temperature, loading rate, specimen size and material structure on the fracture toughness of metal are of principal interest in order to understand the physical nature of the capability of matter to resist the crack propagation.

2. Fracture toughness

The fracture toughness of material is the mechanical property which, according to the standards, is usually related to the crack initiation event.

2.1. STRETCHED ZONE FORMATION AND CRACK INITIATION

The main process responsible for a stretched zone formation is the plastic blunting of the crack tip during monotonic loading of a specimen. The crack initiation instant is the

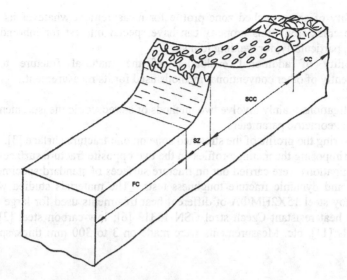

Figure 3. Schematic representation of the fracture surface: FC-zone of the fatigue pre-crack growth; SZ-stretched zone; SCC-zone of the sub-critical crack growth; CC-zone of the catastrophic crack growth.

event interrupting the plastic blunting process being the final stage of the stretched zone formation.

2.1.1. Stretched zone formation

This is why a stretched zone on fracture surface (Fig.3) is generally related to the crack tip plastic blunting process and to the crack initiation when a specimen or a structural component is under monotonically increasing load. Since this process is accompanied by the plastic crack tip opening displacement (CTOD), the stretched zone geometric parameters are expected to correlate with the material fracture toughness [1-4]. In contrast to other methods of metals fracture toughness measurement, the stretched zone as a subject for investigation has a number of unique properties, some of them are as follows:

- a possibility of the material fracture toughness evaluation at any time after the fracture process has been completed;
- a possibility of the direct measurement of the residual (i.e. without elastic component of displacement) critical CTOD related to the actual moment of the crack initiation;
- a possibility of the direct measurement of *local* parameter correlated with the critical crack tip opening displacement (STOD), in contrast to the conventional methods which give the *integral* STOD value;

152

- accessibility of any stretched zone profile for measurement whatever its location along the crack front (this property can have special interest for inhomogeneous materials, particularly, for welding joints);
- a possibility of additional evaluation of the material fracture toughness independently of other conventional methods used for its measurement.

Our investigations mainly involve two methods of stereoscopic measurements of the stretched zone geometric parameters:
(i)   measuring the profile of the stretched zone on one fracture surface [3];
(ii)   superimposing the mating profiles on the two opposite fracture surfaces [5,6]

The investigations were carried out on fracture surfaces of standard specimens used for the static and dynamic fracture toughness tests. The materials studied were: the soviet low-alloy steel 15X2HMФA of different heat treatments used for large pressure vessels [3,4]; heat resistant Czech steel CSN 15313 [6]; low-carbon steel [3]; transit pipe-line steels [11], etc. Measurements were made on 3 to 300 mm thick specimens tested in the.

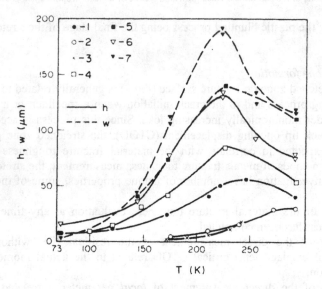

Figure 4. Temperature dependence of the stretched zone geometric parameters - height, *h*, and width, *w*. Legend: steel 15X2HMФA, shell (1,2); CSN 15313 (3); low-carbon steel (4,5); steel 15X2HMФA,plate (6,7). (1,4,5,6,7)-static loading; (2,3)- dynamic loading.

temperature range from 77 to 293 K with the loading rate $\dot{K}_I$ varying from 10^{-2} to 10^6 MPa $\sqrt{m}s^{-1}$ (Figs. 4, 5). The method involves stereo pairs photographed in a scanning electron microscope with their subsequent quantitative analysis in a precision stereocomparator in accordance with the recommendation of photogrammetry

Measurements of the stretched zone profile with the first of the aforementioned methods revealed essential scatter of results associated primarily with the irregular height and width of the stretched zone along the crack front. In this case an average value of the stretched zone height, h, was estimated from the increased number of measurements on the same crack front. To obtain the COD value basing on the assumption of the symmetric stretched zone on both sides of the crack plane which may be argued upon, the measured average stretched zone height was doubled , $2h$. Then this value was compared with the material fracture toughness obtained by conventional methods.

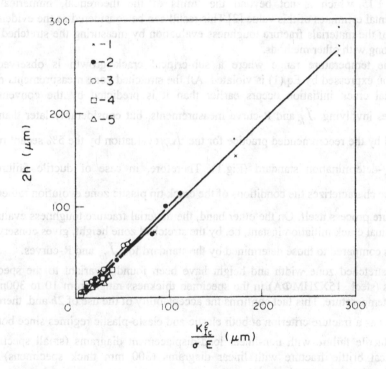

Figure 5. Relationships between the measured stretched zone heigth (2h) and calculated COD. Legend: steel 15X2HMФА, plate (1); steel 15X2HMФА, shell (2,3); CSN 15313 (4); low-carbon steel (5,6). (1,2,5)-static loading; (3,4)- dynamic loading.

When using the second method (superimposing the mating profiles of the two opposite fracture surfaces), it is of importance to find the mating points on the opposite fracture surfaces to ensure the stretched zone profiles along one and the same line. The scatter of the stretched zone measurement results obtained by this method is much smaller relative to the first method. First we shell indicate the main results confirmed by both methods.

1. In the temperature range where no sub-critical crack growth occurs prior to the specimen fracture (Fig.3), a correlation was found between the stretched zone height and the fracture toughness characteristics of steels under static and dynamic loading (Fig. 5). The comparison was made using the formula of linear fracture mechanics:

$$\delta_c = 2h = \alpha \frac{K_{Ic(d)}^2}{\sigma_{y(d)} E} \qquad (1)$$

where the value of $K_{Ic(d)}$ was determined when the condition of plain strain both at static and dynamic (index d) loading was satisfied. The magnitude of the proportionality factor α obtained for all the materials and loading conditions is within the range of 0.56 to 1.15 which is not beyond the limits of the theoretical, numerical and experimental estimates of this value [2].This result can be considered as the evidence of validity of the materials fracture toughness evaluation by measuring the stretched zone height along with other methods.

2. In the temperature range where a sub-critical crack growth is observed the correlation expressed by Eq.(1) is violated. All the stretched zone measurements reveal that actual crack initiation occurs earlier than it is predicted by the conventional techniques involving J_{Ic} and R-curve measurements, but considerably later than it is predicted by the recommended practice for the K_Q evaluation by the 5% secant rule in the K_{Ic}-determination standard (Fig.1). Therefore, in case of ductile failure the K_Q value characterizes the conditions of the crack tip plastic zone evolution rather than the fracture process itself. On the other hand, the material fracture toughness evaluation by the actual crack initiation instant, i.e. by the stretched zone height, gives conservative results as compared to those determined by the standard for J_{Ic} and R-curves.

3. The stretched zone width and height have been found invariant to the specimen thickness (steel 15Х2НМФА) in the specimen thickness range from 10 to 300mm at ambient temperature. This fact confirms the acceptability of the use of $2h$ and, therefore, δ_c values as a fracture criterion at both elastic and elasto-plastic regimes since both the typical ductile failure with non-linear load-displacement diagrams (small specimens) and typical brittle fracture with linear diagrams (300 mm thick specimens) were observed under above conditions.

4. A linear correlation was found between the stretched zone height and width for all the materials studied over a vide temperature range. Only in the region of brittle fracture there is a tendency towards the violation of this correlation. For instance, at the liquid nitrogen temperature the stretched zone height can be determined for some steels but its width is not visible. This can be associated with the character of the initial stages of the plastic zone evolution at the crack tip: first the plastic zone fans are developing normally to the crack plane and consequently the crack faces mutual displacement in this direction is prevalent.

At the same time the method of the stretched zone measurement by superimposing the mating profiles of the two opposite fracture surfaces has a number of specific features which determine its advantages over the other method:

1. The asymmetry of the stretched zone against the crack plane is well pronounced on the opposite mating fracture surfaces. The method made it possible to reveal the fact that the direction of the crack growth during its initiation always deviates from the main direction of the crack growth determined by the design of specimen and its loading conditions. A conclusion was made that the crack initiation occurs into one of two plastic zone fans along the trajectory of the maximum damage accumulation in the material due to plastic deformation. That is why no mirror-symmetric profiles of the stretched zone are generally observed on the opposite fracture surfaces.

2. For all the steels studied the geometrical parameters of the stretched zone on one of the two fracture surfaces depend on the location of the measured profile along the crack front. This is why the one-side fracture surface measurements generally reveal essential scatter of results. On the other hand, the superimposing of the two corresponding opposite profiles always gives close results irrespective of the location of the measured profiles along the crack front.

2.1.2. Crack initiation mechanism

From the results of stereofractographic analysis certain conclusions can be drawn as to both the fracture initiation mechanism in steels and the moment of the crack initiation. As it can be judged from real stretched zone profiles the local direction of the crack initiation makes a considerable angle with the fatigue pre-crack plane. It results in the asymmetry of the mating stretched zones profiles on the opposite fracture surfaces: on one of them the stretched zone is difficult to observe or cannot be found at all. Owing to the fact that stereofractographic analysis allows measurement of fracture surface profiles at any point of the crack front irrespective of its location through the thickness of specimen, the specific feature of the crack initiation mentioned can be observed in the bulk of the specimen. It seems to be associated with the plastic zone evolution features at the crack tip: the crack initiation probably occurs along one of the fans of the zone (namely, along the trajectories of maximum damage of material by plastic deformation) and then, later (i.e., beyond the region of heavy damage of material) the crack make a turn to the plane of maximum hydrostatic tension which coincides with the fatigue pre-crack plane continuation. Owing to the symmetric development of the two fans of the plastic zone there is an equal probability for the crack start along each of them. This assumption is supported by the results of fractographic studies: depending on the location along the crack front of the fracture surface profile being measured, the deviation of the crack initiation direction to one side or another occurs with approximately the same frequency.

2.2. TEMPERATURE - RATE EFFECT ON FRACTURE TOUGHNESS

The fracture toughness property is usually related to the crack initiation event. The typical relationships between fracture toughness and temperature are shown in the Fig.6

156

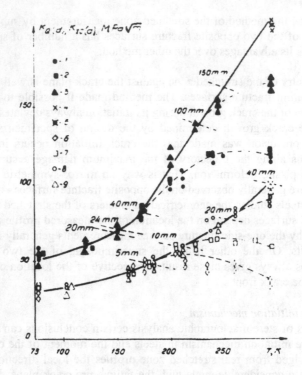

Figure 6. Fracture toughness dependence on temperature for the pressure vessel steel 15X2HMФA. Static loading, specimen thickness, mm: 1-20; 2-24; 3-40; 4-100; 5-150. Dynamic loading, specimen thickness, mm: 6-5; 7-10; 8-20; 9-40.

for static and dynamic loading with the specimens of different thickness. The values of K_Q and K_{Ic} (if valid) are estimated according to the ASTM E399 Standard. The boundaries of maximum temperature for the plane strain condition satisfaction are indicated by dashed lines for each specimen thickness, B, accordingly to the formula

$$K_Q^2 = 0.4 B \sigma_{0.2}^2 \qquad (2)$$

where the values of 0.2% offset yield stress, $\sigma_{0.2}$, were determined with account taken of its temperature and rate dependence. The K_Q and K_{Ic} temperature dependence are similar in shape for static and dynamic loading, but the curves for dynamic loading are located below the static curves and is shifted relative to the latter in the direction of growing temperatures by over 100 K. Similar temperature shift between static and dynamic loading curves for the same steel was independently observed using the stretched zone height as the fracture toughness measure [4].

The results presented in Fig.6 show that for each static and dynamic specimen thickness there is its own specific temperature above which the conditions of plane strain are not valid. This temperature is determined by the point of intersection of the lines representing the plane-strain condition boundary, Eq.(2), and the K_{Ic} or K_{Id} temperature dependence. Obviously, this temperature is related to the brittle-to-ductile transition temperature. Let us designate this temperature by T_{ps} and call it the plane-strain brittle-to-ductile transition temperature. The loading rate effect on the transition temperature is well expressed by a linear relationship between $1/T_{ps}$ and

transition temperature is well expressed by a linear relationship between $1/T_{ps}$ and $\log \dot{K}$ (Fig.7).

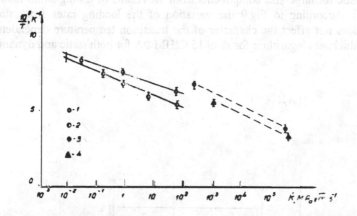

Figure7. Dependence of reciprocal brittle-to-ductile transition temperature on the loading rate logarithm. Low-carbon steel of two different melts, specimen thickness 2.5mm -1,2 [4]. Steel 15X2HMФA, specimen thickness: 3-20mm, 4-40mm.

The comprehensive experimental studies on temperature-rate sensitivity of plastic flow stresses and on fracture toughness have been carried out on both the smooth and pre-cracked specimens of low-carbon steel [7]. The thermodynamic parameters of plastic flow were determined on smooth specimens tested at different temperatures and strain rates. These results were used in order to describe the plastic zone evolution at the crack tip at different temperatures and loading rates. Main conclusion was drawn from this investigation: the fracture toughness dependence on temperature and loading rate can be quantitatively described within the framework of the interaction between thermally activated plastic flow inside of plastic zone and the athermal brittle fracture process involved the micro-crack nucleation at some characteristic distance ahead the crack tip [8]. Two mechanisms of brittle fracture were considered in this work. The first one involves the stage-by-stage decohesion of interatomic bonds at the crack tip, and the second is due to the microcrack coalescence in front of main crack. Quantitative evaluation of the threshold and minimum fracture toughness was made for both mentioned mechanisms corresponding to the cleavage crack propagation through crystalline lattice and to the crack growth trough the polycrystalline structure.

2.3. SPECIMEN THICKNESS EFFECT

Comparison of the stretched-zone geometric parameters values for specimens of different dimensions of steel 15X2HMΦA within the thickness range from 10 to 300 mm (Fig.8) revealed the independence of the stretched zone height, h, and width, w, of the thickness variation [3]. This fact confirms the opinion about the possibility to use the stretched zone height value (as well as COD value) as the fracture toughness characteristic.

On the other hand, the Fig.9 [4,9] shows the dependence of the brittle-to-ductile transition temperature on the specimen thickness within the thickness range from 5 to 300 mm for different low-alloyed steels [4]. This fact opens the possibility of predicting this characteristic for large-size components from the results of testing small laboratory specimens [9]. According to Fig.9 the variation of the loading rates from static to impact ones does not affect the character of the transition temperature dependence on the specimen thickness logarithm: for steel 15X2HMΦA for both static and dynamic

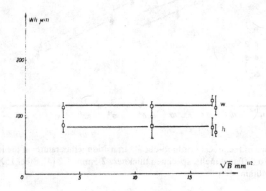

Figure 8.   Stretched zone height and width dependences on specimen thickness at room temperature for steel 15X2HMΦA

loading the log B versus T_{ps} relations remain parallel but shifted with respect to each other by over 100 K as it was previously recognized by both the standard fracture toughness and the stretched zone measurements. The above results of stereofractographic studies of the stretched zone height are in good agreement with the results of fracture toughness evaluation by standard techniques: there is a linear correlation between the both groups of data at the loading rates and temperatures studied up to the temperatures at which the sub-critical crack growth occurs (Fig.5). These data confirm the possibility of evaluating fracture toughness of the material independently by stereofractographic measurement of the stretched zone height.

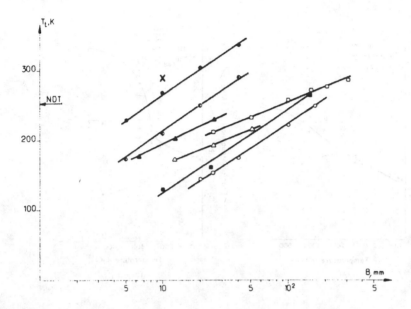

Figure 9. Plot of the transition temperature versus log*B* for various steels.

Some of above statements can be confirmed by the results of fracture toughness evaluation for nodular cast irons of different structure using the specimens of different thickness [10]. The microstructures of three cast irons are characterized by Table 1. The

TABLE 1. Microstructure of nodular cast irons

Type of cast iron	Average diameter of inclusions (μm)	Quantity of inclusions (1/mm2)	Average distance between inclusions (μm)	Pearlite content (%)
I	50	100	35	45-55
II	110	20	92	0
III	40	100	57	45-55

results of fracture toughness estimation under static loading for cast irons I, II and III are shown in Fig. 10, (a), (b) and (c), respectively.

160

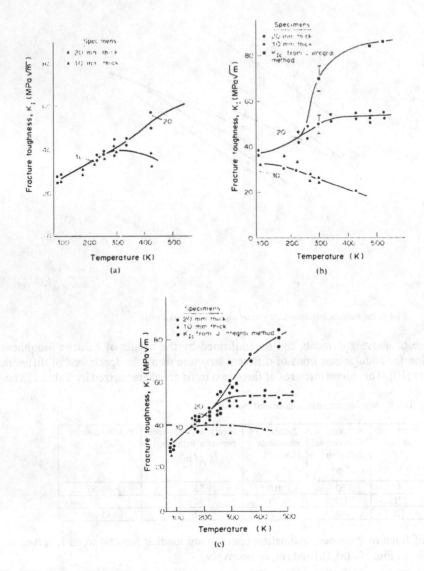

Figure 10.   Fracture toughness dependence on temperature for nodular cast irons (a) type I, (b) type II, (c) type III under static loading. 10 and 20 mm thicknesses are the limits within which equation (1) is satisfied for the corresponding specimen thickness.

For nodular cast iron I under static loading, load versus deflection curves were linear up to fracture almost over the whole temperature range studied. For 20 mm-thick specimens slightly non-linear curves were observed only at temperatures above 500 K. Note that a plane strain condition (2) was not satisfied at temperatures above 443 K. For

10 mm-thick specimens non-linear curves were obtained at temperatures above room temperature. A plane strain condition (2) was not satisfied at T≥210 K.

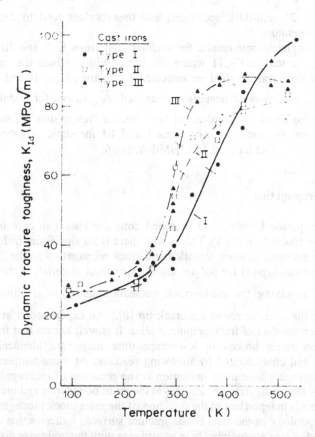

Figure 11.   Fracture toughness as a function of temperature for cast irons I, II and III under impact loading.

For nodular cast iron II load versus deflection curves of two types were obtained. For 20 mm-thick specimens linear curves were observed below room temperature. With an increase in temperature plastic deformation preceded fracture of the specimen and a maximum plastic crack opening displacement was observed at room temperature. For 20 mm-thick specimens equation (2) was not satisfied at T>213 K. For 10 mm-thick specimens essential plasticity was noted even at 173 K and plane strain condition (2) was not satisfied over the whole temperature range.

Cast iron III revealed less plasticity than cast iron II. For 20 mm-thick specimens load versus deflection curves were linear at temperatures below room temperature. Plane strain condition (2) was not satisfied at T>208 K. Ductile failure was observed with 10 mm-thick specimens at temperatures as low as 213 K, while equation (2) was not satisfied at T>150 K.

From Fig.10 it follows that 10 mm-thick specimens fail after more developed plasticity than the 20 mm-thick specimens. In the case where plane strain conditions are satisfied for 10 mm-thick specimens, the same fracture toughness values were obtained as for thicker specimens. In the case of ductile failure, calculated K_Q values were lower than for 20 mm-thick specimens and these values tend to decrease with an increase in temperature.

Fracture toughness test results for nodular cast irons I, II and III under impact loading are presented in Fig.11 where the limits within which the equation (2) for dynamic conditions was satisfied are indicated. Comparing Figs. 10 and 11 one can see that K_{Id} values at elevated temperatures exceed K_{Ic} values for nodular cast irons I and III. It is also should be emphasized here that the "lower shelf " fracture toughness values are about the same for cast irons I and III for static and dynamic loading, whereas it is not the fact for steel 15X2HMФA, Fig.6.

3. Crack propagation

The crack tip plastic blunting and stretched zone are absent in case of the perfectly brittle fracture (point 1 on Fig.1). This is why there is no significant difference between the micro-mechanisms of crack initiation and crack propagation on the "lower shelf" of fracture toughness: typical for polycrystalline structural materials is the "coalescence" mechanism   involving the micro-crack nucleation under local stress σ_f^* on the characteristic distance Xc ahead the crack tip [8]. An experimental evidence for this mechanism are the data of fractographic studies. It is well known that fracture surfaces of low-carbon steels broken in low-temperature range are identical for different temperatures and characterized by following features. At those temperatures, fracture occurs by cleavage whose main indication is the presence of cleavage facets showing river patterns. Generally, fracture initiates at the grain boundaries and the direction of its local extension is independent of the direction of the main crack propagation. Moreover, intricate morphology of the steel brittle fracture surfaces indicates that cleavage facets are distributed in size approximately in accordance with the grain size distribution.

3.1.   BRITTLE-TO-DUCTILE TRANSITION TEMPERATURE

The relation between various fracture transition temperatures and the fracture toughness transition curve has been described in paper [9]. Here we will briefly consider the main results of this work as related to the topic of present paper.

Many different tests were devised to find the transition temperature. Some of them are related to the crack initiation event, the other - to the rack propagation, whereas third group of methods takes into consideration entire process of body's fracture. It is well known that the transition temperatures given by these various tests don't coincide. Moreover, in some cases, there is not a single definition of the transition temperature; for instance, in a Charpy test it can be the temperature for a given level of absorbed energy, or the fracture appearance transition, among other. These tests are only qualitative and they do not provide a possibility to predict quantitatively the fracture of a given structure. Fracture mechanics, on the other hand, lead to the quantitative relation between the loading, the size of a defect and the material fracture toughness which allows to make such a calculation.

The main reason why there is a brittle-ductile transition is the lowering of the yield stress σ_y when the temperature is increased. It leads to a large increase of the plastic zone size at the tip of a crack or at the root of a notch and to a loss of constraint. For a given test the transition temperature should thus be related to the ratio between the thickness, B, and the plastic zone size which is proportional to $(K_{Ic}/\sigma_y)^2$.

Approximating the dependence of fracture toughness, K_{Ic}, on the temperature, T, by the formula

$$K_{Ic} = A\exp(T/T_0)$$

(3)

where A and T_0 are the material parameter, and emanating K_{Ic} between this formula and formula (2) leads to the transition temperature, T_{ps},

$$2T_{PS}/T_0 = \ln[(B/\beta)(\sigma_y/A)^2]$$

(4)

This can be checked by plotting T_{ps} against log B. Fig. 9 shows that a linear relation is indeed observed in such a diagram, for both the static and the dynamic loading. The formula (4) was verified with the experimental data obtained with different steels [9]. This provides us with the possibility of predicting the effect of different variables on transition temperature.

3.1.1. Thickness effect

From Eq.4, the transition temperatures T_t are simply related to the logarithm of the thickness ratio by the expression

$$T_{t1} - T_{t2} = (T_0/2)Ln(B_1/B_2)$$

(5)

where transition temperatures T_{t1} and T_{t2} are corresponding to the thicknesses B_1 and B_2, respectively. The knowledge of T_0, deduced from the K_{Ic} transition curve, is not needed, if the transition temperatures can be determined for two different thicknesses B_1 and B_2. Then transition temperature T_t for any thickness B can be determined as

$$\frac{T_t - T_{t1}}{T_{t2} - T_{t1}} = \frac{LnB/B_1}{LnB_2/B_1}.$$

(6)

3.1.2. Loading rate effect

These effects are more complicated to predict because T_0, A and σ_y, are all strain dependent. However, it was found that, in the instances where it could be checked,

T_0 and A varied only slightly with the strain rate. If this variation is neglected it follows that the transition temperature is inversely proportional to the logarithm of the stress intensity factor rate, $\log \dot{K}_I$, a relation which was indeed found in some cases, Fig. 7.

3.1.3. Transition temperature from different tests

For a given steel the formula (4) yields

$$T_{t1} - T_{t2} = (T_0 / 2) Ln[(B_1 / B_2)(\beta_2 / \beta_1)] \tag{7}$$

In this way it is possible to predict a particular transition temperature from the knowledge of another one. Or again, when the K_{Ic} transition curve is missing one could use

$$\frac{T_t - T_{t1}}{T_{t2} - T_{t1}} = \frac{Ln(B / B_1)(\beta_1 / \beta)}{Ln(B_2 / B_1)(\beta_1 / \beta_2)} \tag{8}$$

to deduce the transition temperature T_t from the knowledge of two other ones, T_{t1} and T_{t2}.

3.1.4. Deducing of the K_{Ic} transition curve from other tests

From formula (4) it appears that the parameters A and T_0 can be calculated when two transition temperatures are known corresponding either to two different thicknesses or to two different kinds of tests (i.e. of β) or both. It is of course also assumed that the temperature variation of σ_y is also known. This can be useful when K_{Ic} tests to access critical defects sizes are difficult or impossible to perform. A somewhat similar proposal was made by other authors who correlated the K_{Ic} transition curve with the Charpy one, using a different procedure based on fracture energy considerations. Present method appears to be based on more direct relations with the plastic zone size and seems to be much more flexible, since it can use types of tests other than the Charpy. Conversely, of course, formula (4) provides a way to estimate any transition temperature from the knowledge of the K_{Ic} transition curve.

3.1.5. Prediction of the irradiation effect

Irradiation effects will change T_0, A and σ_y. It is not feasible to measure K_{Ic} transition curves on irradiated steels for nuclear pressure vessels. However, it is quite possible to obtain the Charpy fracture appearance transition temperature (FATT) in that case. Another transition temperature would be needed, to deduce T_0 and A for irradiated steels. Then formula (4) can be used in order to construct the K_{Ic} transition curve.

4. Conclusions

1. The comprehensive investigation established direct relationship between stretched zone parameters and fracture toughness characteristics of structural steels, in case when no sub-critical crack growth is observed prior to fracture. In particular, if plane-strain conditions are satisfied the stretched zone parameters are independent of the specimen size variation. The stretched zone height being in better correlation with fracture toughness of the material than its width.

2. The method of the stretched zone measurement by superimposing the mating profiles of fracture surface demonstrate the scattering of results comparable or even less than the scattering of conventional fracture toughness measurements.

3. The direction of initial crack propagation differs essentially from the main direction of crack growth, during testing of standard specimens for the fracture toughness measurement. Most likely it can be explained by the fact that the initial crack growth occurs in the direction of one of two plastic zone fans, along the trajectory of the maximum damage of material by plastic deformation.

4. Based on the analysis of experimental data obtained from the investigation of fracture toughness characteristics of low-alloy steels evaluated with specimens of different thickness under static and impact loading, it is proposed to define the critical temperature of brittle-to-ductile transition, T_{ps}, as the upper temperature boundary up to which plane-strain conditions are satisfied at fracture. This temperature is linearly related to the specimen thickness logarithm under both static and impact loading. Simple relationship has been found between the T_{ps} temperature and some other transition temperatures. The advantage of T_{ps} over other transition temperatures is in that it is quantitatively related to the fracture toughness of material.

5. From the analysis of both own and published results on various steels, it appears that the transition temperatures for different tests and for different thickness B are given by formula

$$2T_{PS}/T_0 = \ln[(B/\beta)(\sigma_y/A)^2]$$

where σ_y is the yield stress taken at the strain rate for the test under consideration and A and T_0 are parameters which approximate the K_{Ic} transition curve through the formula

$$K_{Ic} = A\exp(T/T_0).$$

β is the ratio of the thickness to the plastic zone size which takes the value 2.5 for the ASTM plane strain transition and has other values for other transition temperatures. In spite of the fact that from the published data it was not possible to ascertain the above formula with very good accuracy, it was found to be helpful in making various predictions. Thickness and strain rate effects can thus be found with a minimum number of tests, and the various transition temperatures can be deduced from one another. In particular, the K_{Ic} transition curve can be approximately deduced from the knowledge of two transition temperatures obtained with either two thicknesses or two different kinds of tests.

Literature

1. Spitzig, W.A. (1969) Stretched zone observation, in *Electron Fractography*, ASTM STP 453, Philadelphia, pp. 96-110.
2. Krasowsky, A.J. (1988) Stretched zone stereoscopic measurements as an independent control of material fracture toughness, in *Failure Analysis-Theory and Practice.ECF-7.*, Proc. 7th *European Conf. on Fracture*, vol.2, Budapest, pp.796-804.
3. Krasowsky, A.J., and Vainshtok V.A. (1981) On a relationship between stretched zone parameters and fracture toughness of ductile structural steels, *Int. J. Fracture*, **17**, #6, 579-592.
4. Krasowsky, A.J., Kashtalyan, Yu.A., and Krasiko, V.N. (1983) Brittle-to-ductile transition in steels and the critical transition temperature, *Int.J. Fracture*, **23**, 297-315.
5. Krasowsky, A.J., and Stepanenko, V.A. (1979) A quantitative stereoscopic fractographic study of the mechanism of fatigue crack propagation in nickel, *Int. J. Fracture*, **15**, 203-215.
6. Krasowsky, A.J., Krasiko, V.N., Stukaturova, A.S., Bilek,Z., Holzmann, M., and Vlach, B. (1982) Vztah lomove houzevnatosti a kritickeho otevreni trhlini pry dynamickem zatezovani, *Zvaranie*,**31**, 322-326.
7. Pisarenko, G.S., Krasowsky A.J., and Yokobori, T. (1977) Studies on temperature-rate sensitivity of plastic flow stresses and on fracture toughness. *Reports of the Research Institute for Strength and Fracture of Materials*, Tohoku University, Sendai, Japan, **13**, #1, 1-57.
8. Krasowsky, A.J. (1998) Local approach to fracture of structural materials, in G. Cherepanov (ed), *Fracture. A Topical Encyclopedia of Current Knowledge*. Krieger Publ. Co., Malabar, Florida.
9. Francois, D., and Krasowsky, A. (1986) Relation between various fracture transition temperatures and the K_{Ic} fracture toughness transition curve, *Engineering Fracture Mechanics*, **23**, 455-465.
10. Krasowsky, A.J., Kramarenko, I.V., and Kalaida,V.V. (1987) Fracture toughness of nodular graphite cast irons under static, inpact and cyclic loading, *Fatigue, Fract. Engng. Mater. Struct.*, **10**, 3, 223-237.
11. Pluvinage, G., and Krasowsky, A.J. (1992) Dynamic fracture toughness at crack initiation, propagation and arrest for two pipe-line steels, *Engineering Fracture Mechanics*, **43**, #6, 1063-184.

A MODEL OF DAMAGE IN AN AUSTENITIC STAINLESS STEEL BY HIGH TEMPERATURE CREEP

S. K. KANAUN
Instituto Tecnológico y de Estudios Superiores de Monterrey,
Campus Edo de México, División de Ingenieria y Arquitectura.
Apdo. Postal 6-3, Modulo de Servicio Postal, Atizapán,
México 52926.

1. Introduction

In this study the model of quasibrittle fracture developed in [1] is applied for the description of the processes of nucleation and growth of microcracks by high temperature creep in an austenitic stainless steel of the AISI 316 family. Detailed experimental investigations of this steel were undertaken in [2,3].

The model proposed in [1] may be applied to the description of a wide class of fracture processes where tension-induced damage, i.e. microdefects nucleation and growth, precedes the final rupture. In many cases the source of the microdefects nucleation and growth is irreversible deformation (plasticity and creep). A slow (stable) growth of the defects usually takes place until some critical value of the damage density is reached. After that, the microdefects coalesce and form a nucleus of a macroscopic crack which results in the ultimate failure of the specimen. This picture is not universal but in many cases reflects the main features of the fracture processes. This type of fracture is referred as quasibrittle and it is implied that the total strains during all the process are not large and the classic theory of small deformations may be used for the description of the geometry of deformed solids.

The plan of the article is the following. In Section 2 the main equations of the model developed in [1] are revised. In Section 3 the model is applied to the description of damage processes by high temperature creep. The basic experiments for the construction of the material functions used in the model in the case of damage by creep are discussed in this Section. In Section 4 the model is applied to the description of nucleation and growth of intergranular microcracks in an austenitic stainless steel studied in [2,3]. In the Conclusion the possibilities of application of the model to the description of some other damage processes are discussed.

2. The model of quasibrittle damage processes

The central object of the consideration is a representative volume of the damaged

167

E. Bouchaud et al. (eds.), Physical Aspects of Fracture, 167–178.
© 2001 *Kluwer Academic Publishers. Printed in the Netherlands.*

material that consists in an elasto-visco-plastic medium and a set of microcracks that simulate the damage. This volume is subjected to a constant stress field σ. As usual, the properties of the representative volume are ascribed to a point of the continuum that is used as a mathematical model of the specimen. It is assumed that the thermodynamic parameters of state of a representative volume are the stress tensor σ and the damage parameter ρ. The damage parameter is a characteristic of the microdefects distribution in the representative volume. For simplicity, microdefects are simulated by plane circular cuts (cracks) with radii a. The radius of every crack depends only on its orientation $\mathbf{m}$ ($a = a(\mathbf{m})$, $\mathbf{m}$ is the normal vector to the crack plane). In this case the parameter ρ is presented in the form of a product

$$\rho(\mathbf{m}) = \tau(\mathbf{m})\psi(\mathbf{m}), \qquad \tau(\mathbf{m}) = \frac{4}{3}\pi a^3(\mathbf{m})n_0, \tag{1}$$

where n_0 is a numerical concentration of the cracks (their number in a unit volume) and $\psi(\mathbf{m})$ is the distribution function of the crack orientations. Because vectors $\mathbf{m}$ and $-\mathbf{m}$ define the crack of the same orientation, this function is an arbitrary positive function in a unit hemisphere Ω that satisfies the condition of normalization: the integral of $\psi(\mathbf{m})$ over Ω is equal to 1.

For simplicity, only isothermic processes are considered and the temperature T is excluded from the list of parameters of state. Inelastic deformations may also affect the thermodynamic state but that is taken into account via the damage parameter ρ.

Thus, in every point x of the macro continuum the function $\rho = \rho(x, \mathbf{m}, t)$ is defined; this function describes the damage accumulated in this point at a given time t. The main thermodynamic functions of the representative volume depend on damage and are defined as following. The internal energy density U is presented as a sum

$$U = W + \Pi, \tag{2}$$

where W is the elastic energy density, a quadratic form of the stress tensor

$$W = \frac{1}{2}\sigma \cdot \mathbf{B}(\rho) \cdot \sigma. \tag{3}$$

Here $\mathbf{B}(\rho)$ is the elastic compliance tensor that depends on the damage parameter ρ; $\Pi(\rho)$ is the part of internal energy associated with the presence of defects. For microcracks it may be interpreted as a surface energy of new surfaces created in the process of damage growth. Dots in the last equation and below mean the convolution of two tensors with respect to two nearest indexes ($(\mathbf{B}\cdot\sigma)_{ij} = B_{ijkl}\sigma_{kl}$ and here summation with respect to repeating indexes is implied.)

The function $\mathbf{B}(\rho)$ for Poisson's random sets of cracks (the centers of the cracks are distributed independently and homogeneously inside the representative volume) takes the form [4,5]

$$\mathbf{B}(\rho) = \mathbf{B}_0 + \int_\Omega \mathbf{M}(\mathbf{m})\rho(\mathbf{m})d\mathbf{m}, \tag{4}$$

$$\mathbf{M}(\mathbf{m}) = M_1 \mathbf{E}_5(\mathbf{m}) + M_2 \mathbf{E}_6(\mathbf{m}), \tag{5}$$

$$M_1 = \frac{4}{\pi\mu_0} \frac{(1-\nu_0)}{(2-\nu_0)}, \quad M_2 = -2\frac{\nu_0(1-\nu_0)}{\pi\mu_0(2-\nu_0)}$$

$$E_{5ijkl}(m) = [m_i\delta_{ik}m_l]_{(i,j),(k,l)}, \quad E_{6ijkl} = m_i m_j m_k m_l. \tag{6}$$

Here the brackets in indexes mean symmetrization, μ_0 and ν_0 are the shear modulus and Poisson ratio of the undamaged material.

The specific energy Π in Eq.(2) is a sum of the interfacial energies of all defects within the representative volume. This function can be presented in the form of the functional

$$\Pi(\tau, \psi) = (\pi(\tau), \psi) = \int_\Omega \pi(\tau(\mathbf{m}))\psi(\mathbf{m})d\mathbf{m}. \tag{7}$$

where $\pi(\tau)$ stands for the interfacial energy of the defects of the size dependent characteristic $\tau(\mathbf{m})$. Let defects of only one orientation grow. In this case $\Pi(\tau, \psi) = \pi(\tau)$. There are two simple scenarios: a) the number of the defects increases whereas the distribution of their sizes does not change and b) the size of the defects grows but their number stays unchanged. In the case a) Π is proportional to the numerical concentration of the inclusions n_0, i.e., to the first power of τ

$$\pi(\tau) = c_1\tau, \quad c_1 = const. \tag{8}$$

In the case b) the surface energy Π is proportional to the square of a typical size (a^2) of the inclusion, i.e., $\tau^{2/3}$:

$$\pi(\tau) = \pi_0 + c_2\tau^{2/3} \tag{9}$$

and π_0 and c_2 are constants. In general case $\pi(\tau)$ is a monotonically increasing convex function

$$\frac{d\pi}{d\tau} > 0, \quad \frac{d^2\pi}{d\tau^2} \le 0. \tag{10}$$

The balance of energy for the representative volume (the first law of thermodynamics) may be written in the form

$$dU = dA + dQ, \quad dA = \sigma \cdot d\varepsilon. \tag{11}$$

Here dA is the work of stresses σ on the increment of the total deformation $d\varepsilon$ and dQ is the increment of the heat density. The increment $d\varepsilon$ may be decomposed into the sum of the elastic $d\varepsilon^e$ and inelastic $d\varepsilon^p$ parts:

$$d\varepsilon = d\varepsilon^e + d\varepsilon^p. \tag{12}$$

Thus, the density of work is a sum of the work of stresses σ on the elastic deformation $(\sigma \cdot d\varepsilon^e)$ and on the inelastic one $(\sigma \cdot d\varepsilon^p)$. Note that stresses σ applied to the representative volume are considered as a fixed parameter in this analysis.

It is known that a part of the work done on irreversible deformations is converted into heat, and radiated into the thermostat in isothermal processes. Another part of this work is stored inside the body as hidden energy. This energy is associated with the nucleation and growth of the defects

$$\sigma \cdot d\varepsilon^p + dQ = \alpha(\sigma \cdot d\varepsilon^p). \tag{13}$$

Here, the coefficient $\alpha < 1$ characterizes the fraction of the irreversible work which is converted into hidden energy. In general this coefficient is not a constant but depends on the rate and type of loading, temperature, the history of loading, etc. Taking into account Eqs.(12),(13) the energy balance equation (11) may be written in the form

$$\dot{\Pi}(\tau, \psi) - \frac{1}{2}\sigma \cdot \dot{\mathbf{B}}(\rho) \cdot \sigma = \alpha(\sigma \cdot d\varepsilon^p), \quad \dot{f} = \frac{df}{dt}. \tag{14}$$

Here t is time or a time-like parameter (e.g., a parameter of loading), by the derivation of this equation the Hooke's law for the elastic deformations was used in the form $d\varepsilon^e = d\mathbf{B}(\rho) \cdot \sigma$.

The increment of free energy dF of the representative volume takes the following form

$$dF = dU - TdS = dW + \alpha(\sigma \cdot d\varepsilon^p). \tag{15}$$

Here S is the density of entropy of the body; the first equivalence is the definition of the increment of the free energy in isothermic processes. The increment of entropy is the following sum

$$dS = dS^e + dS^i, \quad dS^e = \frac{1}{T}dQ, \tag{16}$$

where dS^e is the increment because of heat transfer and dS^i is the generation of entropy inside the body.

The consequence of Eqs.(3),(15),(16) is the following expression for the entropy generation inside the representative volume

$$\dot{S}^i = \frac{1}{T}\left[\frac{1}{2}\sigma \cdot \dot{\mathbf{B}}(\rho) \cdot \sigma + (1-\alpha)(\sigma \cdot d\varepsilon^p)\right]. \tag{17}$$

As it follows from Eq.(4) for $\mathbf{B}(\rho)$, the derivative $\dot{\mathbf{B}}(\rho)$ may be written in the form

$$\dot{\mathbf{B}}(\rho) = \int_\Omega \mathbf{M}(\mathbf{m})\dot{\rho}(\mathbf{m},t)d\mathbf{m}, \tag{18}$$

where tensor $\mathbf{M}(\mathbf{m})$ is defined in Eq.(5). Thus, the balance equation (14) may be rewritten in the following final form

$$\dot{\Pi}(\tau, \psi) - \frac{1}{2}(\sigma \cdot \mathbf{M}(\mathbf{m}) \cdot \sigma, \dot{\rho}) = \alpha(\sigma \cdot \dot{\varepsilon}^p), \tag{19}$$

$$(\sigma \cdot \mathbf{M}(\mathbf{m}) \cdot \sigma, \dot{\rho}) = \int_\Omega \sigma \cdot \mathbf{M}(\mathbf{m}) \cdot \sigma\dot{\rho}(\mathbf{m},t)d\mathbf{m}, \tag{20}$$

where $\Pi(\tau, \psi)$ is given in Eq.(7). This equation does not define uniquely the function $\rho(\mathbf{m}, t)$ - the parameter of damage. In order to obtain a kinetic equation for ρ we employ the following extremal principle that is an extended version of the second law of thermodynamics.

From all possible variations of the damage parameter $\rho(\mathbf{m}, t)$ that satisfy balance equation (19) the actual variation corresponds to the maximum (supremum) of the entropy production $\dot{S}^i$.

It follows from Eq.(17) for $\dot{S}^i$, that a maximum of the entropy production is achieved together with the maximum of the functional $J(\rho, \dot{\rho})$

$$J(\rho, \dot{\rho}) = \frac{1}{2}\sigma \cdot \mathbf{B}(\rho) \cdot \sigma = \frac{1}{2}(\sigma \cdot \mathbf{M}(\mathbf{m}) \cdot \sigma, \dot{\rho}). \tag{21}$$

Thus, the function $\dot{\rho}(\mathbf{m}, t)$ that corresponds to the actual process of damage growth provides a maximum (supremum) to the functional $J(\rho, \dot{\rho})$ by the constraint of the balance equation (19).

To construct the solution of this variational problem let us select the function $\dot{\rho}(\mathbf{m}, t)$ in the following form

$$\dot{\rho}(\mathbf{m}, t) = \dot{\tau}(\mathbf{m}_*, t)\delta(\mathbf{m} - \mathbf{m}_*), \tag{22}$$

where $\delta(\mathbf{m})$ is Dirac's delta-function on a unit hemisphere. If the velocity rate $\dot{\tau}$ of the parameter τ is chosen in the form

$$\dot{\tau}(\mathbf{m}_*, t) = \frac{\alpha(\sigma \cdot \dot{\varepsilon}^p)}{\pi'(\tau(\mathbf{m}_*)) - \frac{1}{2}\sigma \cdot \mathbf{M}(\mathbf{m}_*) \cdot \sigma}, \qquad \pi'(\tau) = \frac{d\pi}{d\tau}, \tag{23}$$

the condition (19) is satisfied identically and the functional J in Eq.(21) takes the form

$$J(\rho, \dot{\rho}) = \alpha(\sigma \cdot \dot{\varepsilon}^p)\frac{\Phi(\mathbf{m}_*)}{\pi'(\tau(\mathbf{m}_*)) - \Phi(\mathbf{m}_*)}, \qquad \Phi(\mathbf{m}) = \frac{1}{2}\sigma \cdot \mathbf{M}(\mathbf{m}) \cdot \sigma. \tag{24}$$

The direction $\mathbf{m}_*$ in Eqs.(23) should provide a supremum to the functional $J(\rho, \dot{\rho})$ and hence to the function $\Psi(\mathbf{m})$

$$\Psi(\mathbf{m}) = \frac{\Phi(\mathbf{m})}{\pi'(\tau(\mathbf{m})) - \Phi(\mathbf{m})}, \qquad \sup[\Psi(\mathbf{m})] = \Psi(\mathbf{m}_*). \tag{25}$$

If this condition is met for one orientation $\mathbf{m}_*$ only, the solution of Eqs.(19), (21) of the variational problem in question is unique. If there are several orientations where the supremum of $\Psi(\mathbf{m})$ is achieved one may assume that all these orientations have equal chances. This assumption allows us to build a unique solution of the variational problem for this case as well.

3. The damage process by creep under uniaxial loading

Let us consider the process of damage growth by creep when the stress tensor σ is constant and $\sigma_1, \sigma_2, \sigma_3$ are three principal values of this tensor ($\sigma_1 > \sigma_2 > \sigma_3$). We assume that the rate $\dot{\tau}$ of the damage parameter is positive (non negative) in the process. (We do not consider healing of the defects.) Thus the maximum of the function $\Psi(\mathbf{m})$ in Eq.(25) is achieved together with the maximum of the function $\Phi(\mathbf{m})$

$$\Phi(\mathbf{m}) = \frac{1}{2}\sigma \cdot \mathbf{M}(\mathbf{m}) \cdot \sigma. \tag{26}$$

Substituting the expression (5) for the tensor $\mathbf{M}(\mathbf{m})$ in this equation we obtain for $\Phi(\mathbf{m})$

$$\Phi(\mathbf{m}) = \frac{1}{2}\left[M_1(\mathbf{m} \cdot \sigma^2 \cdot \mathbf{m}) + M_2(\mathbf{m} \cdot \sigma \cdot \mathbf{m})^2\right]. \tag{27}$$

Let $\mathbf{e}_1, \mathbf{e}_2, \mathbf{e}_3$ be the principal vectors (eigenvectors) of the stress tensor σ with the corresponding principal values $\sigma_1, \sigma_2, \sigma_3$. It can be shown [1] that extremal values of the function $\Phi(\mathbf{m})$ are achieved for the directions of the principal axes of tensor σ. If $\mathbf{m} = \mathbf{e}_i$ we have

$$\Phi(\mathbf{m}) = \Phi(\mathbf{e}_i) = \frac{1}{2}\left(M_1 + M_2\right)\sigma_i^2 = \frac{(1 - \nu_0)}{\pi\mu_0}\sigma_i^2. \tag{28}$$

Thus, the maximum of $\Phi(\mathbf{m})$ is achieved in the direction of the principal axis that corresponds to the maximal value of $|\sigma_i|$. Let us assume that nucleation and growth of the defects with orientation $\mathbf{m}$ is possible if $(\mathbf{m} \cdot \sigma \cdot \mathbf{m}) > 0$ (only positive normal stresses act across the potential defect). Then the direction of $\mathbf{m}$ coincides with the principal axis that corresponds to the maximal positive principal stress. If all principal values are negative, there is no growth of defects in the model. Thus, the planes of the defects will be orthogonal to the maximal tensile stress σ_1.

Let us consider the fracture process under creep in the stage of a stable creep velocity and uniaxial loading in the direction $\mathbf{e}$ ($\sigma_{ij} = \sigma_0 e_i e_j$). For many metals the velocity of creep strains in this stage is a power function of the second invariant of the stress tensor. Thus, the rate of the work of stresses on irreversible deformations is constant in this case

$$\sigma \cdot \dot{\varepsilon}^p = C \cdot \bar{s}^n = const. \tag{29}$$

$$\bar{s} = \sqrt{\frac{3}{2}s_{ij}s_{ij}}, \qquad s_{ij} = \sigma_{ij} - \frac{1}{3}\sigma_{kk}\delta_{ij}.$$

Here C and n are constants of the material, s is the deviator of stress tensor σ, $\bar{s}$ is the so called equivalent stress.

Since the orientation $\mathbf{m}$ of the defects coincides with the principal direction $\mathbf{e}$ of the stress field, the function $\psi(\mathbf{m})$ is

$$\psi(\mathbf{m}) = \delta(\mathbf{m} - \mathbf{e})$$

during all the process and Eq.(23) for $\dot{\tau}$ takes the form

$$\dot{\tau} = \frac{\alpha(t)(\sigma \cdot \dot{\varepsilon}^p)}{\pi'(\tau) - \sigma_0^2 b}, \qquad b = \frac{(1 - \nu_0)}{\pi\mu_0}. \tag{30}$$

The solution of Eq.(4.11) with the initial condition $\tau(0) = 0$ is

$$\int_0^t \alpha(t)dt = \frac{1}{(\sigma \cdot \dot{\varepsilon}^p)} \left[\pi(\tau) - \sigma_0^2 b\tau \right].$$ (31)

The behavior of the function $\tau(t)$ for various values of σ_0 was analyzed in [1]. It was shown that a stable growth of defects is possible until a moment t_* when the parameter τ achieves its critical value τ_* that depends on applied stress σ_0. When $t = t_*$ the rate of the defects growth increases to infinity. The moment $t = t_*$ may be interpreted as the moment of local failure (the fracture of the representative volume).

The condition of the local failure may be formulated as an infinite velocity of the defects growth $\dot{\tau}$. Thus this condition may be written as the equality to zero of the denominator in the right hand side of Eqs.(23),(30) for $\dot{\tau}$:

$$\frac{d\pi}{d\tau} = \frac{1}{2}\sigma \cdot \mathbf{M(e)} \cdot \sigma \quad \text{or} \quad \frac{d\pi}{d\tau} = \sigma_0^2 b.$$ (32)

Note that this condition is similar to the well known Griffith condition in the linear fracture mechanics. The growth of microdefects in our model is supported by irreversible deformations until some critical value of the parameter τ $(\tau = \tau_*)$. Then the defects may grow spontaneously according to Griffith's scheme. Indeed, the "energy release rate", i.e., δW due to the defects growth

$$\delta W = \frac{1}{2}\sigma \cdot \mathbf{M} \cdot \sigma \delta \tau$$

is equal to the increase of the interfacial energy

$$\delta \pi = \frac{d\pi}{d\tau} \delta \tau.$$

The equality $\delta W = \delta \pi$ gives the necessary condition (32) for local failure.

The critical value τ_* of the parameter τ that corresponds to the moment t_* may be found from the simple equation:

$$\frac{d\pi(\tau_*)}{d\tau} = \sigma_0^2 b = 0.627\frac{(1 - \nu_0^2)}{E_0}\sigma_0^2.$$ (33)

Here b is given in Eq.(30), E_0 is the Young's modulus of the undamaged material. The time of the final fracture t_* is defined from the equation

$$\int_0^{t_*} \alpha(t)dt = \frac{1}{(\sigma \cdot \dot{\varepsilon}^p)} \left[\pi(\tau_*) - \sigma_0^2 b\tau_* \right].$$ (34)

For the application of the model to the analysis of damage processes by complex loading we have to consider the complete system of equations of the macro continuum model of the specimen. In the case of creep, this system consists in the classical equations of the theory of viscoelasticity

$$div(\sigma) = -q, \quad \varepsilon = \varepsilon^e + \varepsilon^p, \quad Rot(\varepsilon) = 0,$$ (35)

$$\varepsilon^e = \mathbf{B}(\rho)\cdot\sigma, \qquad \dot{\varepsilon}^p = C\bar{s}^n s, \qquad (36)$$

where Rot is the operator of incompatibility, the boundary conditions on the surface of the specimen and the kinetic equation for the damage parameter ρ. At a given moment t, the rate of this parameter has the form

$$\dot{\rho}(x, \mathbf{m}, t) = \dot{\tau}(x, \mathbf{m}_*, t)\delta(\mathbf{m} - \mathbf{m}_*), \qquad (37)$$

where

$$\dot{\tau}(x, \mathbf{m}_*, t) = \frac{\alpha\left[\sigma(x) \cdot \dot{\varepsilon}^p(x)\right]}{\pi'(\tau(x, \mathbf{m}_*, t)) - \frac{1}{2}\sigma(x) \cdot \mathbf{M}(\mathbf{m}_*) \cdot \sigma(x)}. \qquad (38)$$

Here the direction $\mathbf{m}_*$ provides supremum to the function $\Psi(x, \mathbf{m})$

$$\sup[\Psi(x, \mathbf{m})] = \Psi(x, \mathbf{m}_*), \qquad \Psi(x, \mathbf{m}) = \frac{\Phi(x, \mathbf{m})}{\pi'(\tau(x, \mathbf{m}, t)) - \Phi(x, \mathbf{m})}, \qquad (39)$$

where function $\Phi(x, \mathbf{m})$ is defined in Eq.(26) and $\sigma = \sigma(x)$.

For the solution of this system one has to point out the initial condition: the value of the damage parameter ρ at initial moment $t = 0$

$$\rho(x, \mathbf{m}, t)|_{t=0} = \rho_0(x, \mathbf{m}). \qquad (40)$$

The two functions that appear in the kinetic equation (38) (α and $\pi'(\tau)$) should be found from experimental observations of damage processes under simplest stress states. For the construction of the function α one has to imply that this function depends on some invariants of the stress-strain state. As it was mentioned above this function depends on creep rate, history of loading, temperature, etc. and one can accept

$$\alpha = \alpha(T, \bar{s}, \lambda), \qquad \lambda = \int_0^t d\bar{\varepsilon}^p, \qquad \bar{\varepsilon}^p = \sqrt{\frac{3}{2}\varepsilon_{ij}^p \varepsilon_{ij}^p}. \qquad (41)$$

Here $\bar{\varepsilon}^p$ is the so called equivalent plastic strain.

From Eq.(30) follows the formula

$$\alpha = \frac{\dot{\tau}}{(\sigma \cdot \dot{\varepsilon}^p)} \left[\pi'(\tau) - \sigma_0^2 b\right] \qquad (42)$$

that may be used for the construction of the function α from direct measurements of damage in the process of creep.

The function $\pi'(\tau)$ should be found from the measurement of the critical values of the parameter τ_* at the moment of final fracture. The necessary equation follows from Eq.(33) and has the form

$$\frac{d\pi(\tau_*)}{d\tau} = 0.627\frac{(1 - \nu_0^2)}{E_0}\sigma_0^2. \qquad (43)$$

Direct measurement of the function $\tau_* = \tau_*(\sigma_0)$ and the construction of the inverse function $\sigma_0(\tau_*)$ give us the dependence $\pi'(\tau)$ that appears in the kinetic equation (38) and in Eq.(42) for α.

4. Kinetic equation for the damage in an austenitic stainless steel

Let us consider the main features of the damage process in an austenitic high creep-resistance steel of the AISI 316 family studied in [2,3]. The creep and damage by the temperature $600°C$ were studied on notched bars for localization of the damage zone. For this steel the power constitutive law for the creep strain velocity may be accepted in the form

$$\dot{\varepsilon}^p = C\bar{s}^n s. \tag{44}$$

Here C and n are constants of the material given in [2,3] ($C \approx 10^{-38.5}$, $n \approx 12.7$, $[\dot{\varepsilon}^p]$ in h^{-1}, $[s]$ in MPa.)

In [2,3] the damage was defined as a set of intergranular cracks developing in the process of creep. Most of these cracks develop perpendicular to the direction of the applied force. As a damage parameter D in the observation the following quantity was used

$$D = \frac{L_c}{L_t}. \tag{45}$$

Here L_c is the cumulative length of cracked grain boundaries (the length of all the cracks divided by the area of observation in arsection of the damaged zone), L_t is the cumulative length of the grain boundaries in the same area. It is known from stereology that when the damage is isotropic, the parameter D coincides with the ratio S_c/S_t, where S_c is the area of intergranular cracks per unit volume and S_t is the area of grain boundaries in the same volume. If the mean size of the cracks has order of the size of the grains d we have for the areas S_c and S_t in a unit volume V ($V = l^3 = 1$) the following equations

$$S_t \approx 3\frac{l^3}{d}, \quad S_c \approx \frac{\pi}{4}d^2 n_0, \quad D = \frac{L_c}{L_t} = \frac{S_c}{S_t} \approx \frac{\pi}{12}d^3 n_0, \tag{46}$$

where n_0 is the numerical concentration of microcracks. Because the parameter τ in Eq.(1) is $\tau = \pi d^3 n_0 / 6$ in this case, the connection between the parameters τ and D takes the form

$$\tau \approx 2D. \tag{47}$$

The empirical kinetic equation for the parameter D was obtained in [2,3] from direct experimental observations of the damage in the process of creep. In [2,3] this equation was proposed in the following incremental form

$$dD = A(\sigma_1)^4 \frac{d\bar{\varepsilon}^p}{\sqrt{\bar{\varepsilon}^p}}. \tag{48}$$

Here $A = 10^{-12}(MPa)^{-4}$, D is undimensional, σ_1 is the maximal principal stress (MPa). From Eqs.(42),(47) and (48) we get the following equation for the function α

$$\alpha \approx 2\frac{A(\sigma_1)^3}{\sqrt{\bar{\varepsilon}^p}} \left[\pi'(\tau) - (\sigma_1)^2 b\right]. \tag{49}$$

176

This equation serves in some region of plastic strains where $\bar{\varepsilon}^p > 0$. Note that the connection of $\bar{\varepsilon}^p$ with the applied stresses may be found from the constitutive equation (44). As it is seen from Eq.(49) the function α that characterizes the portion of the inelastic work connected with the nucleation and growth of microcracks, increases together with the applied stresses. On the other hand with increasing inelastic deformations the number of weak links in the material decreases and the process of microcracks nucleation is being exhausted in the material. As a result, α decreases when $\bar{\varepsilon}^p$ increases.

For the construction of the function $\pi'(\tau)$ let us turn to the condition of the final fracture (43). From the dependences $\sigma(t_*)$ (creep endurance curves, see Fig.2 in [2] and Fig.3 in [3]) and the dependences $D_*(t_*)$ (Fig.7 in [2]) one can reconstruct the dependence $\tau_*(\sigma) = 2D_*(\sigma)$ for time to failure included between 10h and $2 \cdot 10^3$h. Here D_* and τ_* is the values of the damage parameters in the moment of final fracture t_*. For the specimen FLE5 (see [2,3]) the dependence $\tau_*(\sigma)$ is presented in Fig.1. Here square points are experimental data of [2,3], solid line is the accepted approximation of the dependence $\tau_*(\sigma)$.

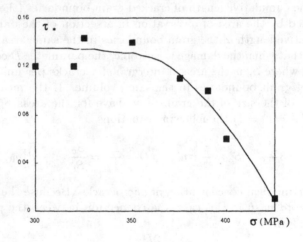

Figure 1. The dependence of critical damage parameter τ_* on applied stress σ.

The function $\pi'(\tau) = \dfrac{d\pi}{d\tau}$ may be constructed now from Eq.(43) using the function $\sigma(\tau_*)$ that is the inverse function with respect to $\tau_*(\sigma)$

$$\pi'(\tau) = 0.627\frac{(1 - \nu_0^2)}{E_0}\sigma^2(\tau). \tag{50}$$

The graph of this function for $E_0 = 144000$ MPa, $\nu_0 = 0.3$ (these are the Young's modulus E_0 and Poisson's ratio ν_0 of the considered steel by T=600°C) is presented in Fig.2.

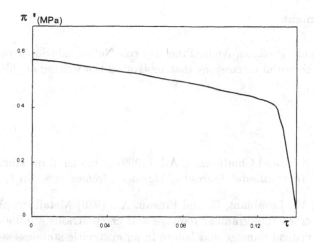

Figure 2. The dependence of π' on damage parameter τ.

Eqs. (44),(49) and (50) give us the solution of the problem of the construction of the function α and $\pi'(\tau)$ for the case of damage processes by creep. Thus, the sufficient system of experimental data for the construction of the functions in the kinetic equations (30) and (38) in the case of creep is the law of evolution of damage in the process of uniaxial creep and the dependence of the critical value of damage parameter on applied stresses.

5. Conclusion

The experimental data presented in [2,3] allow us to build the functions α and π' in the kinetic equation for the damage parameter in the framework of the developed model of damage. When these functions are constructed the model may be applied to the description of damage processes by complex loading histories (see [1]) and to the evaluation of damage in real constructions by high temperature creep.

Another natural area of application of the model is damage processes by low-cycle fatigue and the damage produced by a combination of fatigue and creep in the austenitic stainless steels considered in this study (experimental investigations of these steels by such a loading were undertaken in [6,7]). Some specific difficulties of the application of the model to these cases are connected with a strong dependence of the fatigue damage on the distance from the surface of the specimen. Thus, in the vicinity of the surface the kinetic equation for the damage parameter should differ from the one in the bulk of the material. There is a possibility to take into account this fact in the framework of the proposed model after corresponding corrections of the latter.

Acknowledgment

The author thanks Professor Andre Pineau (Ecole Nationnale Supérieure des Mines de Paris) for the useful discussions that motivated the writing of this work.

References

1. Kanaun, S.K. and Chudnovsky, A.I. (1999) A model of quasibrittle fracture of solids, *International Journal of Damage Mechanics* **8**, No 1, 18-40.

2. Yoshida, M., Levaillant, C. and Pineau, A. (1986) Metallographic measurement of creep intergranular damage and creep strains - influence of stress state on critical damage and failure in an austenitic stainless steel, *International Conference on Creep, April 14-18, 1986, Tokyo, Proceedings*, 327-332.

3. Yoshida, M., Levaillant, C., Piques, R., and Pineau, A. (1990) Quantitative study of intergranular damage in an austenitic stainless steel on smooth and notched bars, *High Temperature Fracture Mechanisms and Mechanics, ESF6* (Ed. by P.Bensussan), Mechanical Engineering Publications, London, 3-21.

4. Kanaun, S.K. (1980) Poisson's set of cracks in a homogeneous elastic medium, *Applied Mathematics and Mechanics (PMM)* **44**, No 6, 1129–1139.

5. Kanaun, S.K. (1993) Effective elastic properties of the medium with an array of thin inclusions, *Revista Mexicana de Fisica* **39**, No 6, 850–869.

6. Argence, D., Weiss, J., and Pineau, A. (1996) Observation and modelling of transgranular and intergranular multiaxial low-cycle fatigue damage of austenitic stainless steel, *Multiaxial and Fatigue Design, ESIS 21* (Edited by A.Pineau, G. Cailletaud and T.C. Lindley), Mechanical Engineering Publications, London, 209-227.

7. Weiss, J., and Pineau, A. (1993) Continuous and sequential multiaxial low-cycle fatigue damage in 316 stainless steel, *Advances in Multiaxial Fatigue, ASTM STP 1191* (Edited by D.C. MacDowell and R. Ellis), American Society for Testing and Materials, Philadelphia, 183-203.

INTERRELATION BETWEEN CONSTITUTIVE LAWS AND FRACTURE IN THE VICINITY OF FRICTION SURFACES

S. ALEXANDROV
Alcoa Technical Center
100 Technical Drive
Alcoa Center, PA 15069-0001, USA

and

Institute for Problems in Mechanics
Russian Academy of Science
101 Prospect Vernadskogo
117526, Moscow, Russia

Abstract - It is well known that cracks can produce stress singularities in linear elastic materials. One common type of stress singularity is $\sigma \propto K/\sqrt{r}$ where r is the distance from the crack tip and K is the stress intensity factor. The stress intensity factor approach is very effective in structural design. In particular, a popular fracture criterion is formulated in the form $K = K_c$ where K_c is a critical value of the stress intensity factor, a material constant. In the present paper the general concept of the stress intensity factor is adopted for fracture analysis in the vicinity of maximum friction surfaces (surfaces where the frictional stress attains its maximum possible value). The basic constitutive law is the rigid/perfectly plastic solid. In this case the equivalent strain rate, ξ_{eq}, is describable by nondifferentiable functions in the vicinity of maximum friction surfaces and its behavior is given by $\xi_{eq} \propto D/\sqrt{x_3}$ where x_3 is the distance from the friction surface. Most fracture criteria in metal forming involve the equivalent strain rate in such a manner that they would immediately predict fracture initiation at the maximum friction surface since the behavior of a fracture parameter is singular and its value approaches infinity at the friction surface. Therefore, by analogy with the mechanics of cracks, it is natural to assume that the intensity of the strain rate singularity can be used to predict fracture in the vicinity of maximum friction surfaces. In the present paper the equivalent strain is used as the fracture parameter. The strain intensity factor E, which controls the magnitude of the equivalent strain in the vicinity of a maximum friction surface, is introduced and then a fracture criterion is postulated in the form $E = E_r$ where E_r is assumed to be a material constant. The effects of temperature, pressure dependence of the yield condition, viscosity, and strain hardening

E. Bouchaud et al. (eds.), Physical Aspects of Fracture, 179–190.
© *2001 Kluwer Academic Publishers. Printed in the Netherlands.*

on the behavior of solution in the vicinity of the maximum friction surface are discussed. A few examples are proposed to illustrate the approach.

1. Preliminary remarks

The stress intensity factor is one of basic concepts in linear elastic fracture mechanics. This parameter appears in asymptotic analyses performed in the vicinity of a sharp crack-tip and is the coefficient of the singular term. In spite of the fact that the assumptions under which the stress intensity factor is determined are not satisfied in real materials (the crack-tip is not sharp and a region of inelastic deformation exists in its vicinity), this approach is effective in structural design ([1, 2] among numerous textbooks and monographs on the subject). In particular, a fracture criterion is formulated in the form $K = K_c$ where K is the stress intensity factor and K_c is its critical value, a material parameter. The stress intensity factor approach has been extended to a class of nonlinear elastic materials in [3, 4] and to the estimation of butt joint strength (free – edge stress intensity factor) [5, 6]. The aim of the present paper is to apply this concept to quite a different field, fracture prediction in the vicinity of frictional surfaces in metal forming processes. Usually, rigid/plastic (in· the broad sense that the elastic portion of the strain tensor is negligible) constitutive laws are very suitable for such applications. Of such laws, the rigid/perfectly plastic model is the simplest which reflects the basic features of material behavior. An analogy to this model in fracture mechanics might be the linear elastic material. Of frictional laws conventionally applied in the modeling of metal forming processes, Tresca's frictional law is most popular and appropriate. An extreme case of this law is the maximum friction law. At sliding this law may be written in the form

$$\tau_f = k \tag{1}$$

where τ_f is the frictional stress and k is the local shear yield stress which may depend on the temperature, the equivalent strain rate, the equivalent strain and other internal variables. An analogy to the maximum friction interface in fracture mechanics might be the sharp crack tip. The remainder of this section shows that the analogy between cracks and friction surfaces is quite complete. Since, in the case of perfectly/plastic materials the velocity fields are singular in the vicinity of friction surfaces where (1) holds. It is interesting, although not very important, that the order of singularity is the same as in crack problems. The singularity is preserved for slight extensions of the model (thermoplasticity, pressure-dependent plasticity), but is lost in the presence of such phenomena as hardening and viscosity.

For the rigid/perfectly plastic material, a few analytical solutions for specific problems involving surfaces with maximum friction are known (vortical flow in a plastic mass [7], compression of a block between platens [8], flow through a plane converging channel [8], flow through an axisymmetric converging channel [9-12], axisymmetric flow of a plastic mass along a rigid fiber [13]). These solutions have been based on different yield conditions and all of them have shown that the equivalent strain rate follows an inverse square law near the friction surface and, hence, approaches infinity at the surface. General study of solution behavior in the vicinity of maximum friction surfaces was initiated in [14] where plane strain deformation was considered. Three-dimensional flow was investigated in [15] for the Mises yield condition and in [16] for an arbitrary smooth yield condition. Axisymmetric flow of the Tresca material was studied in [17]. These results have demonstrated that the aforementioned behavior of the equivalent strain rate is a very general feature of rigid/perfectly plastic solutions. Possible exceptions to this rule have been mentioned in [16] where also the term 'strain rate intensity factor' has been introduced.

The general results for rigid/perfectly plastic material have been extended to thermoplasticity [18]. In this case the strain rate intensity factor enters in the asymptotic representation of the temperature in the vicinity of the maximum friction surface. There are particular solutions which illustrate singular behavior of the velocity field for pressure-dependent incompressible materials [19, 20]. For such materials, however, the order of singularity depends on the constitutive law chosen [21]. In the case of the double-shearing model [22] the order of singularity coincides with that for rigid/perfectly plastic materials.

The behavior of the solution in the vicinity of a maximum friction surface for other constitutive laws suitable for the modeling of metal forming processes (viscous, viscoplastic and rigid/plastic hardening materials) has been studied in [23 – 25]. It has been shown that no singular velocity field exists in these materials, and that the maximum friction law implies sticking conditions. Therefore, a boundary layer theory proposed in [26] for Bingham solids may be combined with the singular behavior of velocity field found in [14–17] to develop an approach to the modeling of fracture. Also, rigid/plastic, hardening materials with a saturation stress have been excluded from consideration in [25]. However, particular examples [27, 28] show that the solution behavior for such materials has features inherent in both rigid/perfectly plastic and rigid/plastic hardening (with no saturation stress) materials. Finally, particular analytical solutions illustrating the evolution of a damage parameter in the vicinity of a friction surface with the sticking conditions have been given in [29, 30]. No general study on the subject is available.

2. Strain rate intensity factor. Perfectly/plastic materials

A particular rigid/perfectly plastic material may be defined by Mises' yield criterion

$$s_{ij}s_{ij} = 2k^2 \tag{2}$$

and the normality rule

$$\xi_{ij} = \left(\sqrt{3}\xi_{eq}/2k\right)s_{ij} \tag{3}$$

where s_{ij} are the deviatoric portions of the stress tensor, ξ_{ij} are the components of the strain rate tensor, $k = const$ and ξ_{eq} is the equivalent strain rate,

$$\xi_{eq} = \sqrt{(2/3)\xi_{ij}\xi_{ij}} \tag{4}$$

Let ω be a tool surface where condition (1) is satisfied, and consider equations (2) and (3) at an arbitrary point, M, on that surface. The tool is regarded as fixed. A Cartesian coordinate system $x_1 x_2 x_3$ is taken to be situated at M with the x_3- axis directed along the normal to ω, away from the rigid tool and toward the plastic material. The x_1- axis is directed along the velocity vector at M, and the x_2- axis is perpendicular to both x_1 and x_3 forming a right – hand system. Since the friction force at sliding is always directed opposite to the velocity vector, it follows from (1) that

$$s_{13} = k \tag{5}$$

at the point M. Substituting (5) into (2) gives

$$s_{ij} = 0 \quad (\text{except } s_{13}) \tag{6}$$

Now if we assume $\xi_{eq} < \infty$ equations (3) combined with (6) result in

$$\xi_{ij} = 0 \quad (\text{except } \xi_{13}) \tag{7}$$

But (7) imposes severe restrictions on the flow and in general is not satisfied. A good example is a steady axisymmetric flow. In this case the circumferential strain rate, which is equal to ξ_{22} in our nomenclature, is defined by the radial velocity and not by derivatives. Thus, if sliding occurs this velocity is not zero and, the circumferential strain rate is not zero. Hence, it is necessary to take

$$\xi_{eq} \rightarrow \infty \quad \text{as} \quad x_3 \rightarrow 0 \tag{8}$$

Using the equilibrium equations it may be shown [16] that the specific form of (8) is

$$\xi_{eq} = D/\sqrt{x_3} + o\left(\sqrt{x_3}\right) \tag{9}$$

where D is the strain rate intensity factor and x_3 is the distance from the friction surface.

The singular behavior of the velocity field has been also found in thermoplasticity [18]. Particular solutions are known for different models of pressure-dependent materials [19 – 21]. In the latter case the order of singularity depends on the constitutive law chosen.

3. Effects of viscosity and hardening

Viscosity and hardening drastically change the solution behavior in the vicinity of a maximum friction surface. As a particular example, consider the compression of an infinite layer between two rough parallel platens. A solution for a rigid/perfectly plastic material is given in [8, 31], and it was found [8] that under certain conditions it is in good agreement with experiment. Extensions of this solution for rigid/hardening and viscoplastic materials have been proposed in [32] and [33] respectively. In both of these cases, however, it can be shown that no solution exists if the maximum friction law is applied. On the other hand, the overall load per unit length is unrealistically high if the friction stress is slightly below the local shear stress. The reason for such behavior of the global solutions originates from different asymptotic behavior in the vicinity of the friction surface. In particular, sliding was forced to occur at the frictional interface whereas the general theory outlined below requires sticking.

A class of viscoplastic materials is defined by (2), (3) and the condition that k is a function of ξ_{eq} such that

$$k \to \infty \quad \text{as} \quad \xi_{eq} \to \infty \tag{10}$$

Equation (1) leads to (5) – (7) and, excluding the special case (7), to (8). Then, combining (8) and (10) shows that stresses are infinite at the maximum friction surface at sliding which has no physical meaning. Thus, one should conclude that sticking occurs at such surfaces. Axisymmetric and planar flows of viscous and viscoplastic materials in the vicinity of maximum friction surfaces have been studied in detail in [23] and [24] respectively. A boundary layer theory for viscoplastic materials has been developed in [26].

A particular class of rigid/plastic hardening materials is defined by (2), (3) and the condition that k is a function of the equivalent strain, ε_{eq}, such that

$$k \to \infty \quad \text{as} \quad \varepsilon_{eq} \to \infty \tag{11}$$

The equivalent strain is given by the following equation

$$d\varepsilon_{eq}/dt = \xi_{eq} \tag{12}$$

where d/dt is the material rate. As before, equation (1) leads to (8). If a material point remains on a maximum friction surface for a finite time interval, then it follows from (8) and (12) that $\varepsilon_{eq} \to \infty$ at this surface. This condition and (11) show that stresses approach infinity at the surface. Axisymmetric and planar flows of such materials have been considered in [25].

Returning to the analogy between fracture mechanics and the subject of the present paper, it is possible to say that viscosity and hardening in the latter case play the same role as plasticity in the former.

4. Basic concept of the proposed approach

Reviews of ductile fracture criteria in metal forming have been given in [34 – 36]. All of these criteria would predict fracture immediately where (8) and, as a consequence, (12) are satisfied. It is similar to the fact that fracture criteria from strength of materials would predict fracture at any crack tip. Extending the concept of the stress intensity factor, it is possible to assume that the strain rate intensity factor introduced in equation (9) controls fracture in the vicinity of maximum friction surfaces. In metal forming, one of the most popular fracture parameters is the strain to fracture, ε_f, so that the fracture criterion is written in the form

$$\varepsilon_{eq} = \varepsilon_f \tag{13}$$

Substituting (9) into (12) gives

$$d\varepsilon_{eq}/dt = D/\sqrt{x_3} + o\left(\sqrt{x_3}\right) \tag{14}$$

in the vicinity of the friction surface. This equation shows that the strain rate intensity factor controls the magnitude of the equivalent strain in the vicinity of the friction surface. It is necessary to mention that the structure of solution to (14) strongly depends on the type of problem. In steady flows, $\partial \varepsilon_{eq}/\partial t = 0$ and, therefore, a space derivative of ε_{eq} must be infinite at $x_3 = 0$. In nonsteady processes, ε_{eq} is given at the initial instant (usually, $\varepsilon_{eq} = 0$) and, in general, the derivative $\partial \varepsilon_{eq}/\partial t$ must approach infinity at this instant. A similar behavior of solution has been found for the heat conduction equation [18]. It is clear that the friction surface is singular for ε_{eq} and that D enters the coefficient, E, of the singular term. This coefficient may be named the strain intensity factor. Then, in place of (13) a possible fracture criterion may be formulated as

$$E = E_c \tag{15}$$

where E_c, for a given material, can depend on various factors, but most important of these is probably the hydrostatic stress. The form of (15) is similar to that of the fracture criterion based on the concept of the stress intensity factor.

5. Modified Couette flow

It is very difficult to find a realistic problem involving friction which would admit an analytical/semianalytical solution for a class of constitutive laws. Therefore, a somewhat artificial problem, modified Couette flow, has been chosen to demonstrate the approach. Consider a circular hollow cylinder of initial internal radius a_0 and initial external radius b_0 and assume that a frictional stress and a pressure are applied to the inner surface and that the circumferential velocity vanishes at the outer radius. Due to this system of loading, the deformation of the cylinder is torsion and expansion. Introduce a polar coordinate system $r\theta$ such that $r = 0$ corresponds to the axis of symmetry of the cylinder. The current inner radius of the cylinder will be denoted by a and the outer by b. Then, $\dot{a} > 0$ is the radial velocity at $r = a$ and its value is assumed to be given. Using the sticking conditions, this problem has been solved in [37] for different materials. In the case of sliding, which is of interest in the present study, this solution can be used with insignificant changes. In order to apply the fracture criterion (15), the solution should be found for the rigid/perfectly plastic material ($k = const$). The radial velocity, u, and the shear stress in the polar coordinate system, $s_{r\theta}$, are determined in the following form

$$u = a\dot{a}/r \quad \text{and} \quad s_{r\theta} = k \sin 2\varphi = ka^2/r^2 \tag{16}$$

where φ is an auxiliary variable. This solution satisfies the maximum friction law. The solution for the circumferential velocity, v, satisfying the condition $v = 0$ at $r = b$ is

$$\frac{v}{\dot{a}} = \frac{\sqrt{r^4 - a^4}}{ra} - \frac{r}{a}\sqrt{1 - \frac{a^4}{b^4}} \tag{17}$$

In the vicinity of the maximum friction surface $r = a$ this equation may be rewritten in the form

$$\frac{v}{\dot{a}} = -\sqrt{1 - a^4/b^4} + 2\sqrt{r - a}/\sqrt{a} + o\left(\sqrt{r - a}\right) \tag{18}$$

Using (16) and (18) the equivalent strain rate may be found as

$$\xi_{eq} = \frac{\dot{a}}{\sqrt{3}\sqrt{a}\sqrt{r - a}} + o\left(\frac{1}{\sqrt{r - a}}\right) \tag{19}$$

Thus, the strain rate intensity factor is given by

$$D = \dot{a}/\sqrt{3a} \tag{20}$$

To find the strain intensity factor, it is convenient to use a new independent variable R defined by

$$r^2 = R^2 + a^2 - a_0^2 \tag{21}$$

and a instead of r and t. Then, substituting (20) into (14) with the use of (16) leads to

$$\frac{\partial \varepsilon_{eq}}{\partial a} = \frac{1}{\sqrt{3}\sqrt{a_0}\sqrt{R-a_0}} + o\left(\frac{1}{\sqrt{R-a_0}}\right) \tag{22}$$

This equation can be immediately integrated with the initial condition $\varepsilon_{eq} = 0$ at $a = a_0$ to give

$$\varepsilon_{eq} = \frac{(a-a_0)}{\sqrt{3}\sqrt{a_0}\sqrt{R-a_0}} + o\left(\frac{1}{\sqrt{R-a_0}}\right) \tag{23}$$

Since the distance from the friction surface is $r - a$, combining (15), (21) and (23) the fracture criterion can be written in the form

$$(a-a_0)/\sqrt{3a} = E_c \tag{24}$$

To determine the thickness of the region where the singular term in dominant, introduce an average value of the equivalent strain rate by

$$\xi_{eq}^{(av)} = \frac{2}{(b^2-a^2)}\int_a^b \xi_{eq} r\,dr\,d\theta$$

and the parameter α by $\alpha = (\hat{r}-a)/(b-a)$ where $\hat{r}$ is the radius at which $\xi_{eq} = \xi_{eq}^{(av)}$. Using (16) and (17) it is easy to find that α is in the range $0.33 < \alpha < 0.37$ if $2 \le b/a \le 40$.

6. Approximate solutions

A classical solution in plasticity theory for compression of a layer between two parallel rough platens is due to Prandtl [31] who performed a stress analysis. A velocity solution was proposed by Nadai (see [8]). In the case of maximum friction, using this velocity field the equivalent strain rate may be calculated in the form

$$\xi_{eq} = \frac{2}{\sqrt{3}}\frac{U}{h}\frac{1}{\sqrt{1-(y/h)^2}} \tag{25}$$

where U is the velocity of the platens, h is the instantaneous gap, and y is a Cartesian coordinate such that $y = 0$ corresponds to the plane of symmetry and $y = h$ to the friction surface. It follows from (25) that

$$D = \sqrt{2}U/\sqrt{3}h \qquad (26)$$

Substituting (26) into (14) and integrating, the fracture criterion (15) may be written in the form

$$\sqrt{2/3}\sqrt{h}\ln(h_0/h) = E_c \qquad (27)$$

where h_0 is the initial value of h. Introduce an average value of the equivalent strain rate by

$$\xi_{eq}^{(av)} = \frac{1}{h}\int_0^h \xi_{eq}\,dy$$

and the parameter α by $\alpha = (h-\hat{y})/h$ where $\hat{y}$ is the value of y at which $\xi_{eq} = \xi_{eq}^{(av)}$.

Using (25) it is easy to find that $\alpha = 1 - \sqrt{1 - 4/\pi^2} \approx 0.23$.

Another classical solution, flow of a rigid/plastic material through an infinite wedge shaped channel, has been proposed by Nadai and Hill [8]. The equivalent strain rate in a polar coordinate system $r\theta$ is defined by

$$\xi_{eq} = \frac{2}{\sqrt{3}}\frac{B}{r^2(c - \cos 2\psi)}\sqrt{1 + \frac{\sin^2 2\psi}{(c - \cos 2\psi)^2}\left(\frac{d\psi}{d\theta}\right)^2} \qquad (28)$$

where B and c are constants and ψ is a function of θ. B cannot be found for flow through an infinite channel, c is determined numerically and, then, ψ is given by the following equation

$$\theta = -\psi + \frac{c}{\sqrt{c^2 - 1}}\arctan\left[\sqrt{\frac{c+1}{c-1}}\tan\psi\right] \qquad (29)$$

in implicit form. The friction surface is defined by the condition $\psi = \pi/4$. In a vicinity of this surface (29) may be rewritten as

$$\theta = \alpha - (\pi/4 - \psi)^2/c + o\left[(\pi/4 - \psi)^2\right] \qquad (30)$$

Substituting (30) into (28) gives

$$D = B/(rc)^{3/2} \qquad (31)$$

Using (14) and (31) the fracture criterion (15) becomes

$$\sqrt{r}\,\ln r\big/\sqrt{3c} = E_c \qquad (32)$$

The thickness of the region where the singular term is dominant may be found numerically with no difficulty and it is of the same order as in the previous problem.

7. Conclusions

An equivalent strain rate singularity of type $D\big/\sqrt{x_3}$ and an equivalent strain singularity

of type $E\big/\sqrt{x_3}$ exist at the maximum friction surface in rigid/perfectly plastic materials. The intensity of this strain singularity, referred to here as the strain intensity factor, characterizes the magnitude of the equivalent strain in a region of the maximum friction surface ("E – dominant" neighborhood). Because of the singular nature of the perfectly/plastic strain rate field, there exists a region in the vicinity of the maximum friction surface where any small viscosity and hardening are not negligible and where the processes of void nucleation, growth, and coalescence that constitute ductile fracture occur. Since the fracture criterion proposed is based on a critical value of E, a necessary condition for its applicability is that this region is smaller than the E – dominant region. In the mechanics of cracks, a similar condition is that the inelastic region is confined to the K – dominant region. The examples considered in the present paper have shown that the thickness of the E – dominant region is about 0.2 – 0.4 of a typical length. The thickness of the boundary layer due to viscosity has been evaluated in [26]. To the best of my knowledge, no result applicable for the purpose of the present theory is available for constitutive laws involving hardening and damage parameters.

Although frictional tool surfaces were explicitly investigated here, the results hold for interior surfaces of velocity discontinuity also because the only condition that was assumed was that the shear stress reaches the shear yield stress on the surface.

In addition to the modeling of metal forming processes, the theory of rigid/perfectly plastic solids is useful for structural design since the limit loads for ideal rigid/plastic and ideal elastic/plastic material models coincide [38]. However, in this case the instantaneous velocity field only is of interest and, therefore, the concept of the strain intensity factor is not applicable. In general, there is an opportunity to formulate a fracture criterion in terms of the strain rate intensity factor and its critical value, $D = D_c$. In particular, such a fracture criterion may be combined with other criteria to predict the strength of adhesive joints. The fracture criterion proposed in [5] is applied at an interface corner, the limit load predicts yielding of the adhesive, and the new criterion might apply for predicting failure at the interface. The strain rate intensity factor may be evaluated from the limit load solution which accounts for the strain rate singularity. In the case of adhesive joints of arbitrary simply connected contour subject to tension such a limit load solution has been proposed in [39].

At this stage, the approach proposed in this paper is rather speculative and is solely based on the analogy between the stress intensity factor in the mechanics of cracks and the strain rate intensity factor in plasticity theory. The applicability of the approach can only be determined by additional theoretical and experimental work.

Acknowledgment – I am very grateful to Dr. Owen Richmond for many helpful discussions on the problem described in this paper.

8. References

1. Kanninen, M.F. and Popelar, C.H. (1985) *Advanced Fracture Mechanics*, Oxford University Press, New York.
2. Meguid, S.A. (1989) *Engineering Fracture Mechanics*, Elsevier Applied Science, London.
3. Rice, J.R. and Rosengren, G.F. (1968) Plane strain deformation near a crack tip in a power – law hardening material, *J. Mech. Phys. Solids* **16**, 1–12.
4. Hutchinson, J.W. (1968) Singular behavior at the end of a tensile crack in a hardening material, *J. Mech. Phys. Solids* **16**, 13–31.
5. Reedy, E.D.Jr. and Guess, T.R. (1993) Comparison of butt tensile strength data with interface corner stress intensity factor prediction, *Int. J. Solids Struct.* **30**, 2929–2936.
6. Akisanya, A.R. (1997) On the singular stress field near the edge of bonded joints, *J. Strain Analysis* **32**, 301–311.
7. Nadai, A. (1931) *Plasticity*, McGraw-Hill, New York and London.
8. Hill, R. (1950) *The mathematical theory of plasticity*, Clarendon Press, Oxford.
9. Shield, R.T. (1955) Plastic flow in a converging conical channel, *J. Mech. Phys. Solids* **3**, 246–258.
10. Browman, M.J. (1987) Steady forming processes of plastic materials with their rotation, *Int. J. Mech. Sci.* **29**, 483–489.
11. Alexandrov, S.E. and Barlat, F. (1997) Axisymmetric plastic flow of an F.C.C. lattice metal in an infinite converging channel, *Mech. Solids* **32**(5), 125-131.
12. Alexandrov, S. and Barlat, F. (1999) Modeling axisymmetric flow through a converging channel with an arbitrary yield condition, *Acta Mech.* **133**, 57–68.
13. Spencer, A.J.M. (1965) A theory of the failure of ductile materials reinforced by elastic fibers, *Int. J. Mech. Sci.* **7**, 197–209.
14. Sokolovskii, V.V. (1956) Equations of plastic flow in surface layer, *Prikl. Math. Mech.* **20**, 328–334 [in Russian].
15. Alexandrov, S.E. (1995) Velocity field near its discontinuity in an arbitrary flow of an ideal rigid-plastic material, *Mech. Solids* **30**(5), 111–117.
16. Alexandrov, S. and Richmond, O. (2000) Singular plastic flow fields near surfaces of maximum friction stress, *Int. J. Non-Linear Mech.* **36**, 1–11.
17. Alexandrov, S.E. and Richmond, O. (1998) Asymptotic behavior of velocity field near surfaces of maximum friction for axisymmetric plastic flow of Tresca material, *Dokl. Acad. Nauk* **360**, 480–482 [in Russian].
18. Alexandrov, S. and Richmond, O. (1999) Estimation of thermomechanical fields near maximum shear stress tool/workpiece in metalworking processes, in J.J.Skrzypek and R.B.Hetnarski (eds.), *Proc. 3rd Int. Cong. Thermal Stress*, Cracow University of Technology, Cracow, pp. 153-156.
19. Pemberton, C.S. (1965) Flow of imponderable granular materials in wedge – shaped channels, *J. Mech. Phys. Solids* **13**, 351–360.
20. Marshall, E.A. (1967) The compression of a slab of ideal soil between rough plates, *Acta Mech.* **3**, 82-92.
21. Alexandrov, S. (submitted) Comparison of double – shearing and coaxial models of pressure – dependent plastic flow at frictional boundaries, *J. Appl. Mech.*
22. Spencer, A.J.M. (1982) Deformation of ideal granular materials, in H.G. Hopkins and M.J. Sewell (eds.), *Mechanics of Solids*, Pergamon Press, Oxford, pp. 607–652.
23. Alexandrov, S.E., Danilov, V.L. and Chikanova, N.N. (2000) On the modeling of the sticking zone in metal forming processes under creep conditions, *Mech. Solids* **35**(1), 149-151 [in Russian].
24. Alexandrov, S. and Alexandrova, N. (in press) On the maximum friction law in viscoplasticity, *Mech. Time-Depend. Mater.*
25. Alexandrov, S. and Alexandrova, N. (submitted) On the maximum friction law for rigid/plastic, hardening materials, *Meccanica*

190

26. Oldroyd, J.G. (1947) Two-dimensional plastic flow of a Bingham solid: a plastic boundary-layer theory for slow motion, *Proc. Camb. Phil. Soc.* **43**, 383–395.

27. Alexandrov, S. (1999) A property of equations of rigid/plastic material obeying a Voce-type hardening law, *Meccanica* **34**, 349-356.

28. Alexandrov, S. and Richmond, O. (in press) Couette flows of rigid/plastic solids: analytical examples of the interaction of constitutive and frictional laws, *Int. J. Mech. Sci.*

29. Alexandrov, S.E. and Goldstein, R.V. (1999) Exact analytical solution to a rigid/plastic problem with hardening and damage evolution, *C. R. Acad. Sci, Ser. IIb* **327**, 193-199.

30. Alexandrov, S.E. and Goldstein, R.V. (in press) Exact analytical solution to a plastic problem with damage evolution, *Prikl. Math. Mech.*

31. Prandtl, L. (1923) Practical application of the Henky equation to plastic equilibrium, *ZAMM* **3**, 401-406.

32. Collins, I.F. and Meguid, S.A. (1977) On the influence of hardening and anisotropy on the plane-strain compression of thin metal strip, *J. Appl. Mech.* **44**, 271-278.

33. Adams, M.J., Briscoe, B.J., Corfield, G.M., Lawrence, C.J. and Papathanasiou, T.D. (1997) An analysis of the plane-strain compression of viscoplastic materials, *J. Appl. Mech.* **64**, 420-424.

34. Atkins, A.G. (1996) Fracture in forming, *J. Mater. Process. Technol.* **56**, 609-618.

35. Shabara, M.A., El-Domiaty, A.A. and Kandil, M.A. (1996) Validity assessment of ductile fracture criteria in cold forming, *J. Mater. Engng Perform.* **5**, 478-488.

36. Janicek, L. and Petruska, J. (1999) Ductile fracture criteria in compression and cracking of notched cylindrical specimens, in M. Geiger (ed.), *Proc. 6th ICTP, Vol.3*, Springer, Berlin, pp. 2287-2292.

37. Alexandrov, S. and Richmond, O. (in press) Frictional effects in the modified Couette flow of solids, in *Proc. 8th Int. Conf. Metal Forming 2000*, Cracow, Poland, Sept. 3-7, 2000.

38. Drucker, D.C., Prager, W. and Greensberg, H.J. (1952) Extended limit design theorems for continuous media, *Quart. Appl. Math.* **9**, 381-389.

39. Alexandrov, S. and Richmond, O. (2000) On estimating the tensile strength of an adhesive plastic layer of arbitrary simply connected contour, *Int. J. Solids Struct.* **37**, 669-686.

POLYCRYSTALLINE PLASTICITY UNDER SMALL STRAINS
Toward finer descriptions of microstructures

F. BARBE, S. FOREST, G. CAILLETAUD

Centre des Matériaux/UMR 7633, Ecole des Mines
de Paris/CNRS, BP87, F-91003 EVRY, France

1 Introduction

After the pioneering work due to Taylor [1], crystal plasticity is a classical topic in the literature. A series of models have been elaborated in the past. Single crystal models came first, with a first type of models in the sixties, mainly applied to pure metals [2], and a second generation able also to represent more complex behaviors, for instance superalloys, in the eighties [3–5]. Transition rules from the macroscopic scale to the microscopic scale in the framework of simplified approaches, and the related homogenization theories, have been developed in the same time [6–9], allowing the user to account for strain or stress heterogeneities in a polycrystalline material element, with the assumption of uniform values in each phase. This last assumption is rather strong, since it does not account for the stress/strain redistribution into the grains, and cannot give any idea of important features like surface effect and grain boundary effect.

On the other hand, the present state of the computing power offers now the possibility to reach original results with FE methods, especially in the study of the mechanics of heterogeneous materials. The first studies have been performed by describing a polycrystal with uniform shape of grains -cubic or polyhedral- or even several grain orientations comprised in a single element. One of the issues successfully addressed with FE deals with the prediction of the mean behavior of polycrystals at large strains together with texture evolution [10–16]. Still, for large strains, neither the distribution of orientations inside the structure nor the shape of the grains have been proved to have a large importance for the purpose of texture evolution [14, 16]. Another kind of problem addressed with FE is the description of plasticity inside grains of a multicrystal with realistic grain shapes [17–19]. Such studies can only rely on quasi-3D structures made of a small number of large grains since experimental determination of the real 3D morphology of a polycrystal is still a difficult task. Their constitutive models take into account the evolution of density of dislocations and they resort to a large number of elements per grain to describe the intragranular behavior and the effects of the grain boundaries. An intermediate problem lying between the case of idealized polycrystals and this of real multicrystalline structures consists in representing each grain of a polycrystal with Voronoï cells, either Voronoï cell elements [20] or several elements inside each grain [21, 22]; but these works are usually restricted to 2D structures.

The purpose of this paper is to present the results obtained in the FE simula-

E. Bouchaud et al. (eds.), Physical Aspects of Fracture, 191–206.
© 2001 *Kluwer Academic Publishers. Printed in the Netherlands.*

tion of a 3D polycrystalline aggregate, with a realistic microstructure. Section 2 presents the crystallographic model which can be applied to represent single crystal behavior. An original transition rule, the β-model [23], is also briefly discussed, in relation with a simpler model [9], which will be used as a reference in our computations. Then, in Section 3, we present computations on a 3D polycrystal with grains having the shapes of Voronoï polyhedra, as was first proposed in [24] and then in [25, 26]. In particular, in Section 3.2, a varying number of elements is used in order to determine how many full integration 20-node brick elements per grain are required, in average, to get a valid response on such or such a scale. The method to generate 3D microstructures is first recalled in Section 3.1.

2 Model for the Single- and Polycrystal

2.1 Single Crystal

It is assumed that slip is the predominant deformation mechanism and that Schmid's law is valid. The resolved shear stress can then be used as a critical variable to evaluate the inelastic flow. A viscoplastic framework is chosen, in order to avoid the problems related to the determination of the active slip systems in plastic models. A threshold is introduced both in positive and negative directions on each slip system : twelve octahedral slip systems will be used for FCC materials. Two variables are defined for each slip system s, r^s and x^s, corresponding respectively to isotropic hardening (expansion of the elastic domain), and kinematic hardening (translation of the elastic domain [27]). A system will be active provided its resolved shear stress τ^s is greater than $x^s + r^s$ or less than $x^s - r^s$ and the slip rate will be known as long as stress and the hardening variables are known. The state variables used to define the evolution of r^s and x^s are the accumulated slip v^s for isotropic hardening and the variable α^s for kinematic hardening. Knowing the stress tensor $\underset{\sim}{\sigma}^g$ applied to the grain g, the resolved shear stress for system s can be classically written according to (1), $\vec{n}^s$ and $\vec{m}^s$ being respectively, for the system s, the normal to the slip plane and the slip direction in this plane. The hardening variables x^s and r^s can then be expressed as a function of α^s and v^s following (2), their actual values allowing then to compute the viscoplastic slip rate $\dot{\gamma}^s$, the viscoplastic strain rate tensor $\underset{\sim}{\dot{\varepsilon}}^g$ (3), and the hardening rules ((4) and (5)). The present formulation gives a saturation of the hardening in both monotonic and cyclic loading, and takes into account the interactions between the slip systems, through matrix h_{rs}, as in [28]. Nine material-dependent coefficients are involved in the model $(E, \nu, K, n, c, d, R_0, Q, b)$.

$$\tau^s = \underset{\sim}{\sigma}^g : \underset{\sim}{m}^s = \frac{1}{2} \underset{\sim}{\sigma}^g : (\vec{n}^s \otimes \vec{m}^s + \vec{m}^s \otimes \vec{n}^s) \tag{1}$$

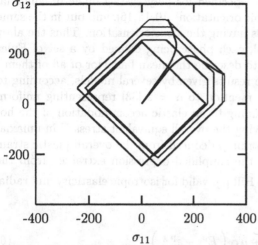

Coefficients (MPa, s)			
τ_0	100	c	10000
Q	20	d	200
b	10	K	20
h_{ij}	1	n	2

Upper path : $\varepsilon_{12}/\varepsilon_{11} = 0.525$
Lower path : $\varepsilon_{12}/\varepsilon_{11} = 0.475$

Figure 1: Illustration of the bifurcation in the stress plane observed for prescribed similar strain paths in the case of a FCC single crystal

$$x^s = c \alpha^s \quad ; \quad r^s = R_0 + Q \sum_r h_{rs} \left\{ 1 - e^{-b \, v^r} \right\} \tag{2}$$

$$\dot{\gamma}^s = v^s \, sign \left(\tau^s - x^s \right) \quad ; \quad \dot{\varepsilon}^g = \sum_s m^s \, \dot{\gamma}^s \tag{3}$$

$$v^s = \left\langle \frac{|\tau^s - x^s| - r^s}{K} \right\rangle^n \quad \text{with} \quad \langle x \rangle = Max(x, 0) \quad \text{and} \quad v^s \, (t = t_0) = 0 \tag{4}$$

$$\dot{\alpha}^s = \dot{\gamma}^s - d \alpha^s \, v^s \quad \text{with} \quad \alpha^s \, (t = t_0) = 0 \tag{5}$$

Such a formulation [4, 29] is an extension of the classical crystallographic approach for single crystal modeling in plasticity or in viscoplasticity (see for instance [1,2,30]). It has been extensively used for single crystal modeling including Finite Element simulations [31, 32]. For this type of approach, as illustrated in fig.1, the presence of corners in the yield surface may produce strong variations of the flow direction for a given loading. This fact can be important in aggregate computations, since complex tridimensional loading paths will be applied to each integration point.

2.2 Transition rule for the polycrystal

In a polycrystalline aggregate, one phase may be characterized by its shape, size, crystallographic orientation, location with respect to the surface of the material,

194

etc ... Most of the models usually specified for polycrystals made of equiaxial grains retain only the crystallographic orientation [10,15,16], and put in the same *crystallographic phase* all the grains having the same orientation. Thus the alloy is considered as a n–phase material, each phase being defined by a set of Euler angles, and then the model is used to describe the mean behavior of all of them.

The relation (6) summarizes the results given by several models, according to the definition of α, with a specific mention to $\alpha = 2$ [33] representing uniform total strain, to $\alpha = 1$ [6], corresponding to an elastic accommodation of the homogeneous medium, or to (7) involving the overall equivalent stress Σ in uniaxial tension and the equivalent plastic strain p deduced from the overall plastic strain tensor $\underset{\sim}{E}^\mathrm{p}$ [9]. This last model is the simplified expression extracted from the general self-consistent model due to Hill [7], valid for isotropic elasticity and radial loading paths.

$$\underset{\sim}{\sigma}^\mathrm{g} = \underset{\sim}{\sigma} + \mu\alpha\left(\underset{\sim}{E}^\mathrm{p} - \underset{\sim}{\varepsilon}^\mathrm{p\,g}\right) \tag{6}$$

with: $\quad \dfrac{1}{\alpha} = 1 + \dfrac{3\mu p}{2\Sigma}\;, \quad p = \left(\dfrac{2}{3}\,\underset{\sim}{E}^\mathrm{p} : \underset{\sim}{E}^\mathrm{p}\right)^{\frac{1}{2}} \quad$ and $\quad \underset{\sim}{E}^\mathrm{p} = <\underset{\sim}{\varepsilon}^\mathrm{p\,g}> \tag{7}$

From a physical point of view, the previous rules simply show that a local plastic strain decreases the local stress, whereas the stress redistribution related to plastic accommodation tends to decrease for larger plastic strains. An alternative formulation, the "β-model", introducing a non linear accommodation, has also been proposed [23]. It can be calibrated from finite element computations, using either an inclusion embedded in a homogeneous medium [34] or a 3D FE polycrystal [35]:

$$\underset{\sim}{\sigma}^\mathrm{g} = \underset{\sim}{\sigma} + \mu\left(\underset{\sim}{\beta} - \underset{\sim}{\beta}^g\right) \tag{8}$$

$$\text{with } \underset{\sim}{\beta} = \sum_g f^g \underset{\sim}{\beta}^g \tag{9}$$

Various expressions can be used for the evolution of the variable β^g in each grain [36]. The model can be used for any kind of loadings, especially cyclic loadings.

Since the loading path in the present paper is just a tension, the Berveiller–Zaoui's (BZ) evaluation of the self-consistent scheme will be considered as a reference for the polycrystal behavior. The following coefficients are then used:
- isotropic elastic behavior: $\quad E = 196000 MPa, \quad \nu = 0.3$
- viscous effect in (4) and (5): $\quad K = 10 MPa.s^{1/n}, \quad n = 25$
- kinematic hardening in (2): $\quad c = 1600 MPa, \quad d = 40$
- isotropic hardening in (2) and (3): $\quad R_0 = 111 MPa, \quad Q = 35 MPa, \quad b = 7$

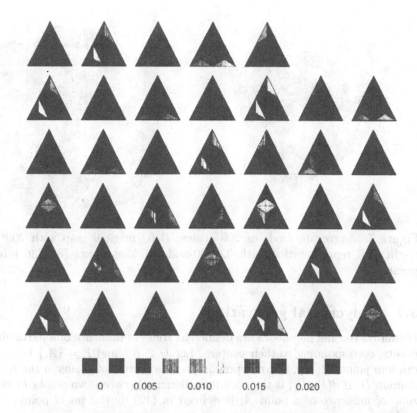

Figure 2: Values of the accumulated slip on each slip system for a 1% tension on a 40-grain aggregate with the BZ model

With such a kind of approach, one has an access to local quantities, like stresses or strains, but assuming that their values are uniform for each crystallographic phase. This is illustrated for example in fig.2, showing the amount of slip for each slip system in a FCC crystal presenting octahedral slip. In each "grain", conventionnally represented by an unfolded Thomson tetrahedron, the twelve slip systems are distributed on four large triangles figuring the slip planes, with 3 slip directions on each. One can observe that, even with a viscoplastic formulation, the number of active slip systems remains low, due to the presence of a threshold.

3 Computing on cubic polycrystals

As an example, we consider a single *cubic polycrystal* made of 200 grains. The original polycrystalline medium is not built from measures on real microstructures but generated in the form of a 3D voxel map representing Voronoï polyhedra (fig.3a). The procedure is explained below. Each grain in the FE structure is defined by a set of integration points rather than a set of elements. The elements are full-integration 20-node bricks -thus containing 27 integration points.

Figure 3: Aggregate made of 200 grains; (left) original map with 200^3 voxels, (right) FE representation with 32^3 20-node-brick elements (884736 integration points)

3.1 Polycrystal generation

Formally, Voronoï polyhedra are defined as zones of influence of a particular set of points, corresponding to their centers. Let $D \subset R^3$, and $E = \{A_i\}$ be a set of N random points $P(x, y, z)$ corresponding to the centers of grains in the continuous domain D. If $d(P_1, P_2)$ is the euclidean distance between two points P_1 et P_2, the zone of influence of a point A_i is defined in (10) by the set of points $P(x, y, z)$ with:

$$iz(A_i) = \{P(x, y, z) \in D \quad | \quad d(P, A_i) < d(P, A_j) \quad \forall j \neq i\} \tag{10}$$

In more physical terms, a point P belongs to the zone of influence of germ A_i if it is closer to A_i than to any other germ. By construction, this zone of influence generates the Voronoï polyhedron centered in A_i. The set of zones of influence $\{iz(A_i)\}$ builds a random tesselation of the domain D into N classes, every A_i being the germ of one class (fig.3a).

A specific procedure was developed to build Voronoï polyhedra inside a discrete domain, made of a 3D voxel map [37]. Each voxel is assigned the number (or label) of the polyhedron to which it belongs. Then, each integration point of the mesh is assigned the number -label- of the voxel at the same location in space. In the case proposed here (fig.3), there are 200^3 voxels vs 96^3 integration points in the most refined mesh.

3.2 Computations and effects of mesh refinement

Now we present a comparison of the results of the same simple tensile test on the aggregate of fig.3 computed with varying number of elements: at the macroscopic scale, at the scale of the mean response per grain and at the intragranular scale.

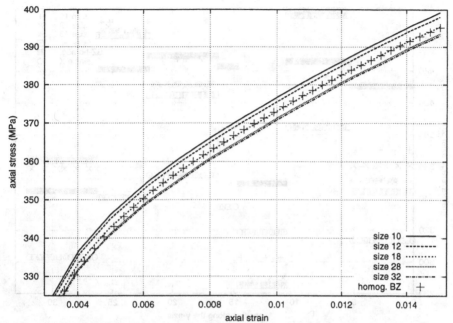

Figure 4: Mean axial stress-strain curves with five different sizes of mesh and with homogenization model

Macroscopic scale

Each structure, made respectively of 10^3, 12^3, 18^3, 32^3 elements, is loaded to $E_{zz} = 1.5\%$ with homogeneous strain boundary conditions. The components of the loading strain tensor result from a preliminary homogenization simulation where the lateral stresses σ_{xx} and σ_{yy} were imposed to remain null. On fig.4, the mean stress-strain curve of the 200-grain-polycrystal is well reproduced whatever our conditions of modeling are: would it be homogenization or FE modeling with 10^3 elements, the response of the 32^3-mesh is respected with less than 1.5% of deviation in stress. The curves also illustrate the effect of the stiffness of the mesh: for a lower number of elements, the response becomes harder.

One may think that, imposing a same strain tensor to all the nodes of the contours, the structures have less freedom to behave on their own than in the case where only axial displacements are imposed to top and bottom faces of the cubes. This latter case with four free lateral faces have been presented in [26], for the same material and the same aggregate; in that case, the mean stress response was 7% lower than that with homogeneous strain boundary conditions. Considering the stress-strain curves of the grains, it has been shown that freeing lateral faces tends to promote the dispersion in strain whereas imposing homogeneous boundary conditions rises the dispersion in stress. So whether the mean stress-strain curves of fig.4 would be more dispersed with free faces, or not, cannot be guessed without further investigation.

198

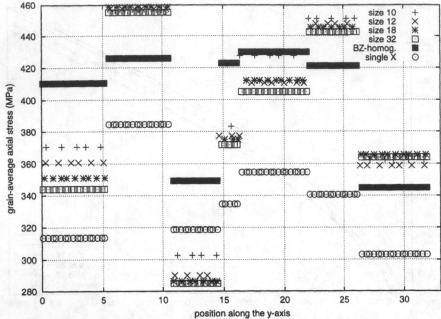

Figure 5: Mean axial stress in grains located along a line crossing the aggregate: mean stress per grain from four different sizes of mesh (10^3, 12^3, 18^3, 32^3 elements), stress from the homogenization model, stress in the corresponding single crystals

Mean grain scale

The average response for each grain can be computed, and compared to the results of the homogenization (BZ) model. It has been made for this 200-grain aggregate in [26], for the same material. It appeared that BZ model underestimates the scatter in stress and strain between the different grains. This is due to the absence of local grain-to-grain interaction for the models with uniform stress/strain fields. Nevertheless, as long as 200 grains are considered on the same plot of stress-strain curves, it is difficult to identify an FE grain and its homogenization correspondent. So here we chose to show a comparison on grains located along a line of integration points crossing the structure (fig.5). This line is parallel to the lateral direction y and is located in the middle of the structure. Since we deal with discretized domains and since refinement varies from 10^3 elements to 32^3, the location of the line varies slightly from a mesh to another. For the 10^3-mesh it has the coordinates $x = z = 14.40$, for 12^3 $x = z = 14.42$, for 18^3 $x = z = 14.42$ and for 32^3 $x = z = 14.50$. These coordinates have been brought on the scale of the 32^3-mesh (32 elements of size 1 in each direction, $0 < x, y, z < 32$). The volume of each grain along the line, divided by the average volume of a grain, gives respectively (from the left to the right): 1.28, 0.96, 1.42, 1.26, 0.76, 2.02, 1.87. Note that the number of points per grain appearing along the line is not related to the volume of the grain, since the line may pass through the middle of a grain or near its boundary. As a matter of fact, the 5th grain has the largest number of points intercepted by the line and is also the smallest grain.

Fig.5 shows the axial stress obtained from 4 FE computations (10, 12, 18 and 32 elements along each edge of the cube), together with the self-consistent (BZ) response, and the response of the single crystal having the same orientation. This last value allows us to check if a given grain behaves simply according to its orientation or is strongly influenced by its neighborhood. Generally speaking, all the grains appear stronger in the polycrystal than as single crystals. This is due to the development of internal stresses. This effect can be classicaly observed, even for the self-consistent approach. Nevertheless, this fact is not necessary true for the FE computations, since stresses can be higher or smaller than stresses of the corresponding single crystals. The difference are even more pronounced ($> 10\%$) in three of the grains (from the left, 1st, 3rd and 4th). As 1st grain can be thought to be strongly influenced by the loading conditions in FE computations, 3rd and 4th cannot. Moreover, these grains have volumes higher than the average volume of a grain. So neither the volume nor the location in the structure are responsible for the difference between BZ and FE. The actual neighborhood of a grain can then be said to influence the mean stress response of a grain by more than 10% in stress, which may result in more than 60% in axial strain.

As the number of elements is increased, the accuracy of the mean response per grain is acceptable for all of the 200 grains with a minimum of 18^3 elements. The 12^3-mesh could also be considered as acceptable except that one can observe a difference in 1st grain: there, the loading conditions obviously appear to influence the behavior. On the other hand, any kind of decrease of the elements number below 12^3 produces non acceptable variations of the results. For instance, the mean behavior per grain provided by the 10^3-mesh really departs from those of the other computations, especially in 3rd and 5th grains. 3rd grain, though big its volume, is the only grain where the stress is lower than the one of the corresponding single crystal, which means that it is greatly influenced by its neighboring grains. 5th grain is the smallest along the line. So it is not surprising to get the largest differences in these grains. Finally, as the discretization of the 12^3-mesh corresponds to about 8 quadratic elements per grain, we will consider that $2 \times 2 \times 2$ quadratic elements and $6 \times 6 \times 6$ Gauss points are the critical numbers in a grain to get a decent estimation of its average stress.

Intragranular scale
As a first step toward the study of the local effect of mesh refinement, we propose to compare the local responses along the line used in the previous analysis, obtained from three sizes of mesh: 12^3, 18^3 and 32^3 elements (fig.6). The volume of each grain has already been defined and the number of integration points on each intersection with a grain can be counted from fig.5.

A von Mises stress plot is considered in fig.6a, and the cumulated slip on all the active slip systems in fig.6b (it is defined as the sum of the slips over the systems, $\sum_s |\dot{h}^s|$). Surprisingly, the von Mises stress profiles are very similar for the three reported mesh sizes. The 12^3-mesh nevertheless fails at reproducing the peak values in the 1st and 5th grains. On the other hand, for the case of the amount of

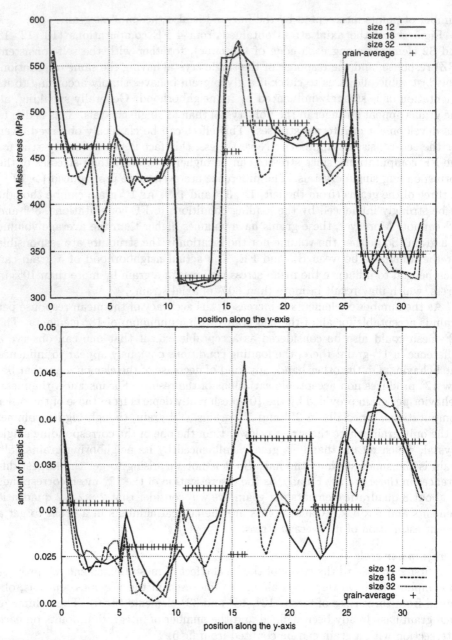

Figure 6: Plots along a line crossing the structure computed with three sizes of mesh (12^3, 18^3, 32^3 elements): von Mises stress (upper), amount of plastic slip (lower). The mean response per grain obtained with 32^3 elements is also figured

plastic slip the lack of discretization becomes obvious: in 1st, 5th and 6th grains, it simply forgets or considerably attenuates peaks figured in 18^3 and 32^3-meshes. By 'peak' we mean an abrupt change in the direction of evolution, which could result in smoothing or hardening of the corresponding region. Such a region may then become a critical place where a crack initiation can preferentially occur under fatigue loading or a region of localized deformation resulting in slip bands. That is why with the refinement of the 12^3-mesh, *i.e.* with about $2 \times 2 \times 2$ elements per grain, one cannot expect a correct representation of gradients inside grains. Fig.7a,c, showing the distribution of von Mises stress on the slices containing the lines of fig.6, allows to observe the poor intragranular resolution of the 12^3-mesh.

Using a 18^3-mesh, a good approximation of the local behavior is obtained, since nearly all the peaks are reproduced. The slight differences, in the amounts reached by the peaks or in the positions of the peaks cannot be said to come from a lack of discretization or from the difference in the location of the line of integration points. However, the comparison would probably not remain so good for any kind of line. This is demonstrated by fig.7 which show respectively the von Mises equivalent on a 12^3-mesh (a), on a 18^3-mesh (b), on a 32^3-mesh (c), then the grain map (d), and finally the amount of plastic slip for the 18^3-mesh (e) and for the 32^3-mesh (f). The von Mises stress is well predicted only far from the grain boundaries since it is rather uniform inside each grain. But it is not the case for the amount of plastic slip: there are some grains where high-slip-activity-structures form near grain boundaries and where the rest of the grain remains unaffected by slip activity. Such features are strongly dependent on the resolution of the grain boundary: if the 18^3-mesh only is considered, one may observe high-slip-activity-regions spreading over grain boundaries. But with a higher resolution, these structures appear to be disconnected at the grain boundaries, thus illustrating the fact that the slip activity is mainly due to the gradients of stress at the grain boundaries and not related to a kind of propagation of slip across boundaries.

Hence, having about $3 \times 3 \times 3$ elements per grain (case of 18^3-mesh) may lead to a first good estimation of the gradients of the fields inside grains. Yet, for a systematic treatment aiming at describing the effect of grain boundaries (*e.g.* plotting field variables *vs* the distance to the grain boundary and averaging over the grains of the aggregate [38]), the uncertainty resulting from the fact that grain boundaries pass inside elements instead of between elements (so-called multiphase elements) could alterate the observations made in the vicinity of the boundaries. Such a systematic treatment requires to have at least $4 \times 4 \times 4$ elements per grain so that one is assured that there are $2 \times 2 \times 2$ single-phase elements inside each grain, *i.e.* enough elements unaffected by the numerical construction of grain boundaries.

4 Conclusion

Comparisons of computations with a varying number of elements have been performed. For each scale of interest -macroscopic, mean grain scale, intragranular- a minimum number of elements per grain has been defined. Note that the elements

202

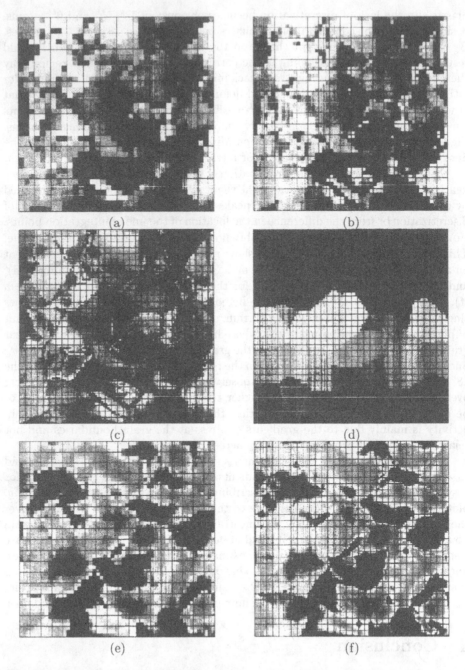

Figure 7: Contour plots on slices of integration points containing the lines investigated previously: (a) von Mises stress in 12^3-mesh ($x = 14.66$), (b) von Mises stress in 18^3-mesh ($x = 14.42$), (c) von Mises stress in 32^3-mesh ($x = 14.50$), (d) grain map in 32^3-mesh, (e) amount of plastic slip in 18^3-mesh, (f) amount of plastic slip in 32^3-mesh

contain 27 integration points. So if 8-integration-points elements were to be used, our results could be useful as long as one adapts them by counting integration points rather than elements.

By combining those full-integration elements to the description of grains with sets of integration points (instead of sets of elements), one obtains an improved resolution of the grain boundaries. Analyzing the intragranular stress-strain curves of each integration point inside a grain, it has been found that the tremendous dispersion of the curves were not due to the multiphase elements. Further, studied as a function of the distance to the grain boundary, the variables featured a continuous spreading from the center to the grain boundary, without any visible step of variation in multiphase elements [38]. So the use of multiphase elements seems to be adequate.

The computations of 3D polycrystals require a huge amount of degrees of freedom but their use is not limited to the study of classical crystal plasticity. A direct application of such computations has been proposed recently: the calibration of a homogenization model with the mean behavior of an FE aggregate [35]. The method presented here has also been extended to size-sensitive crystal plasticity [39].

References

[1] G.I. Taylor. Plastic Strain in Metals. *J. Inst. Metals*, 62:307–324, 1938.

[2] Jean Mandel. Une généralisation de la théorie de la plasticité de W.T. Koiter. *Int J Sol Struct*, 1:273–295, 1965.

[3] E. Jordan and K.P. Walker. Biaxial Constitutive Modeling and testing of Single Crystals at Elevated Temperature. In *Conference on Multiaxial Fatigue*, Sheffield (UK), 1984.

[4] Laurent Méric, Philippe Poubanne, and Georges Cailletaud. Single Crystal Modeling for Structural Calculations. Part 1: Model Presentation. *J Engng Mat Technol*, 113:162–170, 1991.

[5] E.P. Busso and F.A. McClintock. A dislocation mechanics-based crystallographic model of a b2-type intermetallic alloy. *Int J Plasticity*, 12:1–28, 1996.

[6] E. Kröner. Zur plastischen Verformung des Vielkristalls. *Acta Metall*, 9:155–161, 1961.

[7] R. Hill. Continuum Micro–Mechanisms of Elastoplastic Polycrystals. *J Mech Phys Sol*, 13:89–101, 1965.

[8] J.W. Hutchinson. Elastic–Plastic Behaviour of Polycrystalline Metals and Composites. *Proc R Soc London*, A319:247–272, 1966.

[9] M. Berveiller and A. Zaoui. An Extension of the Self–Consistent Scheme to Plastically Flowing Polycrystal. *J Mech Phys Sol*, 26:325–344, 1979.

[10] A.J. Beaudoin, K.K. Mathur, P.R. Dawson, and G.C. Johnson. Three-dimensional deformation process simulation with explicit use of polycrystal plasticity models. *Int J Plasticity*, 9:833–860, 1993.

[11] R. Becker and S. Panchanadeeswaran. Effects of grain interactions on deformation and local texture in polycrystals. *Acta metall mater*, 43(7):2701–2719, 1995.

[12] H. Takahashi, H. Motohashi, M. Tokuda, and T. Abe. Elastic-plastic finite element polycrystal model. *Int J Plasticity*, 10(1):63–80, 1994.

[13] Gorti B. Sarma and Paul R. Dawson. Texture predictions using a polycrystal plasticity model incorporating neighbor interactions. *Int J Plasticity*, 12(8):1023–1054, 1996.

[14] Venugopal Bachu and Surya R. Kalidindi. On the accuracy of the predictions of texture evolution by the finite element technique for fcc polycrystals. *Mat Sc Engng*, A257:108–117, 1998.

[15] A. Staroselski and L. Anand. Inelastic deformation of polycrystalline face centered cubic materials by slip and twinning. *J Mech Phys Sol*, 46(4):671–696, 1998.

[16] D.P. Mika and P.R. Dawson. Polycrystal plasticity modeling of intracrystalline boundary textures. *Acta mater*, 47(4):1355–1369, 1999.

[17] C. Teodosiu, J.L. Raphanel, and L. Tabourot. Finite element simulation of the large elastoplastic deformation of multicrystals. In C. Teodosiu, J.L. Raphanel, and F. Sidoroff, editors, *Proc. of the int. seminar Mecamat'91: Large Plastic Deformations, Fundamental Aspects and Applications to Metal Forming*, Fontainebleau, France:153–168, August, 7–9 1991.

[18] Jörn Harder. A crystallographic model for the study of local deformation processes in polycrystals. *Int J Plasticity*, 15(6):605–624, 1999.

[19] F. Delaire, J.L. Raphanel, and C. Rey. Plastic heterogeneities of a copper multicrystal deformed in uniaxial tension: experimental study and finite element simulations. *Acta mater*, 48:1075–1087, 2000.

[20] Somnath Ghosh and Suresh Moorthy. Elastic-plastic analysis of arbitrary heterogeneous materials with the Voronoï Cell finite element method. *Comput Meth Appl Mech Engng*, 121:373–409, 1995.

[21] Osamu Watanabe, Hussein M. Zbib, and Eiji Takenoushi. Crystal plasticity: micro-shear banding in polycrystals using Voronoï tessalation. *Int J Plasticity*, 14(8):771–788, 1998.

[22] E. Nakamachi, K. Hiraiwa, H. Morimoto, and M. Harimoto. Elastic/crystalline viscoplastic finite element analyses of single- and poly-crystal sheet deformations and their experimental verification. *Int J Plasticity*, 16:1419–1441, 2000.

[23] G. Cailletaud and P. Pilvin. Utilisation de modèles polycristallins pour le calcul par éléments finis. *Revue Européenne des Éléments Finis*, 3(4):515–541, 1994.

[24] S. Quilici and G. Cailletaud. F.E. simulation of macro-, meso- and microscales in polycrystalline plasticity. *Comput Mat Sc*, 16(1–4):383–390, 1999.

[25] F. Barbe, L. Decker, D. Jeulin, and G. Cailletaud. Intergranular and intragranular behavior of polycrystalline aggregates. Part 1: F.E. model. *Int J Plasticity*, in press, 2001.

[26] F. Barbe, S. Forest, and G. Cailletaud. Intergranular and intragranular behavior of polycrystalline aggregates. Part 2: results. *Int J Plasticity*, in press, 2001.

[27] Jean-Louis Chaboche. Constitutive Equations for Cyclic Plasticity and Cyclic Viscoplasticity. *Int J Plasticity*, 5:247–302, 1989.

[28] U.F. Kocks and T.J. Brown. Latent Hardening in Aluminium. *Acta Metall*, 14:87–98, 1966.

[29] Georges Cailletaud. *Une approche micromécanique phénoménologique du comportement inélastique des métaux*. PhD thesis, Université Pierre et Marie Curie, Paris 6, 1987.

[30] Robert J. Asaro. Crystal Plasticity. *J. Appl. Mech.*, 50:921–934, 1983.

[31] L. Méric and G. Cailletaud. Single Crystal Modeling for Structural Calculations. Part 2: Finite Element Implementation. *J Engng Mat Technol*, 113:171–182, 1991.

[32] J. Besson, R. Leriche, R. Foerch, and G. Cailletaud. Object–Oriented Programming Applied to the Finite Element Method. Part II. Application to Material Behaviors. *Revue Européenne des Éléments Finis*, 7(5):567–588, 1998.

[33] T.H. Lin. Analysis of Elastic and Plastic Strains of Face Centered Cubic Crystal. *J Mech Phys Sol*, 5:143–149, 1957.

[34] Philippe Pilvin. The Contribution of Micromechanical Approaches to the Modelling of Inelastic Behaviour of Polycrystals. In André Pineau, Georges Cailletaud, and Trevor Lindley, editors, *Fourth Int. Conf. on Biaxial/multiaxial Fatigue and Design*, pages 3–19, London, 1996. ESIS 21, Mechanical Engineering Publications.

206

[35] F. Barbe, R. Parisot, S. Forest, and G. Cailletaud. Calibrating a homogenization polycrystal model from large scale FE computations of polycrystalline aggregates. *To be presented at 5th Eur Mechanics of Materials Conf*, Delft, The Netherlands, March 5–9 2001. To appear in J Phys IV.

[36] J. Besson, G. Cailletaud, J.L. Chaboche, and S. Forest. *Mécanique des matériaux inélastiques*. To appear, Hermès, 2001.

[37] L. Decker and D. Jeulin. Simulation 3D de matériaux aléatoires polycristallins. *Revue de Métallurgie, CIT/Science et Génie des Matériaux*, February:271–275, 2000.

[38] Fabrice Barbe. *Etude numérique de la plasticité d'agrégats polycristallins*. PhD thesis, Ecole Nationale Supérieure des Mines de Paris, décembre 2000.

[39] S. Forest, F. Barbe, and G. Cailletaud. Cosserat modelling of size effects in crystals. *Int J Sol Struct*, 37(46-47):7105–7126, 2000.

FRACTURE AND MESOSCOPIC PLASTIC DEFORMATION

E. Van der Giessen

Delft University of Technology, Koiter Institute Delft,

Mekelweg 2, 2628 CD Delft, The Netherlands

Abstract

This article discusses some recent advances in the understanding of plastic deformation near a crack tip at a mesoscopic scale. This is a scale that is in between the engineering macroscopic scale and the microscopic scale of the atomic description used by physicists. The mesoscopic scale is where one observes the individual grains or even the dislocation structures. We start out by discussing the near-tip stress fields inside a single crystal, using continuum model of plasticity. Then, we zoom in further and adopt a discrete dislocation dynamics study of a propagating crack. It will be shown that very near the crack tip, the dislocations have a dual role. On the one hand, dislocation activity gives rise to plastic dissipation which increases the crack growth resistance. On the other hand, the local stress concentration associated with discrete dislocations in the vicinity of the crack tip promote fracture.

1 Introduction

For any novice in fracture mechanics it is startling to realize that one can understand or even predict macroscopic fracture even though the mechanism for a crack to propagate operates at the atomic scale. Even though the word 'macroscopic' can mean different things to different people, there is a gap of several orders of magnitude. Intermediate scales can sometimes be neglected for engineering purposes (linear elastic fracture mechanics) but physical insight requires due consideration of the relevant microstructural scales.

This article is concerned in this respect with cracks in ductile metals. Here a number of intermediate, mesoscopic length scales are important because plastic deformation is highly nonuniform. When zooming in from the macroscale, we are first faced with the polycrystalline nature of (most) metals, which renders local plasticity highly anisotropic. Upon further zooming in, we encounter the length scale where dislocation structures become visible.

The objective of the paper is to summarize some recent findings related to plastic deformation at these mesoscopic scales. We first consider the largest of these scales, where a continuum description of plastic flow around the tip of a crack is still valid, but which does account for the anisotropic nature of crystals

E. Bouchaud et al. (eds.), Physical Aspects of Fracture, 207–223.

208

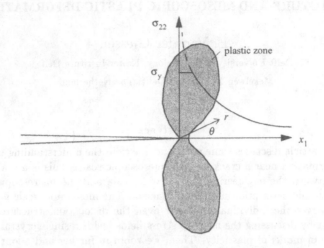

Figure 1: The opening stress in an elastic material is singular, but in a plastic material is limited to a value on the order of the yield stress.

(Sec. 2). Next, we address the role of dislocations in mediating plastic flow near a crack tip (Sec. 3). Section 4 summarizes some very recent results of dislocation dynamics simulations, which show that dislocations have a dual role: they allow for plastic stress relaxation but at the same time their long-range interactions lead to patterns that can locally raise the stresses to very high levels. This seems to resolve the 'paradox' that the next section actually starts out with. The paper closes with preliminary results under cyclic loading conditions.

2   Continuum near-tip plasticity

Imagine a sharp crack in an elastic material subjected to mode I loading. According to linear elasticity, the stresses close to the crack tip vary according to

$$\sigma_{ij} = \frac{K_I}{\sqrt{2\pi r}} f_{ij}(\theta) \qquad (1)$$

where (r, θ) are polar coordinates from the crack tip, K_I is the stress intensity factor and f_{ij} are known functions of θ (see, e.g., Anderson, 1995). Although we know that linear elasticity will not apply to the very tip of the crack, the stresses will grow extremely large when we approach the tip and certainly large enough for atomic debonding to occur, leading to propagation of the crack. If the material can undergo plastic deformation, these high stresses will be relaxed and a plastic zone develops around the tip. Inside this zone, the stresses are on the order of the yield stress, as given for example by the HRR field (Hutchinson, 1968; Rice and Rosengren, 1968) for an isotropic elastoplastic material. This stress level is much lower than the stress for debonding. Hence the paradox that ductile materials would not break — of course, we experience that they do.

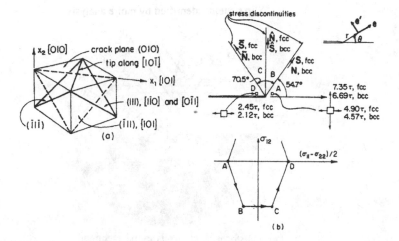

Figure 2: (a) Active slip planes in an fcc crystal for a crack on the (010) plane.
(b) The four sectors around the tip, inside of which the stresses are uniform and
have the values as indicated in the yield surface. From Rice (1987).

The deeper cause for this is that the argument makes use of a description of
plastic flow that makes sense at a macroscopic scale. The stress fields that decide
whether or not the crack propagates, however, operate at much smaller length
scales. When we zoom into the near-tip region from the macroscopic scale, we
first arrive at a scale in which we observe the individual grains that make up a
polycrystalline metal. Upon further zooming-in, we arrive at the grain in which
the crack terminates (in the case of a transgranular crack). The elastic and plastic
behavior of a grain is highly anisotropic, so that the stress fields inside this grain
near the tip of the crack will differ significantly from the HRR solution mentioned
above.

Rice (1987) used a crystal plasticity model to analyze the near-tip fields. Such
a model accounts for the fact that slip occurs on well-defined slip planes and in
particular directions, and is governed by the resolved shear stress. In an fcc crystal,
the slip planes are {111} while the slip directions are ⟨110⟩. When the crack plane
is a (010) with the front along the [10$\bar{1}$]-direction, there are three slip systems that
are potentially active, as illustrated in Fig. 2a. Neglecting any hardening, Rice
(1987) showed that the near-tip stress state has a very particular structure, which
is shown in Fig. 2b. Considering only half of the problem for symmetry reasons,
the solution involves four sectors, with two of the sector boundaries being parallel
to two of the inclined slip planes and the third sector boundary being normal to the
third slip system. Inside each sector, the number of active slip systems is constant
the stress is uniform and the stress jumps discontinuously from one sector to the
adjacent one. The sectors stresses are all on the yield surface, as shown in Fig. 2b
too, so that the material is at yield everywhere.

Saeedvafa and Rice (1988) extended the analysis for fcc and bcc crystals to
account for power-law hardening. The result is an HRR-type solution that is
different from the above-mentioned perfect-plasticity solution in a number of re-

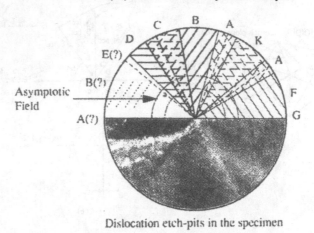

Slip systems identified by moire analysis

Dislocation etch-pits in the specimen

Figure 3: Experimentally determined sectors within a radius of around $500\mu m$ (upper half) and dislocation etch pits (bottom half). From Shield and Kim (1994).

spects. First of the all, the solution involves typically seven sectors (dependent slightly on the hardening exponent) instead of four. The stress states in three sectors is in a corner of the yield surface, but in the remaining four sectors the stress is on a flat segment. In those sectors, only one slip system is active. Furthermore, there are no discontinuities between sectors but thin transition regions. Finally, stresses are no longer constant inside a sector, while the strain field is singular at all angles.

Shield and Kim (1994) performed careful experiments on a pre-cracked bcc single crystal and used Moiré interferometry to measure the strain fields near the crack. From this, they back-constructed the sectors inside which one or more slip systems were activated. As illustrated in Fig. 3, they thus observed eight sectors, which were confirmed by the distribution of dislocation etch-pits. The difference with the analysis of Saeedvafa and Rice (1988) was mainly attributed by elastic unloading that was observed in one of the sectors and subsequent nonproportional loading.

3 Dislocations near crack tips

One of the key features of the HRR-type fields is that they are self-similar. This is a consequence of the fact that it is an asymtotic solution, so that the length of the crack is irrelevant, and the constitutive model is a local theory, i.e. without an internal material length scale. Therefore, the dependence of the radial position from the crack tip is scaled, merely through dimensional arguments, by a length that is determined by the remote loading (i.e. J or K) and the yield strength.

Self-similarity means that the structure of the solution is unchanged as we zoom further and further into the material. However, at some point we arrive at regions of observation that are so small that individual dislocations are being noticeable. At that point, the validity of the continuum plasticity theory breaks down and one needs to adopt a dislocation plasticity description. The remainder of the paper addresses various aspects of this near crack tips.

Roberts (1996) gives an overview of the role of dislocations near crack tips. In particular, he discusses this in relation to the brittle-to-ductile transition. The simple textbook picture is: if there are no dislocations to relieve the high elastic stresses at a crack tip, fracture will be brittle; and if there are dislocations, they will dissipate a portion of the energy and fracture will be ductile. Reality, however, turns out to be more complicated, as we will see in the subsequent section. Nevertheless, it is instructive to see how a dislocation interacts with a crack tip.

Consider a sharp crack with a slip plane emanating from the tip at an angle from the crack plane, Fig. 4. The high stresses at the crack tip can then be effectively reduced by punching a dislocation into the materials, as shown. As the dislocation glides away from the tip, the near-tip stresses increase again and a new dislocation may get punched in. In this schematic, the dislocations are pictured as straight edge dislocations, so that a fully two-dimensional description suffices. Indeed, this is the most efficient way of reducing the near-tip stresses. In keeping with this, Fig. 4 suggests that the dislocations are generated from a straight dislocation source. However, as shown by the detailed studies of Rice and co-workers (e.g. Rice and Beltz, 1994), dislocations are much more likely to nucleate as loops from the crack front. Roberts (1996) argues that dislocation loops nucleated at various locations along the front expand and form a straight dislocation at some distance ahead of the tip, as illustrated in Fig. 5.

The argument above that the motion of dislocations away from the tip leads to a reduction of the stress level near the tip is essentially based on application of the continuum notion of plastic deformation that is caused by the dislocation. From the dislocation point of view, it is a line defect in an otherwise elastic lattice. Modelling this as a line defect in an elastic continuum, the fields caused by a dislocation are singular at the dislocation line (Hirth and Lothe, 1968). Moreover, the fields have long-range effects; the stress field decays as $1/r$ away from the dislocation. The stress field very near the crack tip – so close by that there are no dislocations – remains singular, if crack-tip blunting is neglected. The stress intensity, however, is reduced by the presence of dislocations in the configuration of Fig. 5. The change of the mode I stress intensity at the crack tip due to a dislocation positioned as shown in Fig. 4 is given by (Lin and Thomson, 1986; Lakshmanan and Li, 1988)

$$K_I = -\frac{\mu b}{2\pi(1-\nu)}\sqrt{\frac{2\pi}{r}}\left\{\frac{1}{2}\sin\left(\phi+\frac{\theta}{2}\right) - \frac{3}{4}\sin\left(\frac{\theta}{2}-\phi\right) + \frac{1}{4}\sin\left(\frac{5\theta}{2}-\phi\right)\right\}$$

(2)

where b is the magnitude of the Burgers vector, and μ and ν are the elastic shear modulus and Poisson's ratio, respectively. The sign of the right-hand side

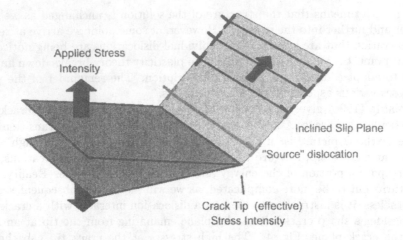

Figure 4: A row of straight edge dislocations on a slip plane emanating from a sharp crack tip [from Roberts (1996)].

immediately shows the shielding effect of a dislocation with the orientation shown in Fig. 6

The effect of a distribution of dislocations around a tip is obtained by straightforward summation of the individual contributions of the type (2). Lin and Thomson (1986), Lakshmanan and Li (1988), Roberts (1996) and several others have used this approach to investigate the transition from a ductile to a brittle mode of fracture. This transition is often presented as a transition with temperature, but there also is a transition in terms of applied loading rate. These studies have shown that this transition is the result of a competition between reduction of the net stress intensity due to distributions of dislocations and their stress-dependent glide motion and nucleation. The latter are quite sensitive to temperature, whereas brittle fracture is largely temperature independent; similarly for the rate dependence. Whether or not the brittle-to-ductile transition is mainly controlled by dislocation nucleation or by the mobility is till under debate (see also Gumbsch *et al.*, 1998). Evidently, these type of calculations have to account for the dynamics of dislocation motion. A somewhat refined approach for this is discussed in the next section.

4 Discrete dislocation dynamics results

4.1 Method of analysis

Most dislocation dynamics studies for cracks are aiming at materials such as silicon, which are initially (almost) dislocation free and where dislocation nucleation indeed happens from the crack tip. Many materials, including most metals, are not

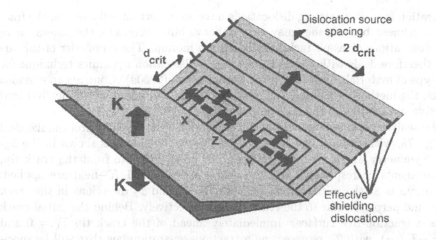

Figure 5: Mechanism for the generation of near-tip edge dislocations from small dislocation loops emitted from the crack front [from Roberts (1996)].

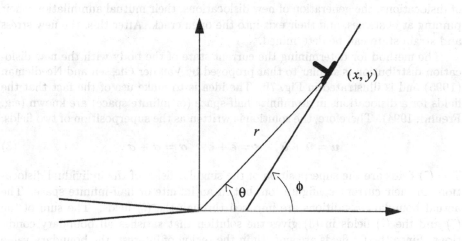

Figure 6: Coordinates of a shielding dislocation near a crack tip on an inclined slip plane.

dislocation free. Pre-existing dislocations have an important influence on the fracture toughness, because they may serve either as bulk sources for the generation of new dislocations or as obstacles to dislocation motion. The remainder of this article therefore deals with a recently developed dislocation dynamics technique for this type of materials developed by Cleveringa *et a.* (2000a). Contrary to previous works, the method accounts for crack-tip blunting and adopts a physically-based cohesive surface to govern crack propagation.

We analyze the two-dimensional plane strain small-scale yielding problem sketched in Fig. 7a, with dislocations restricted to a process window as shown in the figure. Symmetry about the crack plane is assumed. Remote from the crack tip, displacements corresponding to the linear elastic mode I K–field are applied. The origin is at the initial crack tip, with the x_1 and x_2 directions in the crack plane and perpendicular to the crack plane, respectively. Behind the initial crack tip is a traction-free surface. Immediately ahead of the crack tip $T_1 = 0$ and $T_2 = -T_n(u_2)$, with T_n representing a traction–separation law that will be specified later. Further ahead of the initial crack tip, there is no cohesive surface and the condition $\dot{u}_2 = 0$ is applied to enforce symmetry.

At each time step, an increment of the mode I stress intensity factor $\dot{K}_I\Delta t$ is prescribed. At the current instant, the stress and strain state of the body is known, so the forces between dislocations can be calculated. Edge dislocations on multiple slip planes which have an angle of $\pm 30°$ with the x_2-axis are considered. On the basis of these forces we update the dislocation structure, which involves the motion of dislocations, the generation of new dislocations, their mutual annihilation, their pinning at obstacles, and their exit into the open crack. After this, the new stress and strain state can be determined.

The method for determining the current state of the body with the new dislocation distribution is similar to that proposed by Van der Giessen and Needleman (1995) and is illustrated in Fig. 7b. The idea is to make use of the fact that the fields for a dislocations in an infinite half-space (or infinite space) are known (e.g. Freund, 1994). Therefore, the solution is written as the superposition of two fields,

$$\boldsymbol{u} = \tilde{\boldsymbol{u}} + \hat{\boldsymbol{u}}, \quad \boldsymbol{\epsilon} = \tilde{\boldsymbol{\epsilon}} + \hat{\boldsymbol{\epsilon}}, \quad \boldsymbol{\sigma} = \tilde{\boldsymbol{\sigma}} + \hat{\boldsymbol{\sigma}}. \tag{3}$$

The (˜) fields are the superposition of the singular fields of the individual dislocations in their current configuration, but in an infinite or half-infinite space. The actual boundary conditions are imposed through the (ˆ) fields. The sum of the (˜) and the (ˆ) fields in (3) gives the solution that satisfies all boundary conditions. Since the (ˆ) fields are smooth in the region of interest, the boundary value problem for them can conveniently be solved using a finite element method. The calculations to be presented later have used a finite element mesh of 120×100 bilinear quadrilateral elements inside the region analyzed ($1000\mu m \times 500\mu m$). The process window in Fig. 7a is specified by $L_p = 10\mu m$ and $h_p = 12.5\mu m$ and in it there is a fine mesh of 80×80 quadrilateral elements.

Dislocation motion is assumed to occur only by glide with no cross slip, so that dislocations remain on their slip plane. The Peach-Koehler force $f^{(i)}$ acting on

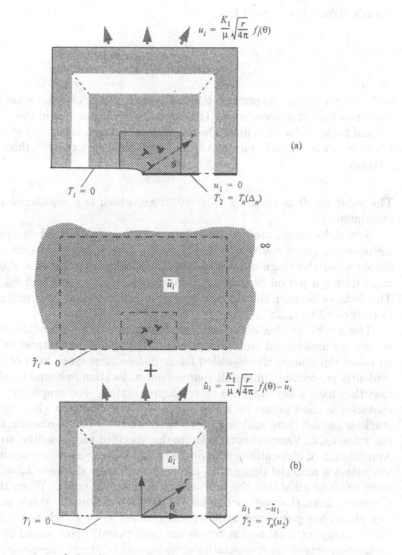

Figure 7: Illustration of the discrete dislocation analysis of the growing mode I crack. (a) Boundary conditions for the symmetric problem. (b) Decomposition into the (~)-solution for the dislocations in half-infinite space and the (^)-solution that enforces the boundary conditions.

the ith dislocation is given by

$$f^{(i)} = n^{(i)} \cdot \left(\hat{\sigma} + \sum_{j \neq i} \sigma^{(j)} \right) \cdot b^{(i)} , \qquad (4)$$

with $n^{(i)}$ the slip plane normal, $b^{(i)}$ the Burgers vector of dislocation i and $\sigma^{(j)}$ is the stress field of dislocation j. The direction of this force is in the slip plane and normal to the dislocation line. The magnitude of the glide velocity v^i of dislocation i is taken to be linearly related to the Peach-Koehler force $f^{(i)}$ through the drag relation

$$f^{(i)} = Bv^{(i)} . \qquad (5)$$

The value for B is taken as $B = 10^{-4} \mathrm{Pa\,s}$, which is a representative value for aluminum.

New dislocations are generated by simulating Frank-Read sources in two dimensions by point sources on a slip plane. The sources generate a dislocation dipole when the magnitude of the Peach-Koehler force exceeds a critical value $\tau_{\mathrm{nuc}} b$ during a period of time t_{nuc}. We choose $t_{\mathrm{nuc}} = 10 \,\mathrm{ns}$ and $\tau_{\mathrm{nuc}} = 50 \,\mathrm{MPa}$. The distance between this dislocation pair is chosen so that their mutual attractive force is equal to $\tau_{\mathrm{nuc}} b$.

The model also features point obstacles that can pin dislocations. These obstacles are understood to represent either very small precipitates or dislocations on other slip planes, the so-called forest dislocations. Such kind of obstacles are evidently pre-existing in metals that are not dislocation free, and we shall see later that they have a key influence on crack propagation. For simplicity, we introduce obstacles as fixed points on a slip plane. A dislocation that glides against an obstacle is pinned there and is only released when its Peach-Koehler force exceeds the value $\tau_{\mathrm{obs}} b$. Various obstacle strengths, specified subsequently, are considered. Annihilation of dislocations with opposite signed Burgers vector occurs when they are within a material dependent, critical annihilation distance $L_e = 6b$. Dislocations can also glide into the free surface of the open crack. When they do, they disappear from the system but leave a lattice step on the crack surface. Since slip planes are positioned symmetrically about $x_2 = 0$, when a dislocation exits the computational region across the (still closed) plane ahead of the crack, a dislocation enters the computational region on the mirror slip plane.

Crack initiation and growth are modeled using a cohesive surface framework (Needleman, 1990). The constitutive relation for the cohesive surface is taken to have the form of the so-called universal binding law,

$$T_n(\Delta_n) = e\sigma_{\max} \frac{\Delta_n}{\delta_n} \exp(-\frac{\Delta_n}{\delta_n}) \qquad (6)$$

where Δ_n is the total separation of the cohesive surface (twice the displacement along $x_2 = 0$) and T_n is the traction normal to the cohesive surface. As the cohesive surface separates, the magnitude of the traction increases, reaches a maximum and then approaches zero with increasing separation. The parameters used in this

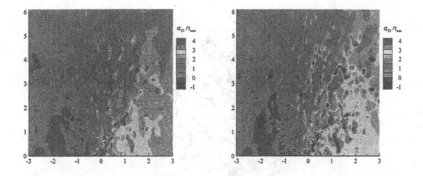

Figure 8: Distributions of the opening stresses σ_{22}, normalized by the nucleation strength τ_{nuc}, for a crystal with three active slip systems, $\phi^{(\alpha)} = (-60°, 0, +60°)$, at load levels of (a) $K_I = 0.45\text{MPa}$; (b) $K_I = 0.6\text{MPa}$. The dislocation distribution is shown in Fig. 9.

study are $\sigma_{\text{max}} = 0.6\,\text{GPa}$ and $\delta_n = 4b$ giving a work of separation, $\phi_n = e\sigma_{max}\delta_n$ of $1.63\ \text{J/m}^2$. The work of separation can be related to a reference stress intensity factor K_0 defined by

$$K_0 = \sqrt{\frac{E\phi_n}{1 - \nu^2}}. \tag{7}$$

With a Young's modulus $E = 70\,\text{MPa}$ and $\nu = 0.33$, representative for aluminum, $K_0 = 0.358\,\text{MPa}\sqrt{\text{m}}$.

4.2   Near-tip fields

We first investigate the stress and strain fields near the crack tip before crack propagation takes place (in fact, we temporarily exclude fracture by using a very high value of the maximum stress σ_{max} in the cohesive surface law). The case shown in Fig. 8 is for a crystal with three slip systems, with the slip planes oriented at angles of $\phi^{(\alpha)} = (-60°, 0, +60°)$ with respect to the crack plane. This configuration is very close to the active slip systems in the fcc crystal in Fig. 2, but just the precise orientation is slightly different. Figure 8 shows how the opening stress field, σ_{22} evolves with increasing applied load. The fields exhibit turbulent stress fluctuations, which are due to the singularities of the individual dislocations, see Fig. 9. In fact, the fluctuations are actually damped in the figure because of the way the contours are plotted on the finite element mesh that was used for the computation (80 by 80 elements in the process window). The figure shows that the stresses very near the crack tip grow very large with increasing K_I over distances that are much larger than the average dislocation spacings. These stresses are high enough to cause atomic debonding and crack advance if this were allowed for in the analysis; we will return to this in the next section.

However, one also observes that further away from the tip, the stresses appear

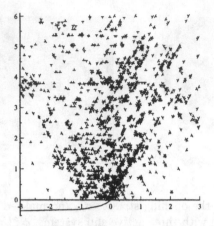

Figure 9: Distributions of dislocations near the tip of a crack in a crystal with three active slip systems, $\phi^{(\alpha)} = (-60°, 0, +60°)$, at a load level of $K_I = 0.6\text{MPa}$, corresponding to Fig. 8b.

to be significantly lower on average. In fact, the stress distribution optically suggests three sectors inside of which the stresses are, on average, rather constant. The boundaries are roughly oriented at 60 and $-60°$ from the crack plane, which is consistent with the A-B and C-D sector boundaries of the continuum solution of Fig. 2. The third sector boundary at $90°$ is not immediately obvious from Fig. 8a. We have computed the average stresses over annular regions of about one micron width around the tip and have checked the levels with the sector stresses following from the Rice (1987) type solution for the present orientation. We find that the σ_{22} component agrees reasonably well, but the comparison is somewhat worse for the other two components.

Similar computations have been carried out for a crystal that has been rotated over $90°$. According the continuum solution, the stress field should have the same sectors as for the orientation above. The discrete dislocation results, however, are not consistent with this. In fact, they suggest different sectors, which do not all need to be yielding on average.

4.3 Crack growth

We mentioned above that very close to the crack tip the stresses do not comply with the average continuum levels, but are very much higher. Figures 10 and 11 illustrates how this gives rise crack growth. These results are for case with two slip systems only, $\phi^{(\alpha)} = (-60°, +60°)$, but they are qualitatively similar with three active slip systems. The first situation shown in Fig. 10a corresponds to the state just before the crack propagates. Prior to this, a dislocation distribution has developed that has reduced the near-tip stress as discussed above and that has given

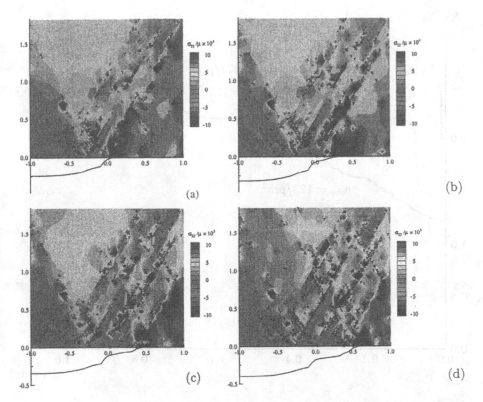

Figure 10: Distribution of dislocations and the opening stress σ_{22} in the immediate neighborhood (2μm $\times$ 2μm) of the crack tip for the case with $\rho_{nuc} = 49/\mu$m^2 and $\rho_{obs} = 98/\mu$m^2 at four different stages of loading (see circles in Fig. 11): $K_I/K_0 = 1.66$ (a); 1.87 (b); 1.94 (c); 2.08 (d). The corresponding crack opening profiles (displacements magnified by a factor of 10) are plotted below the x_1-axis. From Cleveringa *et al.* (2000a).

rise to substantial blunting. Despite shielding of the crack tip by dislocations, the opening stress, σ_{22}, reaches sufficiently high values over a distance of about 0.2μm ahead of the tip to open the cohesive surface. To indicate the stress enhancement involved, the discrepancy between the "continuum" stress level, i.e. the average value in the sector ahead of the crack, and the cohesive strength is a factor of 3 to 4. The stress peak in Fig. 10a is directly due to the presence of the discrete dislocations since quite a few dislocations are located very near the original crack tip.

With continued loading, these stresses cause the crack to propagate into the region of relatively lower stresses. Here, the crack arrests and dislocations on more forward slip planes blunt the tip again. Still, the tractions along the cohesive surface are large enough to cause opening, Fig. 10b. One of the noticeable differences with the state shown in Fig. 10a is that there is a small region of about 0.1μm around the current crack tip where there are no dislocations. Outside this region,

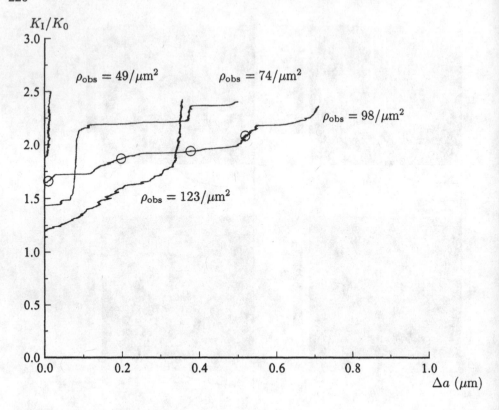

Figure 11: Normalized applied stress intensity factor, K_I/K_0 vs crack extension Δa for various densities of obstacles. The source density is $\rho_{nuc} = 49/\mu m^2$. Dislocation distributions corresponding to the circles are shown in Fig. 10. From Cleveringa *et al.* (2000a).

there are a few dislocations ahead of the crack tip, which have a considerable influence on the near tip stress field. With continued loading, the crack continues to propagate in an almost brittle manner. At the stage shown in Fig. 10c, a tangle of dislocations has formed immediately around the current tip which produces a traction profile ahead of the tip that apparently leads to rapid propagation. Figure 10d shows a stage in which some blunting again accompanies crack growth. As in Fig. 10b, there is a $\approx 0.1\mu$m-sized dislocation-free zone around the tip.

It is of importance to note from Fig. 11 that the response discussed above depends quite sensitively on the density of obstacles. If the obstacle density is only half, $\rho_{\mathrm{obs}} = 49/\mu$m^2, the crack hardly propagates, but continues to blunt. The reason for this is that dislocations are able to move further away from the crack tip and therefore are more able to shield the tip. With increasing obstacle density, the likelihood of dislocations getting trapped near the crack increases, thereby increasing the possibility of development of very high, local opening stresses. The trend is masked somewhat in the toughness curve in Fig. 11 because of statistical effects (Cleveringa et al., 2000a). Since the point obstacles represent also forest dislocations, their density should be expected to increase with plastic deformation, but this is not accounted for in the simulations so far.

4.4   Rate effects

The model involves three time scales: one related to the loading rate, one related to the mobility of the dislocations and the nucleation time t_{nuc}. The time it takes to reach K_0, see eq. (7), is $K_0/\dot{K}_I$. If this time is less than the time to nucleate dislocations, then the material will remain elastic and brittle crack growth will occur. If this time is larger than t_{nuc}, dislocations may get nucleated. Provided that the mobility is high enough, they may glide away from the tip and effectively shield the tip so as to give a ductile response.

Some insight into the transition from brittle to ductile behaviour is provided through Fig. 12 which shows the effect of loading rate on the toughness. The reference loading rate, $\dot{K}_0 = 50$GPa$\sqrt{\mathrm{m}}$/s, is the same as in the previous subsection, and different loading rates differing by roughly two orders of magnitude are considered. As observed above as well, crack initiation is quite sensitive to statistical effects, so that the various resistance curves do not order completely. However, there is a clear tendency that the resistance decreases with increasing loading rate. For the highest loading rate, $\dot{K}_I = 50\dot{K}_0$, there is virtually no increase in resistance once the crack has initiated, indicating that there is insufficient time for dislocation activity to effectively shield the crack tip. For the lower loading rates, crack growth takes place in "spurts" of relatively brittle growth separated by periods in which the resistance increases significantly. The results in Fig. 12 suggest that the spurts are shorter for lower loading rates, although this trend is blurred by statistical effects. Apparently, at the lower loading rates, there is more time available for dislocation tangles to form as the crack approaches.

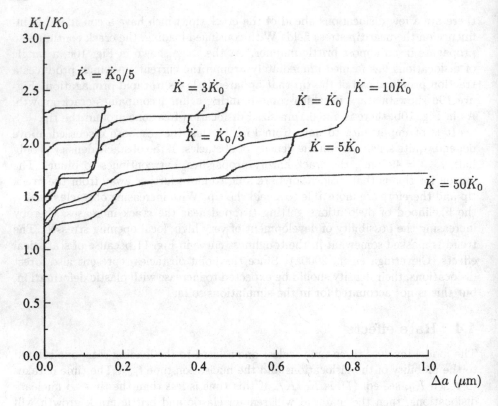

Figure 12: Normalized applied stress intensity factor K_I/K_0 vs crack extension Δa for various loading rates. From Cleveringa *et al.* (2000b).

References

Anderson, T.L. (1995) *Fracture Mechanics*, 2nd ed. CRC Press.

Cleveringa, H.H.M., Van der Giessen, E. and Needleman, A. (2000a) A discrete dislocation analysis of mode I crack growth, *J. Mech. Phys. Solids*, **48**, 1133–1157.

Cleveringa, H.H.M., Van der Giessen, E. and Needleman, A. (2000b) A Discrete Dislocation Analysis of Rate Effects on Mode I Crack Growth, *Mat. Sci. Eng.*, in print.

Freund, L.B. (1994). The mechanics of dislocations in strained-layer semiconductor-materials. *Adv. Appl. Mech.* **30**, 1–66.

Gumbsch, P., Riedle, J., Hartmaier, A. and Fischmeister, H.F. (1998) Controlling factors for the brittle-to-ductile transition in tungsten single crystals. *Science* **282**, 1293–1295.

Hirth, J.P. and Lothe, J. (1968) *Theory of Dislocations*. McGraw-Hill, New York.

Hutchinson, J.W. (1968) Singular behaviour at the end of a tensile crack in a hardening material. *J. Mech. Phys. Solids* **16**, 13–31.

Lakshmanan, V. and Li, J.C.M. (1988) Edge dislocations emitted along inclined planes from a mode I crack. *Mat. Sci. Engrg.* **A104**, 95–104.

Lin, I.H. and Thomson, R. (1986) *Acta Metall.* **34**, 187–206.

Needleman, A. (1990) An analysis of tensile decohesion along an interface. *J. Mech. Phys. Solids* **38**, 289–324.

Rice, J.R. (1987) Tensile crack tip fields in elastic-ideally plastic crystals. *Mech. Mater.* **6**, 317–335.

Rice, J.R. and Beltz, G.E. (1994) The activation energy for dislocation nucleation at a crack. *J. Mech. Phys. Solids* **42**, 333.

Rice, J.R. and Rosengren, G. (1968) Plane strain deformation near a crack tip in a power-law hardening material. *J. Mech. Phys. Solids* **16**, 1–12.

Roberts, S.G. (1996) Modelling the brittle to ductile transition in single crystals. In: *Computer Simulation in Materials Science – nano/meso/macroscopic space and time scales*, ed. by H.O. Kirchner, L.P. Kubin, V. Pontikis, NATO ASI Series E **308**, Kluwer Academic Publishers, 4094–434.

Saeedvafa, M. and Rice, J.R. (1988) Crack tip singular fields in ductile crystals with Taylor power-law hardening. II: plane strain. *J. Mech. Phys. Solids* **37**, 673–691.

Shield, T.W. and Kim, K.-S.(1994) Experimental measurement of the near tip strain field in an iron-silicon single crystal.*J. Mech. Phys. Solids* **42**, 845–873.

Van der Giessen, E., Needleman, A. (1995) Discrete dislocation plasticity: a simple planar model. *Modell. Simul. Mat. Sci. Engin.* **3**, 689–735.

STRAIN LOCALIZATION IN SINGLE CRYSTALS AND POLYCRYSTALS

C. REY, T. HOC, Ph. ERIEAU
Laboratoire de Mécanique des Sols, Structures et Matériaux,
CNRS, UMR 8579,
Ecole Centrale Paris, Grande Voie des Vignes,
92295 Châtenay-Malabry cédex, France

Abstract. Strain localization appears during cold forming processes, leading to softening and to specific deformation texture. This phenomenon has a strong impact on damage and recrystallization. This paper focuses on the description of the localization bands in single and polycrystals submitted to different loading paths. Modelling of localization, based on a crystalline approach using a finite element code, is proposed. The simulation gives some details on the evolution of dislocation densities, on all activated slip sytems and also on the microstructural anisotropy. Relations between localization bands and stored energy inhomogeneities are discussed.

1 Introduction

Localization bands and deformation bands have a strong impact on technological forming processes resulting in softening, damage, rupture, deformation texture and recrystallization. Extensively studied in the last decade, the use of recent experimental techniques as microextensometry and Electron Back Scattering Pattern and of numerical approaches by using finite element software, brought a new interest to these phenomena. Deformation bands were overviewed by D. Kuhlmann-Wilsdorf [1]. According to this author, "Deformation Bands" (DBs) consist of volume elements, typically in sequences slab shaped parallel and presenting alternating average lattice orientation. These DB are due to local simultaneous activation of fewer slip systems than would be required by an homogeneous deformation.

225

E. Bouchaud et al. (eds.), Physical Aspects of Fracture, 225–241.

Relationships between average band width, band length and flow stress at the time of band formation were obtained on the base of low-energy dislocation structure theory (LEDS). It is stated that, among all microstructures in equilibrium with the applied stresses, those formed are the ones which minimize the energy of the system composed of the material and the applied traction. DBs include dipolar or tilted walls constituting domain boundaries, regular deformation bands constituted by slab shaped parallel sequences of volume element with alternating average lattice orientation, transition bands which present strain gradient and lattice orientation change, kink bands and shear bands. At the beginning of the plastic strain, DBs are not yet constituted and the observed intragranular inhomogeneities are mostly composed of domains delimited by intragranular interfaces and presenting different activated slip systems. An example of intragranular interfaces is given on figure 1.

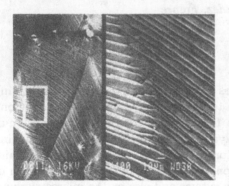 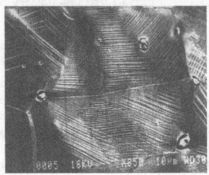

FIG. 1 – Strain inhomogeneities in tensile tested copper polycrystals, E=21%

It is generally assumed that such patterns lower plastic energy. The stability of these interfaces depends on the material stacking fault energy, which rules the hardening properties [2], [3]. Microextensometry techniques by microgrids have pointed out strain localization in bands, at the scale of grains and of aggregates, named mesobands and macrobands. The first ones are mainly associated to ductility, whereas the latter ones are mainly associated to instabilities occuring during sequential tests or for large monotonic deformations. If shear-bands and kink-bands belong to macrobands, it is not established that Deformation Bands correspond to mesobands. The origin of strain localization during monotonic loadings and small plastic deformations has not yet received advanced explanations. By contrast, strain localization which appears for sequential loading paths or for large monotonic strains have been extensively studied [4], [5],[6], [7], [8], [9] [10]. According to some authors [11] such localization, generally named shearbanding, is linked to two main sources of softening, resulting from the evolutions of crystalline textures and intragranular microstructure. For single crystals submitted to large monotonic strain, a bifurcation criterium based on textural softening was proposed by Asaro [12] and Pierce [13]. In steel polycrystals submitted to sequential loading, strain localization was related to the evolution of microstructural anisotropy [14] rather than textural effect. Finite element codes in crystalline plasticity has been extensively used these last decades and some studies focused on prediction of deformation texture in polycrystals [15] and on heterogeneous intragranular and intergranular local strain and rotation fields [16, 5, 17, 14].

This paper is focused on strain localization during monotonic and sequential loading paths in bcc and fcc materials. It includes experimental studies based on microextensometry techniques and numerical works by FEM.

2 Experimental investigations of localization bands

2.1 EXPERIMENTAL TECHNIQUES

Most of the experimental investigations presented in this paper used microgrids and EBSD techniques. The strain field was determined at the grain scale, thanks to microgrids deposited on the surface samples before the mechanical tests. The used square mesh of the grid was about 5 μm for a grain size of 20 μm. The tests were stopped at some intervals and the displacement of the nodes of the grid were recorded. The displacement normal to the surface was obtained by comparison of untilted and tilted images. In all this paper, the axis $\vec{1}$ and $\vec{2}$ correspond to the transverse axis and the tensile or compression axis respectively.

Six components of the gradient tensor $\underset{\sim}{F}$ ($F_{11}, F_{12}, F_{21}, F_{22}, F_{31}, F_{32}$) and three components of the Green-Lagrange tensor $\underset{\sim}{E}$ (E_{11}, E_{12}, E_{22}) were computed from the displacement for each element of the grid at different steps of the plastic deformation. The local crystalline textures of the studied area were simultaneously measured by Electron Back Scattering Diffraction technique(EBSD). More details are given in [18], [19]

2.2 LOCALIZATION BANDS

Mesobands in mild steel polycrystals. Investigations of the intragranular and intergranular strain field have been performed in tensile tested mild steel polycrystals 20 μm grain size, [20]. Figure 2 gives the result of the computation for an aggregate of 114 grains for a macroscopic strain of 25%. The map obtained by reporting the component E_{22} on each element of the undeformed grid, points out a strong strain localization within two families of bands (mesobands), tilted at about 45 degrees of the tensile axis. These mesobands, about two or three grains long, rotate with increasing deformation but stay localized in the same material areas. For an applied strain of 25 % , the local strain reaches 80%. It is worth noting that, in these experiments, the maps were always drawn in the undeformed configuration so that the material rotation is not visible.

Such regular heterogeneities, already observed in steel as early as 5% of macroscopic strain [21], have recently be pointed out in other materials at the very beginning of plastic deformation [22]. The physical and mechanical origins of these

228

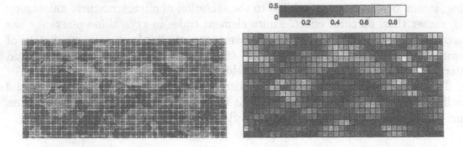

FIG. 2 – Investigated aggregate of 114 grains (left). Inhomogeneities of the E_{22} component of the Green-Lagrange tensor in the reference configuration for applied stretching of 25% (right)

mesobands are not yet clarified. Corresponding to a maximum of the shear stress, mesoband initiation depends on the boundary conditions applied to the sample. Density and morphology of mesobands are rather ruled by microstructure such as grain size and grains misorientation, whereas strain localization amplitude is rather bound to hardening properties. The nucleation of bands in a given point of the aggregate can be the result of local stress effects due to elastic or to plastic incompatibilities at grain boundaries. Few investigations were performed to relate strain localization bands and deformation texture, but Lineau [21], using EBSD technique, showed that the crystalline misorientations within the bands decreased with increasing deformation, leading to preferential orientations.

Iron and copper single crystals. In iron single crystals, oriented in single slip, one macroband (kink band) corresponding to the maximum of the stress-strain curve was observed [18]. The macroband, at 60 degrees of the tensile axis, was limited by two parallel interfaces, one stable, the other one mobile. Large gradients of the strain and rotation fields were computed within the band, the strain maxima being close to the mobile interface. The strain mechanisms within the bands corresponded to the activation of crystalline slip systems. The band (200 μm width at 55% of plastic deformation for a 10 mm x 3mm x 0.7 mm sample) widened by activation of new slip systems near the mobile interface, leading to the existence of an active zone which width remained constant with increasing deformation. The same kinetic was observed in copper single crystal oriented in symmetrical double slip [14]. In every cases the observed localization bands corresponded to a large density of activated slip systems and the computed strain field was obviously not limited only to shearing.

Macrobands in mild steel polycrystals. Macrobands, often called shearbands in the literature, correspond to instability phenomena. Recent works, on mild steel polycrystals submitted to a change of the loading path [14], [19], have brought some new informations on the macrolocalization phenomenon. The mild steel polycrystal was deformed first in plane strain tension in the rolling direction, then stretched in uniaxial tension in the transverse direction (TD test) or in the rolling direction (RD test). The macroscopic results corresponded to classical ones :

1. the strain stress curves corresponding to the second loading, presented a maximum corresponding to the localization, followed by a decrease then a plateau,

2. for TD tests, the maxima of the stress-strain curves were higher than for RD tests,

3. diffuse localization was observed for RD tests and macrobands at 45 degrees of the tensile axis for TD tests,

4. the number of macrobands depended on the prestrain amplitude, as shown on figure 3.

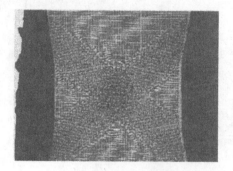 

(a) Macrolocalization, prestrain=12%, tensile strain=4% (TD test)

(b) Macrolocalization, prestrain=24%, tensile strain=6.6% (TD test)

FIG. 3 – Macro-localization in steel polycrystals (initial sample width=3mm).

Computation of the local strain field inside the macroband, at different steps of the test, pointed out a macroband evolution similar to the single crystals one, and the existence of a "mesostructure" not correlated to the grain boundaries. Figure 4 shows that the whole plastic deformation is concentrated within the macrobands, whereas the deformation outside (matrix) is close to zero (elastic deformation). Moreover, a regular pattern of two families of small parallel bands (mesobands) appears within the macrobands making a well defined mesostructure. These mesobands crossing one or two grains exhibit a large value of E_{22}, reaching 150% for a 1.5% applied tensile strain. These mesobands are constituted by bundles of slip sytems, some of them impinging on grain boundaries, whereas some others vanish close to them (figure 5). With increasing deformation, the mesoband traces rotate towards the tensile axis, but as shown by figure 4 b, the maxima of the local strain E_{22} are always localized in the same material points, corresponding to the observed mesobands. It is worth noting that the average deformation within the macroband does not correspond to a pure shear. According to these results, we can say that macrobands in polycrystals behave as a pseudo single crystal, with mesobands playing the role of slip systems. The correlation between mesobands and dislocation microstructure is not yet solved, though bundles of dislocation walls were observed within the macrobands.

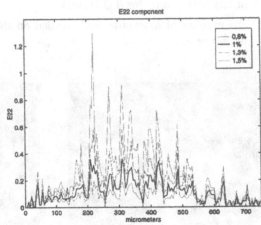

(a) E_{22} component for an applied tensile strain of 1.5%

(b) Evolution of the E_{22} component with increasing applied macroscopic strain as a function of the position within the macroband (prestrain=24%)

FIG. 4 – Component E_{22} of the Green-Lagrange tensor in the reference configuration for the TD test. Tensile axis is vertical. The surface $(250 \times 750\mu m^2)$ is studied by microextensometry (microgrids of $5 \times 5\mu m^2$) for a macroscopic deformation of 1.5% (b) Evolution of E_{22} along a line parallel to the tensile axis for different macroscopic steps

Copper polycrystals. Strong localization was observed in rolled copper polycrystals 500 μm grain size, for 90% applied strain. Figure 6 shows localization in some grains, according to parallel bands at 30 degrees of the rolling direction. Constituted by sinuous glide traces, these bands carried on a large amount of shear, crossed one or two grains and widened by activation of new glide. The most interesting fact is that nucleation of such bands affected only some grains. Though crystalline orientation was unknown in these experiments, we can assumed that texture effect favors activation of parallel short bands in the whole sample. In these experiments the localization bands were named shearbands. Though accurate informations are lacking, it seems that the hereabove phenomena may correspond to macrobands.

231

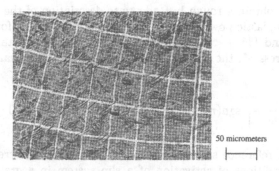

FIG. 5 – Coarse slip systems within the macroband for an applied macroscopic strain of 1.5 %

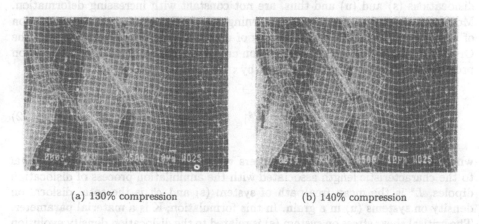

(a) 130% compression   (b) 140% compression

FIG. 6 – Strain localization evolution in rolled copper polycrystal $500\mu m$ grain size (microgrid was deposited at 120%, rolling axis is vertical)

3 Plastic crystal modelling

A crystalline plasticity constitutive model which accounts for deformation by crystallographic slip and for lattice rotation with deformation, was used in the analysis of localization problem and implemented in the Abaqus finite element code by Hoc [23], [24], [25]. Implementation of the constitutive relations as a "User material" subroutine in the Abaqus code has been described by Smelser and Becker [26]. The mechanical modelling follows the framework developed by Peirce [27] *et al* and later modified by Teodosiu *et al* [28] . The kinetics of large transformations used here, are based on the multiplicative decomposition of the deformation gradient into elastic and plastic parts. Plastic anisotropy is described in terms of dislocation densities on each slip system rather than in term of shear strain rates, in order to obtain a rough but adequate description of the microstructure evolution. A viscoplastic power law for the glide on the twenty four slip systems $\{110\} < 111 >$ and $\{112\} < 111 >$ is assumed for the bcc crystals. The applied resolved shear stress τ^s, the critical shear stress τ_c^s and the shear strain rate $\dot{\gamma}^s$ are related by :

$$\dot{\gamma}^s = \dot{\gamma}_0 \left| \frac{\tau^s - \tau_c^s}{\tau_0} \right|^n \mathrm{sgn}(\tau^s) \quad \text{if} \quad | \tau^s | > \tau_c^s \quad \text{and} \quad \dot{\gamma}^s = 0 \quad \text{otherwise} \tag{1}$$

where $\dot{\gamma}_0$ is a reference strain rate, and τ_0 is the friction stress. In this formulation, the conditions of activation of a slip system in a grain are given by Schmid's law. In the phenomenological approach used here, the critical resolved stress is linearly related to slip rate through a hardening matrix. The components of this hardening matrix (24x24) depend on short range interactions between the dislocations (s) and (u) and thus, are not constant with increasing deformation. Moreover, the anisotropy of the hardening matrix increases with the activation of new slip systems. The evolution law of dislocation densities based first on the Orowan's relationship for the dislocation creation and second on the annihilation process of dislocation dipoles, is given by :

$$\dot{\rho}^{sg} = \frac{1}{b} \left(\frac{1}{L^{sg}} - G_c \rho^{sg} \right) |\dot{\gamma}^{sg}| \quad \text{with} \quad L^s = K \left(\sum_{u \neq s} \rho^u \right)^{-\frac{1}{2}} \tag{2}$$

where b is the magnitude of the Burgers vector, G_c is a parameter proportional to the characteristic length associated with the annihilation process of dislocation dipoles, L^s is the mean free path of system (s) and ρ^u is the total dislocation density on systems (u) in a grain. In this formulation, K is a material parameter. The critical shear stress on system (s) is related to the dislocation density evolution on each slip system by the relation :

$$\tau_c^s = \tau_0 + \mu\, b \left(\sum_u a^{su}\, \rho^u \right)^{\frac{1}{2}} \tag{3}$$

where a^{su} are constants characterizing the different kinds of interaction between two families of dislocations (s) and (u). In this paper, six different coefficients a^{su} are considered (see table 1).

Plane	$\{110\} \cap \{110\}$	$\{110\} \cap \{112\}$	$\{112\} \cap \{112\}$
Same	0.2		0.26
Colinear	0.2	0.21	0.26
No colinear	0.23	0.2205	0.299

TAB. 1 – Matrix interaction coefficients

The values of the parameters are determined from mechanical tests (plane tension tests, tensile tests, cyclic shear tests) and crystallographic texture, using an automatic identification procedure given by the software SiDoLo and a polycrystalline model [29], [30]. The computed parameters, given in tables 1 and 2, are close to those proposed in the literature.

$\tau_0(MPa)$	$\rho_0(m^{-2})$	G_c (nm)	K	$\dot{\gamma}_0$ (s^{-1})	n
20	64.10^9	10	20	1.14	15

TAB. 2 – Different parameters of the crystalline model

4   Numerical simulation of strain localization and of associated energy

The polycrystalline structure has to be idealized for the implementation of a finite element code. Between the impossibility of knowing a 3-dimension structure without destroying it and discretizing an arbitrary virtual aggregate, we have chosen to use a part of an actual polycrystal for the definition of the pattern of grains named Finite Element Representative Pattern (FERP) and to build, when it was necessary, the complete structure by repeating this pattern. The FERP used in the hereafter simulations was constructed by discretizing the actual surface of the experimental studied polycrystals, in order to obtain a grid made of 8 noded brick finite elements. Each element was assigned a crystallographic orientation so that the actual grain boundaries were approximately followed. An example of the used FERP is given on figure 7.

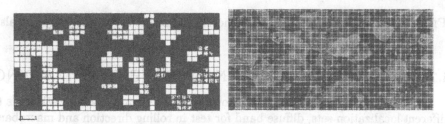

FIG. 7 – Finite Element Representative Pattern and actual aggregate

4.1   STRAIN LOCALIZATION IN MONOTONIC LOADING

Though the main goal of the hereabove crystalline model is to describe the evolution of anisotropy due to activation of slip systems during sequential loading paths, it was used to describe inhomogeneities formation during monotonic loading. To avoid direct effects of boundary conditions on a "numerical sample" composed by a granular medium of about one hundred grains, the FERP was partially embedded in an intermediate medium which behavior was the polycrystal one [20]. The stretch component E_{22} of the Green-Lagrange tensor was computed for a global strain of 2% and of 20% respectively. The results reported on figures 8a,b, point out intra and intergranular heterogeneities constituted by a more or less regular pattern of small bands tilted about 45 degrees to the tensile axis. The length and thickness of numerical bands are of the same order than experimental bands, but the strain intensity is weaker. This means that the equivalent medium is not exactly representative, from strain incompatibilities point of view, of the inner layers of grains.

For monotonic loading, localization was obtained for other materials by different authors [16], [5]. The main difference between this model and the others holds in the description of the hardening matrix composed here of 6 different terms

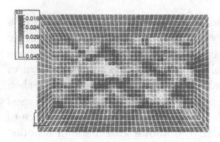 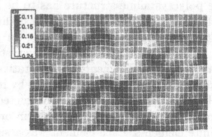

(a) Granular medium at the surface of an equivalent medium E=2%

(b) Granular medium at the surface of an equivalent medium E=20%

FIG. 8 – Simulation of strain heterogeneities in tensile tested steel polycrystals

4.2   STRAIN LOCALIZATION IN SEQUENTIAL LOADING

Experimental investigations have shown that the second loading path leads to different localization sets, diffuse band for test in rolling direction and macroband (defined by two parallel interfaces) in the transverse direction. We assume that the observed differences between the stress-strain curves obtained for DT and DR tests, and also between the bands morphology, come from anisotropy of the microstructure at the end of the first loading path.

In order to simulate qualitatively the strain localization observed after prestraining in the transverse and rolling directions respectively, the densities of dislocation and the orientations of 114 grains computed at the end of the first loading were introduced in the finite element code. Maps of longitudinal deformation for these two numerical samples, are given on figure 9.

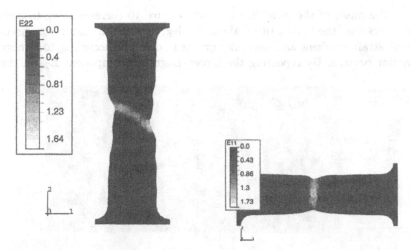

(a) tensile test in the transverse direction (macroband)

(b) tensile test in the rolling direction (diffuse necking)

FIG. 9 – Strain localization for tensile loading (prestrain :18%)

Computed strain localizations (stress-strain curves and band morphology) fit qualitatively well with experimental results. In the model, anisotropy is taken into account through a hardening matrix described in terms of dislocation-dislocation interactions and through a dislocation density evolution law given by equation 2. According to the simulations, localization occurs when the dislocation density on the activated slip systems saturates, whereas the shear amplitude increases.

4.3 STORED ENERGY ASSOCIATED TO STRAIN LOCALIZATION

Strain heterogeneities in polycrystals are known to be preferential locations where nucleation occurs prior to recrystallization. In fact, nucleation has been observed after cold rolling in IF steels next to grain boundaries, in the interior of grains and also in strain localization microbands [31]. Moreover, annealing and recrystallization are related to the amount of energy stored during rolling [32]. As a consequence, nucleation in IF-Ti steels was assumed to be governed by the stored energy.

The crystal model was used to simulate the inhomogeneous deformation of crystal aggregates submitted to plane strain compression (channel die device). Stored energy and plastic energy distribution have been computed in an actual layer of one hundred grains, which orientations were obtained using EBSD technique [33]. Considering a $3\mu m$ mesh for a $25\mu m$ grain size , strain localization bands were

236

observed at the onset of the plasticity domain. Figure 10 corresponding to 30% reduction shows that the bands, tilted about 35 degrees from the extension direction, present strain gradient and share the grains in domains according to a more or less regular pattern. By reporting the Green-Lagrange component E_{22} in the

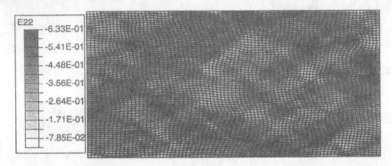

FIG. 10 – Deformed mesh after 30% thickness reduction, E_{22} component isovalues

initial mesh, the localization development has been analyzed at different steps of the test. The bands seen at 30% reduction were already present at the early stages of the simulation, at the onset of plastification, they were tilted of 42 degrees from the compressive axis, widened and rotated with increasing deformation. The model gives access to the crystal rotation fields. Figure 11 shows the Rodrigues angle distribution at 30% reduction in the initial configuration. The map gives for each element,the angle of rotation of the crystal,from its initial position. The main results agree within literature and show that the grain boundaries areas have rotated less than the inner of the grains and that the rotation is inhomogeneous on both sides of a grain boundary.

After cold work, the stored energy is due to the elastic energy of the dislocations arrangements, but according to the review performed by Bever et al [34] , 1 to 15% of the total plastic work spent in straining, remains as stored energy. In order to compare these approaches, two ways of computation of the stored energy were performed. The stored energy was computed by the formula $E_{st}^{\rho} = \frac{1}{2}\mu b^2 \rho$ where ρ was the total dislocation density [32]. The total plastic work used $E_{st}^{p.w.} = \int_t \sigma : \epsilon^p dt$ formula. For 30% reduction figure 12 shows that the ratio $\frac{E_{st}^{\rho}}{E_{st}^{p.w.}}$ is a bit smaller than those reported in the literature. Both stored energy and total plastic work patterns present high values near the grain boundaries but the computed amounts of energy are different in both cases. The E_{st}^{ρ} stored energy and the crystal rotation distribution look similar, which seems to indicate that grain boundaries store a large energy, but display moderate rotations . The total plastic work distribution is close to the strain localization one.

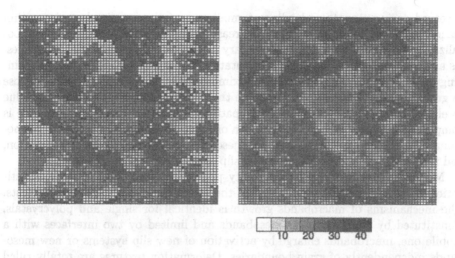

FIG. 11 – Initial position of the grains (left) and crystal Rodrigues rotation angle in degrees (right) for 30% reduction

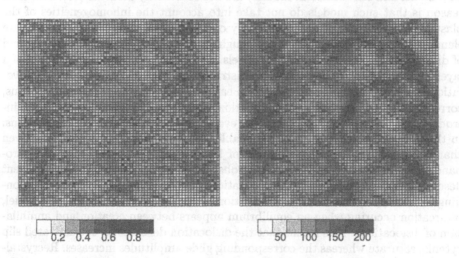

FIG. 12 – Stored energy distributions calculated with : E_{st}^{ρ} (left) and total plastic work $E_{st}^{p.w.}$ $(right) for 30\% reduction (values in \text{MJ}.\text{m}^{-3})$

5 Discussion

Strain localization within bands corresponds, at least for bcc materials, to an increase of ductility or to an instability, and partly rules the deformation texture. Mainly analyzed in bcc materials, a recent work has shown that localization occured at the onset of plastic stage and affected a lot of materials [22]. Mesobands observed during monotonic loading at small and medium strains but also within macrobands, are bound to ductility whereas macrobands deal with instability. Localization phenomena depend on boundary conditions and on material properties. As a matter of fact, the mesoband orientation corresponds, at least at the beginning of the tests, to the macroscopic maximum shear plane. Band nucleation close to grain boundaries (hard zone) may be the consequence of internal stresses due to elastic and (or) plastic strain incompatibility. In other words, a hard zone is compensated by a soft zone. "Soft" grain orientations probably favor such phenomena. Crystalline rotations within the mesobands lead to preferential orientation, and favor activation of only one or two slip sytems.

Macrobands correspond to instability phenomena and the threshold of activation corresponds to the maximum of the stress-strain curves for tensile tests. The mechanisms of macrobands growth is identical for single and polycrystals. Constituted by slip systems or mesobands and limited by two interfaces with a mobile one, macrobands enlarge by activation of new slip systems or new mesobands independently of grain boundaries. Deformation textures are totally ruled by macrobands.

It is now accepted that textures predicted by homogeneization models (Taylor's model and self consistent models) are not sufficiently accurate [35]. The main reason is that such models do not take into account the inhomogeneities of the plastic deformation inside the grains. By contrast, crystalline models using finite element method [17, 23, 24] allow a rough description of localization bands and of dislocation microstructure. Such models are based on an actual description of a layer of grains (rather than a random distribution of orientations) and on an evolution law of the dislocation densities on each slip system, composed of two terms, corresponding to creation and annihilation of dislocations. The length scales introduced in the code correspond to the evolutive mean free paths of dislocations. In this framework, crystalline model enables to describe qualitatively, for a given change of loading path, the threshold of bifurcation, the morphology of macrobands and the texture within the macrobands. Moreover, a simple but sufficient description of the evolution of the dislocation microstructure and of the corresponding anisotropy is obtained. No bifurcation criterium is introduced in the model, bifurcation occuring when an equilibrium appears between creation and annihilation of dislocations. This means that the dislocation densities on the activated slip sytems, saturate whereas the corresponding glide amplitudes increases. Recrystallization of rolled polycrystals depends on the misorientions between grains, on the grains orientation and also on the inhomogeneities inside the grains (intragranular interfaces as well as localization bands). The crystalline model shows that for a significant applied strain, plastic energy pattern is different of stored energy one.

The evolution of the dislocation microstructure in steels submitted to tension, rolling and change of loading path, has been extensively studied [36, 37, 38], but there is a lack of information about correlation with mesobands. We can only add that walls of dislocations observed in grains of polycrystals at large deformation, are often parallel to slip systems in mesobands.

6 Conclusion

Strain localization was analysed through the combination of experimental and numerical techniques. The experimental study called for several observations at different scale, in order to identify the physical mechanisms which control localization. A crystalline model based on a description of dislocation density evolution on each slip system was proposed. The use of finite element method is a direct way to account for actual grain to grain interaction and to set implicitly characteristic lengths, assumed to be the evolutive mean free path of dislocations. Such framework gives an adequate description of the anisotropy evolution during the loading path. Though a coarse description of grain boundaries and a microstructure described by continuous dislocation distributions, the results of the simulation are in qualitative agreement with experiments. The main interest of the model is an access to physical variables which cannot be measured directly. According to the model, plastic instability occurs for a saturation of the dislocation densities on the activated slip systems. Significant improvements of the understanding and modelling of strain localization require to gather more informations at a more microscopic scale, and to correlate localization bands and microstructure.

References

[1] D. KUHLMANN-WILSDORF. Regular deformation bands and the LEDS hypothesis. *Acta mater*, 47(6) :1697–1712, 1999.

[2] A. REY, C.and ZAOUI. slip heterogeneities in deformed aluminium bicrystals. *Acta Metallurgica*, 28 :687–697, 1979.

[3] C. REY and A. ZAOUI. Grain boudary effects in deformed bicrystals. *Acta Metall*, 30 :523–535, 1982.

[4] R.J. ASARO. Crystal plasticity. *Journal of applied mechanics*, 50 :921–934, december 1983.

[5] R. BECKER. Effects of strain localization on surface roughening during sheet forming. *Acta Metal.*, 46(4) :1385–1401, 1998.

[6] A. KORBEL and P. MARTIN. Microstructural events of macroscopic strain localization in prestrained tensile specimen. *Acta Metal.*, 36(9) :2575–2586, 1988.

[7] D. PEIRCE, R.J. ASARO, and A. NEEDLEMAN. Material rate dependence and localized deformation in crystalline solids. *Acta Metal.*, 31(12) :1951–1976, 1983.

[8] H. DEVE, S. HARREN, C. McCULLOUGH, and R.J. ASARO. Micro and macroscopic aspects of shear band formation in internally nitrided single crystals of e-TI-MN alloys. *Acta Metal.*, 36(2) :341–365, 1988.

[9] A. NEEDLEMAN, R.J. ASARO, J. LEMONDS, and D. PEIRCE. Finite element analysis of crystalline solids. In Elsevier Science Publishers B.V., editor, *Computer Methods in Applied Mechanics and Engineering 52*, 1985.

[10] E.F. RAUCH and C. G'SELL. Flow localization induced by a change in strain path in mild steel. *Materials Science and Engineering*, A111 :71–80, 1989.

[11] A. B. LOPES, E.F. RAUCH, and J.J. GRACIO. Textural vs structural plastic instabilities in sheet metal forming. *Acta Metal.*, 47(3) :859–866, 1999.

[12] R.J. ASARO. Geometrical effects in inhomogeneous deformation of ductile single crystal. *Acta metal.*, 27 :445–453, 1979.

[13] D. PEIRCE. Shear band bifurcation in ductile single crystals. *J. Mech. Phys. Solids*, 31 :133–153, 1983.

[14] E. LABBE, T. HOC, and C. REY. A simplified cristallographic approach of bifurcation for single crystals and polycristals. In *Proc. Euromech. Magdeburg*, February 1998.

[15] S.R. KALIDINDI, C.A. BRONKHORST, and L. ANAND. Crystallographic texture evolution in bulk deformation of fcc metals. *J. Mech. Phys.*, 40(3) :537–569, 1992.

[16] J. HARDER. A cristallographic model for the study of local deformation processes in polycrystals. *Int. J. of plasticity*, 15 :605–624, 1998.

[17] F. DELAIRE, J.L. RAPHANEL, and C. REY. Plastic heterogeneities of a copper multicrystal deformed in uniaxial tension : experimental study and finite element simulation. *Acta Metal.*, 48(5) :1075–1087, 2000.

[18] C. REY and P. VIARIS. Experimental analysis of bifurcation and post bifurcation in iron single crystals. *Mater. Sci. Eng. A*, A234-23 :1007, 1997.

[19] T. HOC, C. REY, and P. VIARIS DE LESEGNO. Mesostructure of the localization in prestrained mild steel. *Scripta Met.*, 42 :749–754, 2000.

[20] T. HOC and C. REY. Effect of the free surface on strain localization in mild steel. *Scripta Met.*, 42 :1053–1058, 2000.

[21] J. Zieb et al, editor. *Evolution of the Local Strain Field and local Cristallographic Rotation Field in Grains of polycrystals*, March 1996.

[22] P. DOUMALIN. *Microextensométrie locale par corrélation d'images appliquée aux études micromécaniques en utilisant le microscope électronique à balayage*. PhD thesis, Ecole Polytechnique, 2000.

[23] T. HOC. *Etudes expérimentale et numérique de la localisation lors de changement de trajets dans un acier doux*. PhD thesis, Ecole Centrale Paris, 1999.

[24] T. HOC and C. REY. Experimental analysis and crystallographic model of plastic deformation after a change of loading in steel polycrystals. In A. Gonis et al, editor, *Materials Research Society, symposium proceedings*, volume 578, pages 67–72, 2000.

[25] T. HOC, C. REY, and J.L. RAPHANEL. Experimental and numerical analysis of localization during sequential tests for IF-Ti steel. *Acta metall.*, 2001. in press.

[26] R.E. SMELSER and R. BECKER. In *ABAQUS User subroutines for Material Modeling*, 1989.

[27] D. PEIRCE, R.J. ASARO, and A. NEEDLEMAN. An analysis of nonuniform and localized deformation in ductile single crystals. *Acta Metal.*, 30 :1087–1119, 1982.

[28] C. TEODOSIU, J.L. RAPHANEL, and L. TABOUROT. Finite element simulation of the large elastoplastic deformation of multicrystals. In C. Teodosiu, J.L. Raphanel, and F. Sidoroff, editors, *Large Plastic Deformation, Proc. int. seminar MECAMAT'91*, pages 153–167, August 1993.

[29] P. PILVIN. Une approche inverse pour l'identification d'un modèle polycristallin élastoviscoplastique. In *Colloque national en calcul des structures*, 1997, Giens.

[30] T. HOC and S. FOREST. Polycrystal modelling of IF-Ti steel under complex loading path. *Int. J. of Plasticity*, 17 :65–85, 2001.

[31] M.R. BARNETT and J.J. JONAS. Influence of ferrite rolling temperature on microstructure and texture in deformed low-C and IF steels. *ISIJ International*, 37(7) :697–705, 1997.

[32] F.J. HUMPHREYS and M. HATHERLY. *Recrystallization and related annealing phenomena*. Pergamon, Oxford, 1996.

[33] Ph. ERIEAU, T. HOC, H. BIAUSSER, and C. REY. Modelling heterogeneities and computation of stored energy distribution in an if-Ti steel. In N. Hansen et al, editor, *Proceeding of the 21st RisøInternational Symposium of Material Science : Recrystallization-Fundamantal aspects and Relations to Deformation Microstructure*, pages 339–344, RisøNational Laboratory, Roskilde, Denmark, 2000.

[34] M.B. BEVER, D.L. HOLT, and A.L. TITCHNER. *The stored energy in cold work. Progress in materials science.* Pergamon, Oxford, 1973.

[35] T. LEFFERS. Why we cannot simulate deformation textures. In J.A Szpumar, editor, *Proceeding of the Twelfth International Conference on Textures of Materials*, volume 1, pages 261–266, 1999.

[36] R.A. JAGO and N. HANSEN. Grain size effects in deformation of polycrystalline iron. *Acta metall.*, 34(9) :1711–1720, 1986.

[37] Q. LIU, B.L. LI, W. LIU, and X. HUANG. Deformation structure evolution of IF steel during cold rolling. In N. Hansen et al, editor, *Proceeding of the 21st RisøInternational Symposium of Material Science : Recrystallization-Fundamantal aspects and Relations to Deformation Microstructure*, pages 423–430, RisøNational Laboratory, Roskilde, Denmark, 2000.

[38] E.F. RAUCH. *Etude de l'Ecrouissage des Métaux.* PhD thesis, I.N.P. Grenoble, 1993.

FATIGUE AND STRESS CORROSION

MODELLING IN FATIGUE
REMARKS ON SCALES OF MATERIAL DESCRIPTION:
APPLICATION TO HIGH CYCLE FATIGUE

K. Dang Van
École Polytechnique
Laboratoire de Mécanique des Solides - UMR 7649 C.N.R.S.
91128 Palaiseau

1.Remarks on scales of material description in fatigue

Fatigue failure is the final step of complex physical processes difficult to model: It begins with the appearance of slip bands in some grains which broaden progressively during the following cycles. These phenomena occur simultaneously with the development of localised damage. The processes result in the formation of intragranular microcracks at a stage determined by the imposed load level. After a certain number of cycles, a main crack initiates, then propagates and shields the other defects and, consequently leads to a the final rupture of the mechanical structure.

This description of the fatigue damage phenomenon is schematic, since it does not do refer to the principal types of fatigue which are depending on the imposed stresses and strains level Low cycle fatigue and high cycle fatigue are traditionally distinguished.

Low cycle fatigue involves significant plastic deformation (of the order of few percent or a fraction of percent) and corresponds to short life time (thousand to some ten thousands of cycles). In this loading regime, the metal grains suffer deformation in a homogeneous manner and the initiation of the first microcracks in the persistent slip bands happens quite early in the life of the structure. Most of the lifespan is then spend in propagation.

In the high-cycle fatigue regime, usually no visible irreversible deformation at the macroscopic level can be detected. This type of fatigue is characterised by a large heterogeneity of plastic deformation from grain to grain (only certain misoriented crystals undergo plastic slips) and, by that way, a very heterogeneous distribution of microcracks can be observed even if the macroscopic loading is homogenous. In high-cycle fatigue, the initiation of the first visible cracks takes a large part of the fatigue life of the structure. This is the reason of the great importance to study the conditions governing the crack initiation process. Many researches have thus been undertaken on this topic for many years. They allow a better understanding of

E. Bouchaud et al. (eds.), Physical Aspects of Fracture, 245–258.

some aspects of the high cycle fatigue phenomena. However many unsolved questions remain and even contreverses arise with the development of sophisticated experimental techniques and observation tools. Thus, the defects can be detected more and more early, so that the widely spread current tendency consists in considering that fatigue is essentially a propagation phenomenon and than the fatigue limit of a material is better defined by a non –propagation limit.

Nevertheless, this way to consider the problem is questionable for several reasons.

First of all, the studies on short cracks (that is the ones of very small size, generally inferior to 50μm) show that they can progress, even if they are submitted to the $\Delta K < \Delta K_s$ provided that the amplitude of applied stress is sufficient (larger than the fatigue limit of the material). Kitagawa [1] has besides proposed to use this property for characterising the limit between the initiation phase and the propagation phase. Some authors try to describe this evolution with the help of macroscopic parameters. Thus, one proposes « corrections » to the stress intensity factor to correlate the results, or empirical formulas in which intervene macroscopic parameters such as ΔJ, the CTOD or even the shear stress acting on the plan of the microcracks, the plastic slidings acting on these planes etc. (see D.L.Mcdowell [2]) Nevertheless the way to evaluate these quantities is questionable because, as one will see later, these local parameters differ from the macroscopic quantities. The use of macroscopic parameters at the scale of these defects, does not seem relevant since the material at that scale of description cannot be considered any more as homogenous and, consequently, the mechanical quantities at the microscopic scale (the scale of dislocations or a little beyond) or mesoscopic scale (the scale of grains) differ from those used in these theories. This is because the local redistributions induced by the inhomogeneous incompatible strains are not accounted for. Finally, considering the fatigue problem only as a propagation phenomenon, does not simplify the problem. On the contrary, it introduces a lot of complications in practical applications because the mechanical structures are frequently submitted to multiaxial loadings. Fracture mechanics cannot nowadays treat these problems with predictive capabilities.

This is the reason why, we choose a different way, while trying to bring back the problem of the evaluation of the fatigue resistance to a problem of mechanics of classical continuous media but at a relevant scale, *the mesoscopic scale.*

To validate our proposition, we first recall the general relations between macroscopic quantities which are usually used by the engineers and local quantities.

Because they differ, we will propose an approach to evaluate the local mechanical parameters near the fatigue limit. For that purpose, we will suppose that the material is in an elastic shake-down state before the initiation of fatigue crack. This hypothesis is very natural and from this point of view the considered fatigue limit, consists simply in assuming that it is not possible to find a state where the material continues to remain apparently elastic, at the end of a certain number of cycles.

2. Relations between macroscopic and mesoscopic mechanical parameters

Two different view points can be adopted for the studies of material properties:
- The engineer approach, which is purely macroscopic and phenomenological.
- the physicist approach, which considers the monocristalline grain (mesoscopic scale), or even the dislocations (microscopic scale).

These two research directions are developed in parallel and their quantitative links for applications are often difficult to establish.

Without considering such a fine scale as the physicist scale, it is clear that the global mechanical properties of the materials depend on their constituents and their defects (pores, microcavities, cracks, oxydized zones...). However the macroscopic properties are the only properties which can be evaluated by the engineers with ordinary direct experimental measurements. The constitutive relations are then formulated by means of macroscopic mechanical variables such as the stress Σ and the total strains E or plastic strains E_p as well as their rates. These relations are sufficient for a large number of engineering applications, such as the design of structures. Nevertheless when one is interested in finer scale properties, as for example when trying to characterise material damage and its evolution (fatigue crack initiation), then such approaches are not appropriate: this is because damage is controlled by phenomena at a smaller scale; their direct effects cannot then be detected by macroscopic measurements. What the engineer perceives is already in somehow filtered by the representative volume of the macroelement, and, a fortiori, by the test specimen. At the local scale in which we are interested, it is necessary to take these microeffects into account in a way similar to structural effects which are well known by engineers. Now, one of the principal difficulties in the study of fatigue is the difficulty to discriminate between work hardening, usually beneficial to the material and damage, which is defined here as the general degradation of mechanical properties. This is the reason why direct macroscopic approach, from experiments and phenomenological modelling is insufficient. It is essential to identify clearly the parameters characterising material damage, independently of work hardening effects. These damage parameters are tensorial by nature and need to be averaged to obtain macroscopic effects which are measurable by the engineers. These parameters must be, as much as possible, consistent with the principles of mechanics and thermodynamics.

2.1. MACROSCOPIC SCALE AND REPRESENTATIVE ELEMENTARY VOLUME

The engineer evaluates mechanical parameters (for example the stress field Σ or the strain field E) not at a point of a structure, but over a finite surface S or volume V that define the scale which is used. More precisely, this volume can be of the order of one cubic millimetre in relation with the dimension of usual strain gauges, or it can also be related to the local mesh size in a finite element calculation. In this volume, the mechanical quantities Σ, E are supposed constant. Nevertheless, this volume is heterogeneous and even anisotropic since it is constituted by a large number of monocristalline grains. If the average grain size is about 10μm, V contains roughly one million grains which have different properties, because of their

248

orientations, their initial strain hardening state etc... There is furthermore other origins of heterogeneities, such as grain boundaries, inclusions, precipitates, so that, in this volume, the local parameters actually differ considerably from the macroscopic values. It is precisely these local parameters which are at the origin of fatigue cracks initiation. It is therefore interesting to know the relations between the macroscopic and local quantities which intervene in the fatigue initiation process to critically review different approaches proposed so far. It is necessary for that purpose to define the elementary representative volume, notion that will be noted RVE in the following: this volume must be sufficiently small so that one can distinguish the microscopic heterogeneities and sufficiently large to be representative of the overall macroscopic behaviour. In the theories on polycrystalline agregates it is generally assumed that the geometry of each phase is known in order to propose statistical hypotheses on homogeneity and ergodicity which define precisely the R.V.E. (Kröner, 1980; Willis, 1981; Hashin, 1983; Stolz, 1996). We will limit ourselves to the intuitive definition given above.

We are interested in the following by the characterisation of damaged media which contain defects such as cavities, cracks, oxydized zones... As we try to detect the onset of crack initiation, it is necessary to distinguish defects for which the material cannot be considered as completely separated, (although one can observe discontinuities of material properties as for example persistent slip bands) from the cracks which are really initiated. This is the reason why, we propose to characterise the latter by the fact that, *these defects cannot transmit traction* under the effect of external loadings. Generally, one can characterise the damaged medium only by this property, which is different from work hardening which precedes the onset of damage phenomena in a metal with no defects. In this manner, a line of physical properties discontinuity, on which interatomic cohesion forces corresponding to the Barenblatt's theory (Barenblatt, 1967) is still acting is not considered as a crack, since $\sigma . n \neq 0$ on this line.

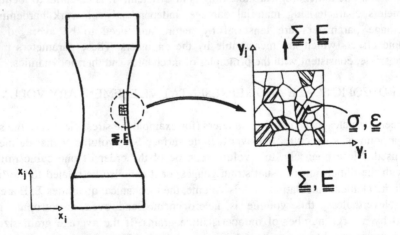

Figure 1: Different scales of material description: the macro and mesoscopic scales

Let us then consider a representative elementary volume element V with boundary dV (figure 1). This "macroscopic" volume element is, for an external observer, apparently homogenous and submitted to homogenous loadings.

In the following, capital letters are reserved for quantities corresponding to V, that we will qualify as "macroscopi"c such as Σ stress tensor, E strain tensor, U displacement vector. *By definition, the quantities Σ and E are constant in V.*

However, *at the local scale, V is inhomogeneous and contains defects*. One will note by $\sigma(x)$, $\varepsilon(x)$, u (x) the corresponding local fields.

According to what was said before, it is natural to suppose that a damaged zone or a cracked zone can be characterised by the fact that the stress vector is zero on his boundary. If at the local scale the volume V contains damaged zones, the volume of which is Z (t), D (t) = V- Z (t) represents the sound material volume at time t and δZ the boundary between D and Z.
In the following, one notes by < f > the average of f (x) in the sound material volume defined by:

$$< f >= \frac{1}{V} \int_{D(t)} f dv$$

Following relations linking up the macroscopic quantities to the local ones can be demonstrated (see for instance Ref. [3]). (The demonstration is essentially based on the virtual works theorem).

2.2 RELATION BETWEEN STRESS TENSORS

Let us impose on the boundary of the RVE, δV, the traction forces $\underline{T}$ (x) = $\underline{\Sigma}$. $\underline{n}$, where $\underline{\Sigma}$ is a constant tensor that corresponds to the macroscopic stress field. The resulting local field is statically admissible with T, i.e. must verify $\underline{\sigma}$. $\underline{n}$ = $\underline{T}$ on δV and $\underline{\sigma}$. $\underline{n}$ = 0 on δ.Z(t).

One can see easily that $\underline{\Sigma}$ = <$\underline{\sigma}$>

The reason is that, whatever the constant strain field $\underline{E}$ deriving from the displacement field $\underline{U}$ = $\underline{E}.\underline{x}$, one has

$$(\Sigma - \sigma).Edv = (\Sigma - \sigma).n.uds - \sigma.n.uds = 0$$

The preceding (classical) result means that *the average of local stress is equals to the macroscopic stress*. This is intuitively evident. Nevertheless, because of local heterogeneities they may differ, then their difference is in average (on the RVE) equal to zero. In fatigue however, one have to characterise this difference since, the cracks initiate in the most critical places relative to the level of loading or to the resistance characteristics. (It is precisely what we propose to do in the fatigue nucleation model proposed hereafter.)

One will see later that *this property cannot be extended to other mechanical quantities*.

2.3. RELATION BETWEEN TOTAL STRAIN TENSORS

In particular, concerning strains, one has the following relation between the total macroscopic strain $\underline{E}$ and the local microscopic strains $\underline{\varepsilon}$.

$$E = <\varepsilon> - \frac{1}{2V}(u \otimes n + n \otimes u)ds$$

Strain rates verify an analogous relation.

These relations are also intuitively evident for they mean that the macroscopic deformation is the sum of the average of local deformations of the material and those induced by holes and cracks contained in V. (Imagine the deformation of a sponge or of a emmenthal cheese!)

It is only when these defects do not exist in the RVE that one has $\underline{E} = <\underline{\varepsilon}>$

2.4 CONSERVATION OF THE TOTAL ENERGY AND THE TOTAL ENERGY RATE

Concerning the energy, as $\underline{\sigma}.\underline{n} = 0$ on the boundary of internal damaged zones, one has

$$<\sigma.\varepsilon> = \sigma.\varepsilon dv = \sigma.n.uds = T.uds = \Sigma.E$$

which corresponds to the conservation of energy.

The same reasoning applies to the work rate.

$$<\sigma \dot{\varepsilon}> = \Sigma \dot{E}$$

Note that these identities also apply to non-associated fields, provided that the stresses are statically and the strains are cinematically admissible.

2.5 RELATIONS BETWEEN THE ELASTIC AND PLASTIC MACROSCOPIC AND MESOSCOPIC STRAIN TENSORS

Let us now consider the elastic and plastic strain tensors, in the case *when the RVE does not contain damaged zones*. Let us recall that in this case the relation between total strains is simply $E = <\varepsilon>$.

If one admits that the local deformation can be split into elastic and plastic strain just as well for the macroscopic part according to the formulas

$$\varepsilon(x) = \varepsilon_e(x) + \varepsilon_p(x) \quad , \quad \varepsilon_e(x) = M(x)\,\sigma(x)$$
$$E(x) = E_e(x) + E_p(x) \qquad E_e = M°.\,\Sigma$$

where $M(x)$ is the local elastic compliance tensor (which depends on x), and $M°$ the macroscopic elastic compliance tensor (which is constant), one can write from the previous formulas

$$E_p = <\varepsilon(x)> - E_e = <\varepsilon_p(x)> + <M(x).\sigma(x)> - M°.\Sigma$$

As usually $<M(x).\sigma(x)> - M°.\Sigma \neq 0$ (mean value of a product is different from the product of the mean values), because the material behaviour at the microscopic scale is heterogeneous, *there is no simple direct relation between the macroscopic plastic deformation and the average value of the microscopic plastic deformation.*
One has in corollary the same property for the elastic parts.

2.6. RELATION BETWEEN MACROSCOPIC AND MESOSCOPIC PLASTIC DISSIPATION

In the same way, *there is no any link between the macroscopic plastic dissipation and the local plastic dissipation.* More precisely, one has

$$\Sigma.\,E_p \neq <\sigma.\varepsilon_p>$$

for on one hand $\underline{\varepsilon}_p(x)$ is not cinematically admissible, on the other hand $\underline{E}_p$ is different from $\underline{\varepsilon}_p$. The right hand side of this relation can be interpreted as the total plastic dissipation, only if the solid corresponds to a perfectly plastic material. If it is not the case, one can see that there is no link with the plastic macroscopic dissipation represented by $\underline{\Sigma}.\underline{E}_p$. Let us recall that a certain number fatigue models are based precisely on the energy dissipated per cycle which cannot be identified by only global measures on classical test specimens. However, the difference between macroscopic and mesoscopic plastic dissipations decreases in relative value with increasing plastic strain because the material behaviour tends toward perfect plasticity. Thus the use of this parameter to modelise fatigue is adequate only if plastic deformation relatively important.

The case where the RVE presents already damaged zones, which is more complex will not be examined here. The reader can find the main results demonstrated by H.D. Bui et al. in reference [3].

3 Conclusion

Most of the existing fatigue models (initiation as well as propagation) are based on « engineering macroscopic parameters », even if they treat of mesoscopic or microscopic phenomena.

In some cases this way is justified: for instance, when one considers the problem of *fatigue propagation of long cracks*. In that case, the *stress field which is governed by stress intensity factors is used as the principal part of asymptotic solution*. (Let us notice that the local field is evaluated under the assumption of small strain, linear elasticity behaviour. In the close vicinity of the crack tip, the corresponding solution is out of the scope of the basic hypothesis on which the obtained solutions are based.) This means that if one suppresses the outer part of some RVE surrounding the crack tip, and replaces the external loading by a tension T= Σ.n, then the material response is not changed. This RVE, is sufficiently large to represent the mean properties of the material so that local fluctuation is attenuated. For instance, if the active stress intensity amplitude range ΔK is of the order of 10 MPa$\sqrt{m}$ for a material presenting a yield strength of about 500 MPa, the radius of the RVE can be estimated by the equation

$$500\sqrt{r} = 10$$

which gives r_0 of the order of 400 μm which is already a quite large dimension in comparison with the grain size. If ΔK is about 5 MPa$\sqrt{m}$, then r_0 is about 100μ so that one can estimate that the corresponding RVE contains at least 1000 grains which is sufficient for a statistical estimation and according to relation between stress (see § 2.2), it is then reasonable to use the macroscopic stress for calculating ΔK.

In many other cases however, it seems not pertinent to use directly the macroscopic mechanical quantities as it is done in many metallurgical models, particularly the models dealing with short cracks propagation, because at that scale the microscopic stress or strain can be very different from the corresponding macroscopic quantities.

4. A multiaxial fatigue criterion based on a multiscale approach

Prediction of high cycle fatigue resistance is of great importance for structural design. In spite of this clear industrial need, modelling of metal behaviour in a high cycle fatigue regime remained until now very often on empirical approaches: Wöhler curve, Goodman – Haigh or Gerber diagrams are still very popular tools to engineers. However, these concepts are not appropriate when studying multiaxial stress cycles that are frequently encountered on modern mechanical components. These multiaxial stresses arise from different factors like: external loadings, geometry of the structure which can induce multiaxiality even if the loading is uniaxial, and finally residual stresses.

4.1. BACKGROUND

In order to derive a multiaxial endurance fatigue criterion, we have proposed an original method of computing, based on a multiscale approach. This method is quite different from existing fatigue approaches since it uses mesoscopic mechanical parameters.

This model arises from the observations recalled at the beginning that generally the first fatigue damage processes begin in grains which have undergone plastic deformation, the appearance of slip bands in some grains which broaden progressively with the applied cycles; this stage is then followed by localised damage corresponding to the formation of intragranular microcracks; these microcracks can be arrested by grain boundaries, but may also propagate. After a certain number of cycles, a main crack initiates, grows in size and shields the other defects and consequently leads to the final rupture of the mechanical structure. In a high cycle fatigue regime, even if it is necessary to have plastic deformation at the micro or the meso scale (corresponding to the grain size), most of the time no visible irreversible deformation at the macroscopic level can be detected. It is thus characterised by a large heterogeneity of plastic deformation from grain to grain: only certain misoriented crystals undergo plastic slips and by this way, a very heterogeneous distribution of microcracks can be observed. In this fatigue regime, the initiation of the first visible cracks takes a large part of the fatigue life of the structure. It demonstrates the importance of studying the conditions governing the crack initiation process.

In most of the existing fatigue models, these conditions are described with the help of *macroscopic parameters* which are evaluated according to different assumptions of homogeneity and isotropy... However, because the phenomena which cause fatigue initiation are microscopic, the local parameters (for example local stress σ) differ from the macroscopic ones. Thus the use of classical macroscopic engineering parameters seems not appropriate since, at that local scale, the material cannot be considered any more as homogeneous. In particular, the local redistributions induced by the inhomogeneous incompatible strains ε_p, and as a consequence the local residual stresses ρ, are not accounted for.

4.2 FORMULATION

The originality in our proposal is precisely the use of local mesoscopic mechanical parameters σ to derive fatigue resistance criteria. These parameters are evaluated from the macroscopic parameters thanks to an hypothesis of elastic shake down. *More precisely, it is postulated that near the fatigue limit threshold, the mechanical structure shakes down elastically at all scales of material description. Under this assumption, the precise knowledge of the local constitutive equations, which is not possible to evaluate, is not necessary.* Physical interpretation of this hypothesis is that after a certain number of loading cycles the response is purely elastic (or at least *plastic dissipation rate becomes negligible*). Then, using shakedown theorems, Melan's theorem and its generalisation by different authors, (cf. for instance Ref. [4]), it is possible to derive a method for estimating the apparent stabilised stress (tensorial) cycle at the macroscopic and mesoscopic level which intervenes in the proposed fatigue criteria, provided that the material is considered as a structure, made of grains of different crystallographic orientations.

Theoretical developments of this theory are presented in Ref. [5]. For practical applications, it is just necessary to remember the way to derive local parameters from macroscopic stress cycles near the fatigue limit which is presented below.

The general relation between macroscopic and local stress tensor is:

$$\sigma(m,t) = A.\Sigma(M,t) + \rho(m,t)$$

This relation is well known in the theory of polycrystalline aggregates. *It is not linear but affine*, so that using macroscopic stress for studying phenomena which occur in grains is incorrect.

In this equation A is an elastic localisation tensor, which depends on the microstructure; $\Sigma(M,t)$ is the macroscopic stress tensor at time t in the representative volume element $V(M)$ surrounding M and $\sigma(m,t)$ and $\rho(m,t)$ are respectively the local stress tensor and the residual stress tensor at any point m of $V(M)$.

For the sake of simplicity, let us assume that A = Identity, (elastic homogeneity); then ρ characterises the local stress fluctuation in $V(M)$.

If elastic shakedown happens, then ρ must become independent of time after a certain number of cycles, so that the local plastic yield criterion f(m) is no longer violated. Assuming that an approximate elastic shakedown occurs if the loading cycles are near the fatigue limit, then there must exist a local fixed (independent of time) residual stress tensor $\rho^*(m)$ and a fixed set of local hardening parameters $\alpha(m)$ such that:

$$\sigma(m,t) = \Sigma(M,t) + \rho^*(m)$$

$$f(\sigma(m,t), \alpha(m)) \equiv f(\Sigma(M,t) + \rho^*(m), \alpha(m)) \leq 0$$

If Mises criterion is chosen, then $\sigma(m,t)$ belongs to the hypershere in 5 dimensional space), representing the limiting value of the Mises norm for which elastic shakedown is possible. Mandel and al. [4] showed that $\rho^*(m)$ can be approximately taken to be the centre of the smallest hypersphere surrounding the loading path in the deviatoric macroscopic stress space represented on Fig.2. In this figure, S (resp. s) represents the deviatoric stress corresponding to Σ (resp. σ). Because of shakedown hypothesis at all scale of material description, S at stabilisation is also elastic and is noted by $S^{el.}$. Finally, by that construction, the local stress state is known at any time t of the apparent stabilised state.

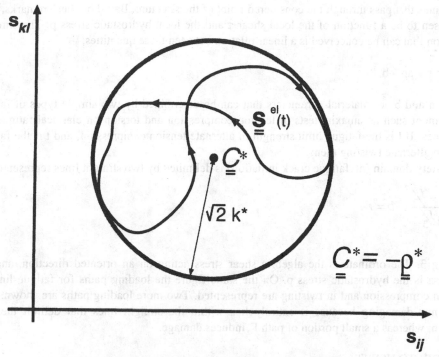

Fig.2: Determination of the local residual stress ρ^* at shakedown state.

It remains to choose a fatigue criterion. As the local stress is approximately known at any time t, it is natural to try to take into account the characteristic of the loading path (as in plasticity). Thus, a reasonable fatigue criterion could be stated as follows:

Crack initiation will occur in a critically oriented locus (most of the time corresponding to a grain) within $V(M)$ that has undergone plastic deformation, if, for at least one time instant t of the stabilised cycle, one has:

$$F[(\sigma(m,t)] \geq 0 \text{ for } m \in V(M)$$

For such a criterion, the current stress is considered, in contrast with most (all?) existing fatigue criteria. Damage arises over a precise portion of the loading path (or equivalently, over a precise time interval of the loading period). As cracks usually occur in transgranular slip bands, the local shear acting on these planes is an important parameter. Moreover, the normal stress acting on these planes accelerates damage formation. However, this quantity is rather difficult to compute in a general case because it depends of the considered plane. For this reason, hydrostatic stress is preferred because it is much easier to use, being an invariant scalar. Furthermore, it can be interpreted as the mean value of the normal stresses acting on all the planes that pass through the considered point of the structure. Based on these remarks, $F(\sigma)$ is chosen to be a function of the local shear τ and the local hydrostatic stress p. The simplest criterion that can be conceived is a linear relation between these quantities,

$$F(\sigma) \equiv \tau + ap - b,$$

where a and b are material parameters that can be determined by two simple types of fatigue experiment such as uniaxial tests of tension-compression and torsion on classical fatigue test machines. If f is the fatigue limit strength in alternate tension compression, and t is the fatigue limit in alternate twisting then:
The safety domain (no fatigue crack initiation) is delimited by two straight lines represented on

$$a = \frac{t - f/2}{f/3}, \quad b = t$$

the Fig 3. The ordinate is the algebric shear stress acting on an oriented direction, and the abscissa is the hydrostatic stress p. On the same figure the loading paths for fatigue limit in tension compression and in twisting are represented. Two more loading paths are shown. Path Γ_1 is non-damaging because it entirely lies within the straight lines that delimit the safe domain, whereas a small portion of path Γ_2 induces damage.

4.3 HOW TO USE IT

To check automatically the fatigue resistance of a structure is a rather difficult task, because at each point one has to consider the plane on which the loading path $(\tau(t), p(t))$ is a "maximum" relative to the criterion. This computation can be simplified as follows. The maximum shear stress according to Tresca's measure is calculated over the cycle period:

$$\tau(t) = \text{Tresca}(\sigma(t)).$$

For this, it is useful to notice that:

$$\text{Tresca}(\sigma(t)) = \text{Tresca}(s(t)) = \text{Max}_{ij} \, |\sigma_i(t) - \sigma_j(t)|/2.$$

The stresses $\sigma_i(t)$ and $\sigma_j(t)$ are principal local stresses at time t. The quantity d that quantifies the danger of fatigue failure defined by,

$$d = \text{Max}_t \frac{\tau(t)}{b - ap(t)}$$

Is calculated over the loading period. The maximum is to be taken over the cycle. If $d > 1$, the fatigue failure will occur.

Working this way, all couples (τ,p) are situated in the positive part of τ. All facets, which could be involved by the crack initiation are automatically reviewed. Couples (τ,p) verifying the condition d>1 are associated with specific facets. Therefore, the criterion also provides the direction of crack initiation.

Another possibility is to use the octahedral shear $J_2(\sigma(t))$ instead of $\tau(t)$. However, this method does not give the critical facets.

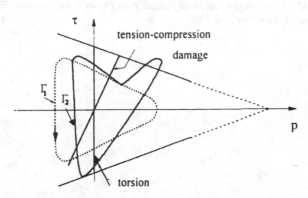

Fig. 3 : Fatigue limit domain and loading paths

Another interesting proposal deriving from the multiscale approach was given by I.V.Papadopoulos: the fatigue limit for a given periodic loading $\lambda Q(t)$ corresponds to the limit of the intensity λ such that elastic shakedown is possible. Beyond this limit, plastic shakedown or ratchet phenomena will induce damage and fracture because of subsequent softening. The limit size k* of the hypersphere surrounding the loading path (as explained previously), is one possible and natural way to characterise this state. If k* is greater than some limit value which depends on the local maximum hydrostatic stress in the cycle, fatigue will occur.

This corresponding fatigue criterion is:

$$k^* + \alpha p_{max} - \beta > 0$$

As previously, the parameters α and β can be identified by two different tests. Using this method, it is no longer necessary to describe the whole loading path once k* is determined. In many cases, the predictions are very similar to the method based on the current stress state presented above.

References

[1] Kitagawa H. and Takahashi S. (1976°, Applicability of Fracture Mechanics to very Small Cracks in the Early Stage, *Proc. 2nd Int. Conf. Mech. Behaviour of Materials (ICM2)*, Boston, Mass. 627-631.

[2] McDowell D.L. , (1996), Basic Issues in the Mechanics of High Cycle Metal Fatigue, *Int. J. of Fracture*, Vol. 80, 103-145

258

[3].Bui H.D., Dang Van K., Stolz C., (1982), Relations entre Grandeurs Microscopiques et Macroscopiques, *C.R. Acad. Sci. Paris*, tome 292,série II, 1155-1158.

[4] Mandel J., Halphen B., Zarka J., (1977), Adaptation d'une Structure Elastoplastique à Ecrouissage Cinematique, *Mech. Res. Comm.* 4, 309-314.

[5] K. Dang Van, Introduction to Fatigue Analysis in Mechanical Design by the Multiscale Approach, *C.I.S.M. Courses and Lectures N° 392 on "High Cycle Metal Fatigue, from Theory to Applications"*, Ed. by K. Dang Van & I.V. Papadopoulos, Springer Wien New York 1999.

THE INFLUENCE OF MICROSTRUCTURE AND MOIST ENVIRONMENT ON FATIGUE CRACK PROPAGATION IN METALLIC ALLOYS

J. PETIT

Laboratoire de Mécanique et de Physique des Matériaux, UMR CNRS N° 6617, ENSMA,

Téléport 2, BP. 109, 86960 FUTUROSCOPE Cedex, France

Abstract

An overview on some aspects of the role of microstructure and of moist environment on fatigue crack propagation is given. The respective influence of several microstructural parameters as grain size, texture, alloy phases, ageing conditions, and of extrinsic factors as crack closure and environment for metallic alloys is illustrated and discussed.

Introduction

Defect tolerance design approach to fatigue is based on the premise that engineering structures are inherently flawed. The useful fatigue life then is the number of cycles to propagate a dominant flaw of an assumed or measured initial size (which can be the largest undetected crack size estimated from the sensitivity of non-destructive inspection methods) to a critical dimension which may be dictated by the fracture toughness, limit load, allowable strain or allowable compliance change. In most metallic materials, catastrophic failure is preceded by a substantial amount of stable crack growth under cyclic loading conditions. The propagation rate of these cracks for given loading conditions depends on several factors including crack length and geometry of the cracked structure, crack closure, temperature, environment, test frequency and propagation mechanisms in relation with material microstructure. Indeed the fatigue fracture behaviour of metals and metallic alloys have been shown widely influenced by material composition and by various microstructural parameters. Numerous articles can be found in the literature on these topics. Some excellent reviews can be mentioned (1-3) but it would be a task beyond the scope of this paper to refer to all of them.

In this overview a selection of demonstrative examples provided by the literature and the works of the present authors will give illustrations of the respective role of grain size, texture, alloy phases and age-hardening. In comparison to the intrinsic propagation behaviour characterised under inert environment and after crack closure correction, an evaluation is made of the role of two dominant extrinsic parameters, i.e. environment and crack closure, since they can deeply affect the fatigue behaviour in different ways (4).

E. Bouchaud et al. (eds.), Physical Aspects of Fracture, 259–269.

INFLUENCE OF MICROSTURAL PARAMETERS

Grain size.

Grain boundaries can constitute high obstacle to the movement of dislocations, thus increasing the flow stress according to the Hall-Petch relation. Fatigue process being basically the result of cumulated plastic deformation, can be affected by hardening processes, and can also interact with grain boundaries. Numerous studies have shown that the decrease of grain size improves the fatigue limit but conversely reduces the resistance against crack propagation. Recently, Haberz et al. (5) have clearly demonstrated in ARMCO iron with different grain sizes ranging between 3 and 3000 µm, that the nominal stress intensity threshold is strongly influenced by the grain size (Fig. 1). But after correction for closure, the effective threshold is not significantly affected. Consequently, the change of the threshold is mainly caused by the increase of the roughness induced closure effect (6,7) with increasing grain size. Such observations are in accordance with many others, but the range of grain size here explored is exceptional. Finally, it comes out from this work that there is no substantial influence of grain size on the effective behaviour of a stage II crack. But when the near-threshold propagation is characterised by a strong localisation of the plastic deformation (8,9), the resulting crack growth behaviour can be strongly affected by the microstructure even after closure correction as illustrated in figure 2 for a Ti-6Al-4V alloy in different thermo-mechanical conditions.

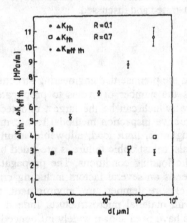

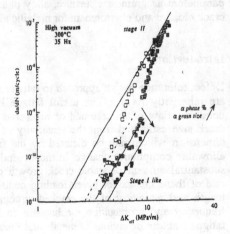

Fig. 1. Threshold and effective threshold as functions of the grain size in ARMCO steel (Haberz et al., 1993).

Fig. 2. Intrinsic fatigue crack propagation Ti-6Al-4V at 300°C :
☐ bimodal 40% α_p (ϕ=20µm)
- ▨ globular 80% α_p ((ϕ=8µm +lamellae of 50 µm)
- ■ globular 75% α_p ((ϕ=8µm)

Texture

The texture can modify the nominal crack propagation (R = 0.1 in air) as illustrated in figure 3 for a 2090 - T8X Aluminium Lithium alloy (10). Faster crack growth and lower threshold are observed in the 1.6 mm T83 thin sheet compared to the 12.7 mm T81 thick plate for tests performed in the LT orientation. But after closure correction both materials present the same effective behaviour. The more zig-zaging crack path in the thick plate only increases the contribution of the roughness induced closure (or non closure) as illustrated in figure 3, but does not affect the crack growth mechanism in itself.

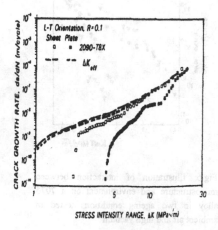

Fig. 3: Fatigue crack propagation in 1.6 mm-thin sheet and 12.7 mm-thick plate of 2090 T8X Al-Li alloy. (Vankateswara et al. 1991).

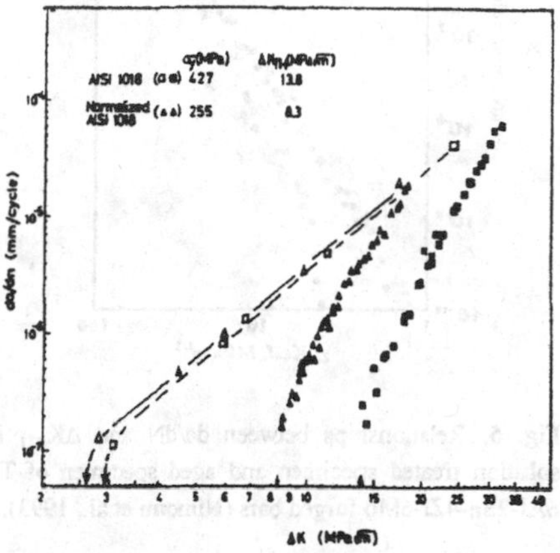

Fig. 4 : Rate of growth vs ΔK for AISI 1018 in Duplex and normalized conditions. Effective data (open symbols) are plotted using crack opening measurements from Minakawa et al., 1982

Alloy phases

Two examples have been selected to illustrate the influence of alloy phases on crack propagation data in duplex (iced water quenched after 1 h à 760°C) and normalised AISI-1018 are plotted in figure 4 (11). Obviously the resistance of the duplex structure against crack propagation for a fatigue test performed at a R ratio of 0.05, is very much higher than that of the normalised material. It could be attractive to relate such performance and high threshold level to the increase in the yield stress (427 MPa and 255 MPa respectively for duplex and normalised conditions). But after closure correction, the effective propagation in both microstructures is identical. Hence the differences in the nominal curves must be attributed to the difference existing in the contribution of crack closure. Here again the microstructure does not affect the effective stage II propagation.

The next example (figure 5) shows the effect of retained metastable β phase on fatigue crack propagation characteristics of forged bars of a Ti 6246 Titanium alloy. The crack propagation rates in the aged material (6 h at 863 K of solutionizing) are substantially lower than in the assolutionized microstructure, with a threshold range decreased of more than 50 %. But in this case, even after closure correction, there is still a large difference between the two microstructures. This example shows that when the propagation mechanism in itself is changed from one microstructure to the other, the effective behaviour is also modified.

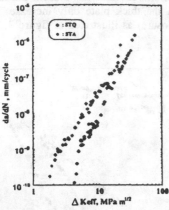

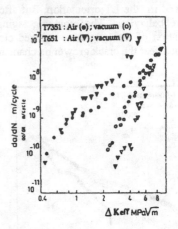

Fig. 5. Relationships between da/dN and ΔK_{eff} in solution treated specimen and aged specimen of Ti-6Al-2Sn-4Zr-6Mo forged bars (Niinomi et al., 1993).

Fig.6: Illustration of interaction between microstructure and environment on a 7075 alloy in two ageing conditions tested in ambient air and high vacuum

Aged conditions

An illustration of the coupled influence of microstructure and environment still existing after closure correction is given in figure 6 on a 7075 alloy in two aged conditions tested in ambient air and high vacuum (4). The peak-aged matrix contains shareable Guinier-Preston zones and shareable precipitates which promote a localisation of the plastic deformation within a single slip system in each individual grain along the crack front. The over-aged matrix contains larger and less coherent precipitates which favour a wavy slip mechanism (4,12,13). In vacuum, the peak-aged T651 condition leads to a highly retarded crystallographic propagation (so called stageI-like regime) while the over-aged T7351 condition gives a conventional stage II propagation. In ambient air, the single slip mechanism which is still operative in the peak-aged alloy, is assumed to offer a preferential path for hydrogen embrittlement which leads to a strongly accelerated propagation. Conversely ambient air has little influence on the stage II regime as observed in the over-aged alloy. It can be noticed that the influence of ambient air has inverted the ranking of the propagation curves with respect to the microstructures. It can be underlined that a crystallographic propagation is faster than stage II in air, but is highly retarded in vacuum. These results indicate a high sensitivity to environment of the slip mechanisms, specially near the grain boundaries, and hence a large influence of environment on the microstructural barrier effects.

INTRINSIC FATIGUE CRACK GROWTH

The above examples have shown that if one intends to analyse the specific role of microstructure it would be useful to analyse the crack propagation behaviour of the material in conditions where the influence of crack closure and environment are eliminated, that is to say to examine the intrinsic fatigue crack growth behaviour.

On the basis of numerous experimental data obtained in high vacuum on technical Aluminium alloys with various ageing conditions, on Aluminium based single crystals and on steels and Titanium alloys, it has been shown (4) that the intrinsic fatigue crack growth can be described according to three characteristic regimes (see example of Al alloys in figure 7):

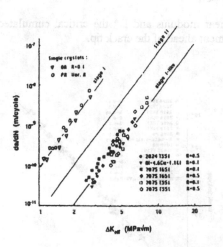

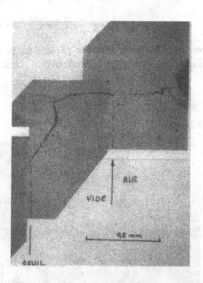

Fig.7 : Illustration of the three intrinsinc propagation regimes for Al alloys .

Fig. 8 : Stage I to stage II transition in a peak aged single crystal of Al-4.5 % Zn - 12.5 % Mg pre-oriented for single slip (high vacuum, R = 0.1, 35 Hz).

- The faster intrinsic stage I, has been identified on single crystals of Al-Zn-Mg alloys (4) with a peak-aged microstructure which favours crystallographic propagation along a PSB (persistent slip bands) which develops in a {111} plane pre-oriented for single slip. This regime is also observed on various materials in the early growth of microstructural short cracks (14).

- The intermediate intrinsic stage II is commonly observed on polycrystals and single crystals when crack propagation proceeds at macroscopic scale along planes normal to the loading direction. Such propagation is induced by microstructures which promote homogenous deformation and wavy slip, as large or non coherent precipitates or small grains size. Figure 8 illustrates a typical change from a near-threshold stage I to a mid ΔK stage II propagation in an Al-Zn-Mg single crystal.

264

The slowest regime, or intrinsic stage I-like propagation corresponds to a crystallographic crack growth observed near the threshold in polycrystals or in the early stage of growth of naturally initiated micro-cracks, when ageing conditions or low stacking fault energy generate heterogeneous deformation along single slip systems within individual grains (see example in figure 9). Crack branching and crack deviation mechanisms (7) and barrier effect of grain boundaries (4), are assumed to lower the stress intensity factor at the crack tip of the main crack.

The stage II regime is in accordance with a propagation law derived from the models initially proposed by Mc Clintock (15), Rice (16) or Weertman (17):

$$da/dN = A/D^* \ (\Delta K_{eff}/\mu)^4 \tag{1}$$

where A is a dimensionless parameter, μ the shear modulus and D^* the critical cumulated displacement leading to rupture over a crack increment ahead of the crack tip.

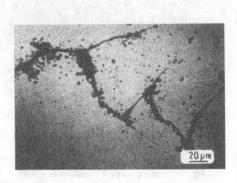

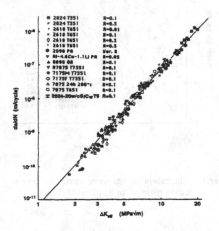

Fig. 9. Stage I-like propagation in 2024T351 tested in high vacuum (R = 0.5, da/dN = 2.10^{-11} m/cycle).

Fig. 10. Intrinsic state II propagation. Al based alloys.

Intrinsic data for well identified stage II propagation are plotted in figure 10 in a da/dN vs $\Delta K_{eff}/E$ (E : Young Modulus) diagram for a wide selection of Al alloys, and in figure 11 for a selection of steels and a TA6V Ti alloy compared to the mean curve for Al alloys. This diagram constitutes an excellent validation of the above relation and confirm that the LEFM concept is very well adapted to describe the intrinsic growth of a stage II crack which clearly appears to be nearly independent on the alloy composition, the microstructure (when it does not introduce a change in the deformation mechanism), the grain size, and hence the yield stress. The predominant factor is the Young modulus of the matrix, and the slight differences existing between the three base metals can be interpreted as some limited change in D* according to the alloy ductility (4).

The stage I-like regime cannot be rationalised using the above relation (Fig.3 and 12). The retardation is highly sensitive to the microstructure; it is well marked when the number of available slip systems is limited (Ti alloys) or can be nearly absent when some secondary slip systems can be activated near the boundaries as observed in Al-Li alloys (20).

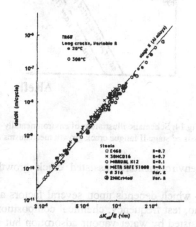

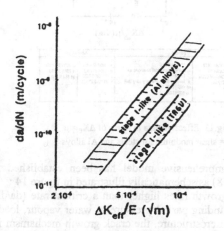

Fig. 11. Intrinsic stage II propagation : Steels and TA6V alloy compared to mean curves for Al alloys after rationalisation in term of $\Delta K_{eff}/E$.

Fig. 12. Comparison of intrinsic stage I-like propagation for Al alloys and TA6V alloys in a da/dN vs $\Delta K_{eff}/E$ diagram

ENVIRONMENTALLY ASSISTED PROPAGATION

Following the rationalisation of intrinsic stage II propagation as presented above, some similar rationalisation of FCG in air could be expected after correction for crack closure and temperature effects ($\Delta K_{eff}/E$). Figure 13 presents a compilation of stage II propagation data obtained in ambient air for almost the same alloys as in vacuum (see Fig. 10 and 11). Obviously rationalisation does not exist in air. The sensitivity to air environment is shown strongly dependent as well on base metals, addition elements, and microstructures (see 7075 alloy in three different conditions) as on the R ratio and the growth rate range. However a typical common critical rate range can be pointed out at about 10^{-8} m/cycle for all materials. This critical step is associated to stress intensity factor ranges at which the plastic zone size at the crack tip is of the same order as grain or sub-grain diameters. In addition there is a general agreement to consider that, for growth rates lower than this critical range, crack propagation results from a step-by-step advance mechanism instead of a cycle-by-cycle progression as generally observed in the Paris regime in air.

266

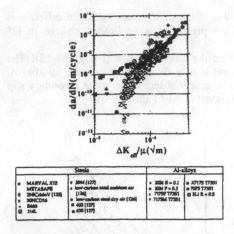

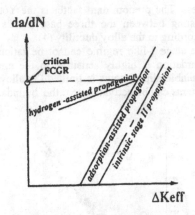

	Steels		Al-alloys	
• MARVAL X12	▼ 30M [127]	□ 30M R = 0.1	● X7175 17351
● METASAFE	• low-carbon steel ambient air	■ 30M P = 0.5	■ 7075 17351
◆ 2NiCrMoV [125]	[126]	● 7175F 17351	▣ 3Li R = 0.5
x 30NCD16	• low-carbon steel dry air [126]	▼ 7175M 17351	
+ B460	■ 40B [127]		
▣ 316L	▲ 430 [127]		

Fig.13. Effective data in terms of $\Delta K_{eff}/\mu$
(μ = shear modulus) for steels and Al alloys

Fig.14. Schematic illustration of environmentally-assisted stage II fatigue crack growth mechanisms

A comprehensive model has been established for environmentally assisted crack growth (4,8,18) as schematically illustrated in figure 14 :
- at growth rates higher than a critical rate $(da/dN)_{cr}$ which depends upon several factors as surrounding partial pressure of water vapour, load ratio, test frequency, chemical composition and microstructure, the crack growth mechanism is assisted by water vapour adsorption but it is still controlled by plasticity as in vacuum.
- at growth rates lower than $(da/dN)_{cr}$, an Hydrogen assisted crack growth mechanism becomes operative, Hydrogen being provided by adsorbed water vapour when some critical conditions are fulfilled.
At room temperature and for conventional test frequencies of 20 to 50 Hz, $(da/dN)_{cr}$ is about 10^{-8} m/cycle as pointed out in figure 15. As recently described (18), the adsorption assisted stage II propagation verifies relation (1), adsorption being just assumed to reduce the cumulated displacement D* in accordance with Lynch approach (19). The modelling of the hydrogen-assisted propagation has to be developed by the introduction of the coupled effect of two concurrent mechanisms, i.e. Hydrogen action and plastic cyclic accumulation, which can be strongly affected by environment and temperature.

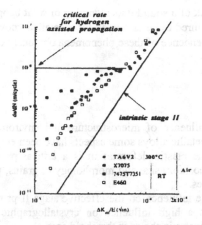

Fig. 15: Critical rate for enhanced near-threshold environmental effect

Up to now no fatigue crack growth law is available which takes into account the different processes described above. Theoretical models do not account for environmental effects and therefore they are merely valid for fatigue crack propagation in inert atmospheres. Similarly fatigue crack growth laws considering the strained material at the crack tip as a low-cycle fatigue micro-sample indirectly integrates environmental effects mainly through the use of a Manson-Coffin law.

The occurrence of hydrogen-assisted crack growth is associated to a typical change in the slope of the propagation curves which becomes close to 2 to 1 at low rates, and the transition from one regime (adsorption assisted) to the other (hydrogen-assisted) often corresponds to a more or less well marked plateau range (Fig. 14 and 15).

The present authors have proposed the following relationship based on a superposition principle to describe the propagation in an active moist environment:

$$\frac{da}{dN} = \frac{A}{D_i^2}\left(\frac{\Delta K_{eff}}{\mu}\right)^4 + \frac{B}{\sigma\mu}(\Delta K_{eff}^2 - \Delta K_{eff,th}^2) \qquad (2)$$

where $\Delta K_{eff,th}$ denotes the threshold obtained under closure-free conditions, B is a dimensionless constant and σ a strength parameter.

The first term accounts for the adsorption-saturating regime previously described. The second one was subsequently added in an attempt to describe the hydrogen-assisted propagation regime. The ΔK_{eff}^2 dependence might be viewed as a coarse description of the dislocation dragging via the CTOD. $\Delta K_{eff,th}$ would thus denote a threshold value of this sweep-in mechanism to enable the attainment of a critical hydrogen concentration at the crack tip. However such a formulation for hydrogen-assisted propagation is still highly empirical.

Problems arise from the lack of a sound understanding of what happens ahead of the crack tip. Some critical issues are required to be answered for a detailed knowledge to achieve this goal. Finally the temperature dependence of these phenomena obviously constitutes a prime issue for investigation.

CONCLUSION

In this overview of the influence of microstructure and environment on the fatigue crack propagation behaviour of metallic alloys some aspects have been emphasised:
- the homogeneity of the plastic deformation is a critical process crack propagation. It depends on alloy composition, size and morphology of grains, phase distribution, size and coherence of precipitates.
- microstructure has little influence on the effective stage II propagation for a given based matrix metal, but has a high influence on crystallographic propagation (stage I-like regime) which predominates in the near-threshold area
- microstructure can strongly modify the closure contribution.
- ambient air and moist environments can strongly affect the propagation mechanisms. An important role of water vapour has been underlined. The behaviour in moist environments has been described by superimposing two distinct processes: adsorption of water vapour molecules which promotes the growth process without altering the basis intrinsic mechanism of damage accumulation, and Hydrogen-assisted propagation which is operative at rates below a critical range depending on several factors including test frequency.
- the water vapour assistance favours the activation of multiple slip systems and hence promotes the stage II regime.
- three intrinsic crack propagation regimes have been clearly identified.
- constitutive laws are proposed for the intrinsic propagation and for water-vapour assisted propagation.

REFERENCES

1) Cazaud, R., Pomey, G., Rabbe, P., Janssen, Ch. : in "La fatigue des métaux", Dunod, Paris, 1969.

2) Fine, J. : in ASM "Materials Science Seminar", St Louis, USA, 1979.

3) François, D.: Advances in Fatigue Science and Technology, C. Maura Branco et al. eds., Kluwer Acad. Pub., 1989, p. 23-76.

4) Petit, J., de Fouquet, J. and Henaff, G.(1994): Handbook of Fatigue Crack Propagation in Metallic Structures, A. Carpinterie ed. Elsevier pub., 1994, vol. 2, p. 1159-1204.

5) Haberz, K., Pippan, R. and Stüwe, H.P. (1993): in Proceeding of Fatigue 93, Vol. 1, J.P. Baïlon and J.I. Dickson eds., EMAS pub., 1994, p.525-530

6) Suresh, S. and Ritchie, R.O. : Fatigue Crack Growth Threshold Concepts, D. Davidson et al. eds., TMS AIME pub., 1984, p. 227-262.

7) Suresh, S. (1985), Metall. Trans. A., 1985, vol. 16A, 249-260.

8) Petit, J.: Fatigue crack growth threshold concepts, D. Davidson et al. eds., TMS AIME pub., 1984, p. 3-24.

9) Sarrazin, C., Lesterlin, S. and Petit, J.: ASTM STP 1296, in press.

10) Vankateswara, K.T. and Ritchie, R.O.: in report L.B.L. 30176, Center for Advanced Materials, Univ. of California, 1991.

11) Minakawa, K., Matsuo, K., Y. and Mc Evily A.J.: Metall. Trans. A., 1982, Vol.13A, p.439-445.

12) Lindigkeit, J., Terlinde, G., Gysler, A. and Lütjering, G.: Acta Metal., 1979, vol. 27, N° 11, p. 1717-1726.

13) Kirby, B.R. and Beevers, C.J.: Fatigue Engng. Mater. Struct., 1979, Vol.1, p. 203-215.

14) Turnbull, A. and de los Rios, E.R. : Fatigue Fract. Engng. Mat. Struct., 1995, Vol. 18, n°11, p. 1355-1366.

15) Mc Clintock, F.A.: Fracture of Solids, Inter science pub.,1963, p. 65-102.

16) Rice, J.P. ASTM STP 416., 1965

17) Weertman, J.: Int. J. Fract. Mech., 2, 1966, p. 460-467

18) Hénaff, G., Marchal, K. and Petit, J.: Acta Metall. Mater., Vol. 43, n° 8,1995, p. 2931-2942.

19) Lynch, S.P.: Acta metall., 36, 1988, p. 2639.

20) Xu, Y.B., Wang, L., Zhang, Y., Wang, Z.G. and Hu, Q.Z.: Metal Trans., 22A, 1991, p. 723-729

CYCLIC STRAIN LOCALIZATION IN FATIGUED METALS

HAEL MUGHRABI
Universität Erlangen-Nürnberg
Institut für Werkstoffwissenschaften
Martensstr. 5, D-91058 Erlangen, Germany

Abstract

Strain localization in ductile metals subjected to cyclic deformation is a commonly observed feature which can be regarded as the first sign of fatigue damage. Cyclic strain localization can occur in different forms, e.g. in persistent slip band (PSBs), at notches at the free surface, at the tips of propagating cracks, at precipitate-free zones (PFZs), at grain boundaries or at pores at or just below the surface. In this paper, these forms of cyclic strain localization will be discussed with reference to their thresholds and the responsible microstructural mechanisms in terms of the slip modes in single- and multiphase mono- and polycrystals. In particular, the irreversible cyclic slip mechanisms responsible for the evolution of fatigue damage will be considered.

1. Introduction

The fatigue of metals during cyclic deformation evolves in a sequence of events, as indicated in Fig. 1 [1]. The initial stages of cyclic deformation are characterized by bulk

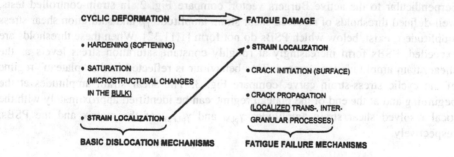

Figure 1. Cyclic deformation and fatigue damage: sequence of events. After [1].

271

E. Bouchaud et al. (eds.), Physical Aspects of Fracture, 271–281.

microstructural changes during cyclic hardening or softening and, in many cases, subsequent cyclic saturation. The onset of the development of fatigue damage is characterized by some form of *localized* cyclic strain enhancement. Subsequently, cyclic strain is strongly localized in few active sites, leading to local trans- or intergranular crack initiation, which occurs at the surface in most cases, followed by the spreading of these cracks and, ultimately, fatigue failure. In this brief article, the different forms of cyclic strain localization will be outlined with respect to different mono- and polycrystalline single- and multi-phase materials, and the responsible microstructural mechanisms will be discussed. Pertinent reviews of related studies to which reference will be made, can be found in refs. [1-7].

2. Cyclic Shear Strain Localization in Persistent Slip Bands

2.1 PERSISTENT SLIP BANDS IN SINGLE-PHASE MATERIALS

Since their first discovery by Thompson et al. [8], persistent slip bands (PSBs) are the most well-known form of cyclic strain localization [1-7]. They have in the past been studied in most detail on copper single crystals (mainly at room temperature), compare [1-8], but, in particular in more recent times, also in much detail on nickel single crystals, compare [9]. In this article, we shall only be considering studies of PSBs at room temperature, although interesting effects are indeed observed at other (higher and lower) temperatures, compare [6,9,10]. In single-phase materials such as, in particular, pure face-centred cubic (fcc) metals and weakly alloyed substitutional solid solutions, PSBs form after strong initial cyclic hardening as a consequence of local instabilities of the dense work-hardened dislocation microstructure. These PSBs have, in their classical form, the so-called "ladder" structure and consist of thin lamellae ($\lesssim 1\mu m$) embedded in the so-called matrix structure. The PSBs lie parallel to the active glide plane in which the dislocations are distributed in the form of an edge di-/multipole wall structure perpendicular to the active Burgers vector, compare Fig. 2. In strain-controlled tests, well-defined thresholds of resolved shear strain amplitude γ_{pl} and saturation shear stress amplitude τ_s exist, below which PSBs do not form [1,11,12]. When these thresholds are exceeded, PSBs form increasingly at roughly constant saturation stress levels as the shear strain amplitude is increased. This behaviour is reflected in the "plateau" regime of the cyclic stress-strain curve, compare Fig. 3. The shear strain amplitudes at the beginning and at the end of the plateau regime can be identified approximately with the local resolved shear strain amplitudes $\gamma_{pl,M}$ and $\gamma_{pl,PSB}$ of the matrix and the PSBs, respectively.

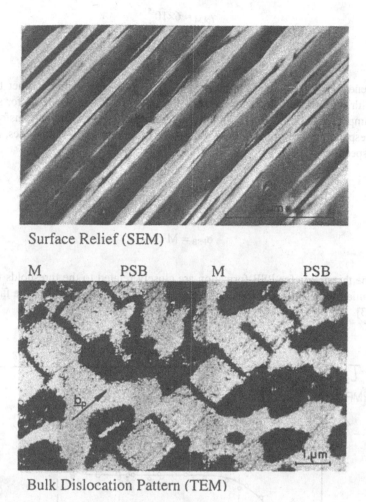

Surface Relief (SEM)

M   PSB   M   PSB

Bulk Dislocation Pattern (TEM)

Figure 2. Persistent slip bands in fatigued copper. Top picture: surface (SEM). Bottom picture: bulk (TEM). M: matrix, PSB: persistent slip band. After [1].

Quite typically, one finds, as a measure for the degree of strain localization in single-phase materials:

$$\gamma_{pl,PSB} \approx 10^2 \cdot \gamma_{pl,M} \tag{1}$$

The threshold values of the amplitudes for PSB-formation are, for example, in the case of copper single crystals at room temperature:

274

$$\gamma_{pl,M} \approx 6 \times 10^{-5} \qquad (2)$$

and
$$\tau_{PSB} \approx 28 \text{ MPa}. \qquad (3)$$

Quite generally, $\gamma_{pl,M}$ is found to lie around 10^{-4} in the case of many other fcc metals [12]. With appropriate orientation factors M (e.g. Sachs factor M = 2.24 for single slip at low amplitudes), compare [1,13], the above threshold values can be transformed into the corresponding thresholds of the axial plastic strain and stress amplitudes, $\Delta\varepsilon_{pl}/2$ and σ_{PSB}, respectively, according to

$$\Delta\varepsilon_{pl}/2 = \frac{\gamma_{pl}}{M} \qquad (4)$$

and
$$\sigma_{PSB} = M \cdot \tau_{PSB}. \qquad (5)$$

Since the thresholds for PSB-formation are closely related to the thresholds for fatigue crack initiation, the above thresholds lie close to the threshold values of the fatigue limit [1,12,13].

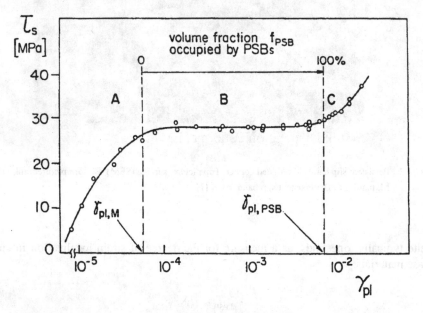

Figure 3. Cyclic stress-strain curve, τ_s versus γ_{pl}, of fatigued copper single crystals. After [1,11].

The reason for the formation of PSBs in fcc single-phase material lies in the metastable nature of the work-hardened dislocation microstructure which consists of patches (veins) of dense (hard) di- and multipolar elongated edge dislocation loops separated by dislocation-poor (soft) regions. A critical situation is reached, when the dense veins are hardened to such a degree that the imposed plastic strain can no longer be accommodated compatibly in the soft and hard regions. Then, local instabilities lead to the formation of PSBs which are able to accommodate much larger local strains than the work-hardened matrix structure.

Simple arguments show that, because of the high local strain amplitude in PSBs, dislocation glide in PSBs is highly irreversible in the sense that profuse mutual annihilation of unlike dislocations occurs and that cyclic saturation in the PSBs is the result of a dynamic equilibrium between dislocation annihilation and multiplication [1,11,12]. On the other hand, in the matrix structure, the dislocations glide to-and-fro over only small distances, and dislocation glide is hence quasi-reversible [11,12].

In low alloy-content solid solutions such as Cu-Zn or Cu-Al alloys, the cyclic slip mode is wavy (easy cross slip) as in pure copper, nickel, etc., and similar PSBs with the ladder structure, as discussed for copper, form. Thus, in the case of Cu-2at.% Al and Cu-2at.%Co solid solution single crystals, the cyclic stress-strain curve shows a well-developed plateau, as shown by Wilhelm and Everwin [14]. At higher alloy contents, e.g. in Cu-16 at.% Al [15] or in Cu-22 at.% Zn [16], the cyclic slip mode is much more planar (difficult cross slip), and very sharp PSB-like slip bands develop, which Laird and co-workers [15] referred to as persistent Lüders bands (PLBs). These PLBs, which are not "permanent" but decay quickly as new ones form [16], lead to a plateau in the cyclic stress-strain curve. This plateau is, however, much shorter than in pure copper single crystals.

The early more detailed observations of strain localization and slip step heights in PSBs relied mostly on interferometric surface observations, pioneered by Laird and his research group [2,5,15,17]. An interesting feature of PSBs is that, as small irreversible microstructural changes accumulate in the PSBs, a cell structure gradually forms in the PSBs and the activity of the old PSBs decays, and new PSBs form. At the same time, a so-called "secondary" cyclic hardening stage is observed [18]. In more recent studies, the sharp-corner technique of the Basinskis [6] provided valuable additional information. Most recently, atomic force microscopy, applied by Holste's research group in Dresden [19,20], has revealed many finer details of the activity, irreversibility and persistence of PSBs in nickel single crystals.

In passing, we note that in pure body-centred cubic (bcc) metals, PSBs in their strict meaning usually do not form. However, in the presence of dynamic strain ageing, e.g. in α-Fe with some carbon or in low carbon steels deformed cyclically at slightly elevated temperature (ca. 100° C) and/or at low strain rates ($\approx 10^{-5}$ s^{-1}), cyclic strain localization in PSB-like bands is observed [11,21]. In this case, however, these "PSBs" usually do not contain the ladder structure but are more or less dislocation-poor channels containing sometimes a weakly developed cell structure. In one exceptional case, PSBs with the ladder structure have been observed in surface grains of a fatigued low carbon steel [22].

2.2 PERSISTENT SLIP BANDS IN MULTI-PHASE MATERIALS

The most pertinent studies on multi-phase alloys are those of Laird and co-workers on precipitation-hardened Al-alloys, compare [2,5,15,17,23,24] and of Wilhelm on Al-Zn-Mg and on Cu-Co alloys [25], as reviewed in [26]. An important difference exists between peak-aged alloys with shearable precipitates and over-aged alloys. In the former case, either PSBs which are thinner than in copper (but thicker than the particle diameter) develop in materials with basically wavy slip and rather small particles, or very sharp, even thinner PSBs which are much thinner than the particle size form in materials with basically planar slip and rather large particles. These two extreme cases are illustrated in Figs. 4a,b. In both cases, cyclic strain localization is much more severe than in single-phase materials (by one or two orders of magnitude) and is frequently reflected in extreme cyclic softening till failure without attainment of saturation. This unstable cyclic deformation behaviour is in marked contrast to the case of stable cyclic deformation (without strain localization) of over-aged alloys in which the particles are not cut but circumvented (Orowan bowing), as demonstrated by Calabrese and Laird [23,24] for Al-Cu and by Stoltz and Pineau [27] for the nickel-base superalloy Waspaloy.

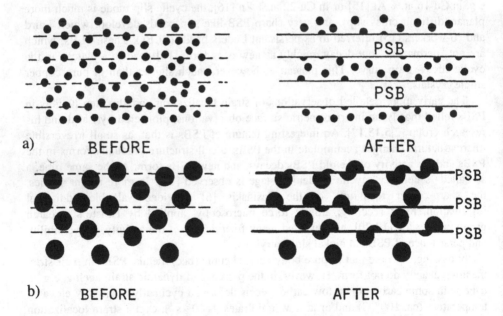

Figure 4. Schematic illustration of two typical types of PSBs in fatigued alloys with shearable precipitates.
a) Wavy slip, small particles. b) Planar slip, larger particles. After [1].

In the case of PSBs in monocrystalline precipitation-hardened alloys with shearable particles, Wilhelm [25] has shown that the PSBs lead to a plateau in a "pseudo" cyclic stress-strain curve in which the peak stresses attained before cyclic softening occurs are plotted (in lack of saturation stresses). There has been much discussion about how and why PSBs form in alloys with shearable precipitates. In all cases, the main reason lies, of course, in an instability of the cyclically hardened precipitate structure, and, in many cases, the details are as follows [26]. In PSBs of the type shown in Fig. 4a, the more or less random to-and-fro slip "chops" the particles up into more or less well separated slices of thicknesses down to atomic dimensions (compare Fig. 5) which can go into solid solution, provided there is enough atomic mobility (diffusion).

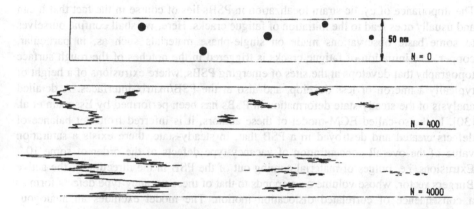

Figure 5. Computer simulation of "chopping-up" of shearable precipitates by random to-and-fro slip for different numbers of cycles N. After [26].

The net result is a PSB which is rather soft, and no longer precipitation- but only solid-solution-hardened. A plausible criterion for the number of cycles, N, until this state is reached is as follows. Assuming random to-and-fro slip with a slip irreversibility p (p ≤ 1) and allowing for dislocations to be arranged in groups of n dislocations, the (irreversible) root mean square (r.m.s.) displacement $\sqrt{<x>^2}$ between adjacent atomic planes, spaced a distance a apart, can be calculated easily. Then, the criterion is that if the r.m.s. displacement exceeds, on the average, the particle diameter d, the precipitates no longer contribute effectively to precipitation hardening. Thus, the criterion is

$$\sqrt{<x^2>} = \sqrt{4N \cdot \gamma_{pl,loc} \cdot n \cdot b \cdot a} > d \,, \tag{6}$$

where $\gamma_{pl,loc}$ denotes the mean local shear strain amplitude in the PSB and b the magnitude of the Burgers vector. As shown elsewhere [26], the number of cycles necessary until this condition is fulfilled can vary significantly, depending on the particular alloy considered. However, one obtains reasonable numbers in those cases in which the necessary data are available. This situation corresponds closely to the case of Cu-2 at% Co, precipitation-hardened by spherical pure Co precipitates [25,26]. In the case of PSBs of the type illustrated in fig. 4b, eq. (6) can be fulfilled (locally) long before the particles are completely chopped up, because severe cyclic slip is confined to rather few discrete very sharp PSBs. Examples of PSBs of this type are found in particular in γ'-precipitation-hardened superalloys, fatigued at not too high temperatures, in which the ordered shearable γ' particles are cut and induce an extremely planar slip mode [27,28,29].

2.3 FATIGUE CRACK INITIATION IN PSBs

The importance of cyclic strain localization in PSBs lies of course in the fact that it can and usually does lead to the initiation of fatigue cracks. Here, we shall confine ourselves to some basic observations made on single-phase materials such as, in particular, copper. The initiation of fatigue cracks is triggered in the notches of the rough surface topography that develops at the sites of emerging PSBs, where extrusions of a height of typically a micron or less develop, and also at the PSB/matrix-interfaces. A detailed analysis of the steady-state deformation in PSBs has been performed by Essmann et al. [30]. In the so-called EGM-model of these authors, it is inferred from net balance of defects created and destroyed in a PSB that, in steady-state, there exists a saturation value of the overall concentration of vacancy-type defects of the order of some 10^{-4}. Extrusions, i.e. bulges of material coming out of the PSB in the direction of the active Burgers vector, whose volume corresponds to that of these vacancy-type defects form as a consequence of correlated dislocation motion. The model excludes an analogous mechanism of intrusion formation.

Many authors speak of extrusion-intrusion pairs, compare [6]. The author's view, however, based on detailed studies on fatigued copper crystals and as elaborated in [1,12,30], is as follows. Close inspection shows, in most cases, that initially only extrusions form, as predicted by the EGM-model, and, in their wake, small microcracks develop, frequently at the PSB-matrix interfaces. These microcracks are probably what many authors then refer to as intrusions.

Furthermore, there are a number of reports in the literature showing that, when fatigue cracks are introduced, PSBs are often observed at the tips of these cracks, as they propagate. In addition to transgranular (mode II) shear cracks forming at emerging PSBs, intercrystalline cracks can be initiated by PSBs (in which the active Burgers vector has a large component parallel to the surface) which impinge at grain boundaries (GBs), compare [1,12,30,31]. These PSB-GB cracks are observed mainly in the range of small to intermediate amplitudes [12,31]. On the other hand, Neumann and Tönnessen [32], using a sharp-corner technique, have noted that, at low amplitudes, many fatigue cracks form at PSBs located at twin interfaces. They made the interesting observation that such cracks formed only on every second twin interface and explained this in terms of the elastic anisotropy of copper.

3. Other Forms of Cyclic Strain Localization and Crack Initiation

In recent years, there has been an increasing interest in ultrafine-grained materials of extraordinary high strength, produced by severe deformation via the equal channel angular (ECA) extrusion technique, compare [33]. In these severely deformed materials, the microstructure is rather unstable under cyclic loading. While cyclic strain localization in PSBs of the classical type cannot occur in the very small grains (≈ 200 nm), a new type of strain localization, extending over many grains, occurs in the form of macroscopic shear bands and frequently leads to early crack initiation [34].

In addition to the effects discussed so far, cyclic strain localization frequently also occurs as a consequence of local stress enhancements at notches, sites of surface roughness, pores or heterogeneities such as inclusions. Quite recently, an interesting form of fatigue crack initiation at pores at or just beneath the surface of cast alloys has found interest, as documented, for example for the case of the fatigued magnesium-base cast alloy AZ 91 [35].

An explanation for subsurface fatigue crack initiation has been offered, based on FEM-simulations of the local stresses and plastic strains in the vicinity of pores lying just beneath the surface of a tensile-stressed material [36]. Figure 6 shows an example of the distribution of the equivalent plastic strain around a near-surface pore. For reasons of symmetry, the figure shows only one quarter of the pore. It is evident that, in

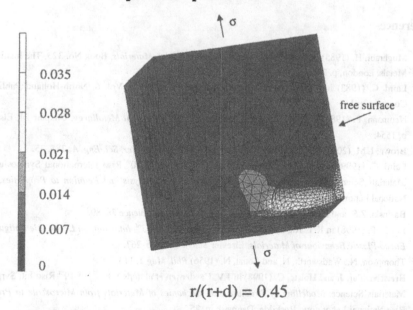

Figure 6. FEM-simulation of local equivalent plastic strains in a tensile-stressed material in the vicinity of a spherical pore lying just underneath the surface (r: radius of pore, d: distance from centre of pore to surface). After Borbély et al. [36].

the narrow "bridge" of material between the pore and the surface, very high local strains occur. It is thus easy to imagine that, under cyclic stressing, a severe cyclic strain localization can occur, leading to crack initiation.

4. Conclusions

The examples discussed above illustrate that cyclic strain localization is usually a consequence of the occurrence of local microstructural instabilities under cyclic loading. Such instabilities can occur in rather homogeneous cyclically hardened materials. Moreover, they can also be triggered by microstructural heterogeneities and by surface roughness and notch effects. Fatigue crack initiation is usually a consequence of cyclic strain localization. Microstructurally induced cyclic strain localization can be reduced or avoided by small particles (dispersoids) which serve to disperse slip, rendering it more homogeneous.

Acknowledgements

Sincere thanks are expressed to Mrs. Renate Graham, Mrs. Waltraud Kränzlein and Dr. Heinz Werner Höppel for their able support during the preparation of this paper.

References

1. Mughrabi, H. (1985) in *Dislocations and Properties of Real Materials*, Book No. 323, The Institute of Metals, London, p. 244.
2. Laird, C. (1983) in F.R.N. Nabarro (ed.), *Dislocations in Solids*, Vol. 6, North-Holland Publishing Company, p.1.
3. Neumann, P. (1983) in R.W. Cahn and P. Haasen (1983), *Physical Metallurgy*, Chapter 24, Elsevier, p. 1554.
4. Brown, L.M. (2000) Special Issue in Honour of Prof. T. Mori, *Mater. Sci. Eng. A*, **285**, 35.
5. Laird, C. (1999) in J.B. Bilde-Sørensen et al. (eds.), Proc. of 20th Risø International Symposium on Materials Science: *Deformation-Induced Microstructures: Analysis and Relation to Properties*, Risø National Laboratory, Roskilde, Denmark, p. 85.
6. Basinski, Z.S. and Basinski S.J. (1992) *Progress in Materials Science* **36**, 89.
7. Lukáš, P. (1998) in K.T. Rie and P.D. Portella (eds.), Proc. of 3[rd.] *Int. Conf. on Low Cycle Fatigue and Elasto-Plastic Behaviour of Materials*, Elsevier Science Ltd., p. 267.
8. Thompson, N., Wadsworth, N. and Louat, N. (1956) *Phil. Mag* **1**, 113.
9. Bretschneider, J. and Holste, C. (1998) in J.V. Carstensen et al. (eds.), Proc. of 19th Risø Int. Symp. on Materials Science: *Modelling of Structure and Mechanics of Materials from Microscale to Product*, Risø National Laboratory, Roskilde, Denmark, p. 25.
10. Holzwarth, U. and Essmann, U. (1994) *Appl. Phys.* **A58**, 197.

11   Mughrabi, H., Ackermann, F. and Herz, K. (1979) in J.T. Fong (ed.), *Fatigue Mechanisms*, ASTM STP 675, p. 69.

12.   Mughrabi, H., Wang, R., Differt, K. and Essmann, U. (1983), in J. Lankford et al. (eds.), *Fatigue Mechanisms: Advances in Quantitative Measurement of Physical Damage*, ASTM STP 811, p.5.

13.   Mughrabi, H. and Wang, R. (1981) in N. Hansen et al. (eds.), Proc. of 2nd Risø Int. Symposium on Metallurgy and Materials Science: *Deformation of Polycrystals, Mechanisms and Microstructures*, Risø National Laboratory, Roskilde, Denmark, p. 87.

14.   Wilhelm, M. and Everwin, P. (1979) in P. Haasen et al. (eds.), Proc. of 5th Int. Conf. on the Strength of Metals and Alloys (ICSMA 5), Vol. 2, Pergamon Press, p. 1089.

15.   Hong, S.I. and Laird, C. (1991) *Fatigue Fract. Engng. Mater. Struct.* **14**, 143.

16.   Lukáš, P., Kunz, L. and Krejči, J. (1992) *Mater. Sci. Eng. A*, **158**, 177.

17.   Lee, J.-K. and Laird, C. (1983) *Phil. Mag. A*, **47**, 579.

18.   Wang, R. and Mughrabi, H. (1984) *Mater. Sci. Eng.* **63**, 147.

19.   Schwab, A., Meissner, O. and Holste, C. (1998) *Phil. Mag. Letters* **77**, 23.

20.   Hollmann, M., Bretschneider, J. and Holste, C. (2000) *Cryst. Res. Technol.* **35**, 479.

21.   Wilson, D.V. and Tromans, J.K. (1970) *Acta metall.* **18**, 1197.

22.   Pohl, K., Mayr, P. and Macherauch, E. (1980) *Scripta metall.* **14**, 1167.

23.   Calabrese, C. and Laird, C. (1974a) *Mater. Sci. Eng.* **13**, 141.

24.   Calabrese, C. and Laird, C. (1974b) *Mater. Sci. Eng.* **13**, 159.

25.   Wilhelm, M. (1981) *Mater. Sci. Eng.* **48**, 91.

26.   Mughrabi, H. (1983) in J.B. Bilde-Sørensen et al. (eds.), Proc. of 4th Risø Int. Symp. on Metallurgy and Materials Science: *Deformation of Multi-Phase and Particle-Containing Materials*, Risø National Laboratory, Roskilde, Denmark, p. 65.

27.   Stoltz, R.E. and Pineau, A.G. (1978) *Mater. Sci. Eng.* **34**, 275.

28.   Clavel, M. and Pineau, A. (1982) *Mater. Sci. Eng.* **55**, 157.

29.   Obrtlik, K., Lukáš, P. and Polák, J. (1998) in K.T. Rie and P.D. Portella (eds.), *Proc. of 3rd Int. Conf. on Low Cycle Fatigue and Elasto-Plastic Behaviour of Materials*, Elsevier Science Ltd., p. 33.

30.   Essmann, U., Goesele, U. and Mughrabi, H. (1981) *Phil. Mag. A*, **44**, 405.

31.   Bayerlein, M. and Mughrabi, H. (1992) in K.J. Miller and E.R. de los Rios (eds.), *Short Fatigue Cracks*, ESIS 13, Mechanical Engineering Publications, London, p. 55.

32.   Neumann, P. and Tönnessen, A. (1988) in P.O. Kettunen et al. (eds.), *Proc. of 8th Int. Conf. on the Strength of Metals and Alloys (ICSMA 8)*, Vol. 1, Pergamon Press, p. 743.

33.   Lowe, T.C. and Valiev, R.Z., eds. (2000) *Investigations and Applications of Severe Plastic Deformation*, Kluwer Academic Publishers.

34.   Vinogradov, A., Kaneko, Y., Kitagawa, K., Hashimoto, S. and Valiev R. (1998) *Mater. Sci. Forum* **269-272**, 987.

35.   Eisenmeier, G., Ottmüller, M., Höppel, H.W. and Mughrabi H. (1999) in *FATIGUE '99: Proc. of Seventh Int. Fatigue Congress*, Vol. 1, Higher Education Press, Beijing, P.R. China, EMAS Ltd., West Midlands, UK, p. 253.

36.   Borbély, A., Mughrabi, H., Eisenmeier, G. and Höppel, H.W. (2000). Publication in preparation, to be submitted to *Int. J. Fracture*.

ENVIRONMENTAL EFFECTS ON FATIGUE IN METALS

T. Magnin

Ecole Nationale Supérieure des Mines de St-Etienne, URA CNRS 1884,
158 Cours Fauriel, 42023 ST ETIENNE CEDEX 02

ABSTRACT

Anodic dissolution and hydrogen effects are known to play an essential role on the fatigue behaviour of metallic materials in aqueous environments. The aims of the paper are :
- to give some pedagogical examples of those effects and to relate such examples to industrial problems.
- to precise the main electrochemical, metallurgical and mechanical parameters which must be taken into account to model corrosion fatigue.
- to detail crack initiation and crack propagation mechanisms in terms of localised corrosion-deformation interactions.
Finally, the possible relation between stress corrosion cracking and corrosion fatigue is shown for both crack initiation and crack propagation.

INTRODUCTION

The deleterious effect of aqueous environment on fatigue crack initiation and propagation in metals and alloys has been observed since a long time [1-5]. The applied electrochemical potential has a large influence on localised corrosion reactions and, consequently, on corrosion fatigue (CF) damage. These complex effects are used for cathodic protection in offshore structure. Nevertheless CF is in fact influenced by various mechanical, chemical and microstructural parameters which locally interact. Even if anodic dissolution versus hydrogen effects are known to occur during CF, the damage mechanisms are still under controversies and fatigue lifetime predictions are still empirical.

The application of linear elastic fracture mechanics (LEFM) led to the determination of threshold stress intensity factors for long crack growth, which are quite useful for engineering material/environment systems. Nevertheless, it is now well established [6,7] that short cracks can develop and grow from smooth surfaces at crack tip stress intensities below the previous threshold, resulting from localised corrosion-deformation interactions. CF damage models of short crack growth are then necessary to improve the fatigue damage predictions in aqueous solutions.

283

E. Bouchaud et al. (eds.), Physical Aspects of Fracture, 283–304.

In the first part of this chapter, crack initiation mechanisms are reviewed and quantitative approaches given, through limitations mainly related to the difficult evaluation of localised corrosion-deformation interactions. A particularly interesting example of electrochemical and mechanical coupling effects during CF in duplex stainless steels will illustrate the complexity of the interactions which lead to crack initiation.

Crack propagation modellings are then presented through these behaviour of both short and long cracks. Finally the possible relation between stress corrosion cracking and CF is shown for crack initiation and crack propagation processes.

CORROSION FATIGUE CRACK INITIATION

It is well known that slip bands, twins, interphases, grain boundaries and constituent particles are classical sites for crack initiation. Moreover, persistant slip bands (PSB)/grain boundaries interactions are often observed to be preferential crack initiation sites during CF, as well as localised pits around metallurgical heterogeneities.

The main need in the fatigue crack initiation modelling is related to the quantitative approach of local synergetic effects between environment and cyclic plasticity. In this section, quantitative approaches of corrosion fatigue crack initiation from different electrochemical conditions are presented through their interests and their limits. Then improvement of such models are given through corrosion-deformation interaction effects recently analysed. Finally, an interesting example is given about the coupling effects between cyclic plasticity and corrosion which must be taken into account to improve the crack initiation resistance of duplex stainless steels in chloride solutions.

Classical Approaches of Corrosion Fatigue Damage

Electrochemical corrosion can be schematised as an 'electronic pump' or an electronic circuit' related to oxidation and reduction reactions :

$$M \rightarrow M^{n+} + ne^- \; : \; \text{anodic dissolution}$$

$$\left. \begin{array}{l} 2H_2O + O_2 + 4e^- \rightarrow 4OH^- \\ 2H^+ + 2e^- \rightarrow H_2 \end{array} \right\} \text{cathodic reactions}$$

together with cation hydrolysis reaction : $M^{n+} + nH_2O \rightarrow M(OH)_n + nH^+$.

M^{n+} is a solvated ion, e^- is an electron and n represents the ion state of charge. The electrons, liberated by the oxidation, must flow through the material M to be consumed in an appropriate cathodic reaction. Beyond a solubility limit, precipitates of hydroxide or hydrated oxide are formed, and this surface film can provide a barrier to further dissolution. Some of them are named 'passive', for stainless steels or aluminium alloys for instance. Theses film will play an important role in environment sensitive crack initiation and fracture. Under equilibrium conditions, film stability is referred to $E = f(pH)$ diagrams, where E is the electrical potential related to the chemical free energy G by the relation : $G = -n \, EF$, and F is the Faraday's number. The electrical potential also affects the anodic dissolution reaction. At equilibrium, one can define the 'electrode potential' (related to ΔG) and the current density I (I $\sim e^{-\Delta G^*/RT}$ where ΔG^* is the activation enthalpy of dissolution).

Thus, the relation $E = f(I)$ gives the different corrosion domains for a given metal in a given solution. Figure 1 schematises such a relation (polarisation curve) in the case of an austenitic stainless steel in an acidic Cl⁻ solution. Five domains can be considered for corrosion and corrosion fatigue damage :

(i)   In Zone 1, $E > E_r$, pitting occurs by destabilisation of the passive film. Pits will act as stress concentrators during fatigue.

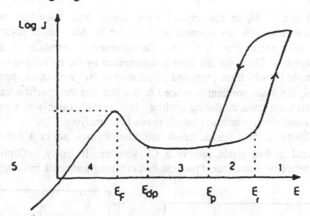

FIGURE 1 General polarization curve for an austenitic stainless steel in acidic Cl⁻ solutions.

During corrosion fatigue under pitting conditions, pits grow into the material. If such a pit reaches a critical depth d_{CL}, a fatigue crack can develop. The critical depth is then a function of the applied stress range [8].

Let us suppose the following conditions :
• constant corrosion conditions (pH, concentration of bulk solution)
• constant alternating load, $d\Delta P/dt = 0$
• constant loading frequency $dv/dt = 0$

It is well established that growth kinetics of corrosion pits are determined by a simple power law :

$$d_L(t) = C(t - t_0)^\beta, \, t > t_0 \tag{1}$$

where t_0 is the incubation time for pit nucleation. If the pit depth reaches the critical value :

$$d_L(t) = d_{CL} \tag{2}$$

corrosion fatigue crack initiation occurs. The critical pit depth d_{CL} depends on the applied stress range $\Delta\sigma_0$, cyclic yield strength σ_{FC}, fatigue crack growth threshold ΔK_0, and the geometry of the specimen, expressed in terms of a geometrical factor G. It can be calculated by elastic-plastic fracture mechanisms based on the Dugdale model. d_{CL} is then given by the following equation :

$$dCL = \frac{\cos\left(\pi\Delta\sigma_0/4\sigma FC\right)\cdot\pi\Delta K_0^2}{32\,G2\sigma_{FC}^2\left[1-\cos\left(\pi\Delta\sigma_0/4\sigma FC\right)\right]} \tag{3}$$

The number of cycles to initiate a corrosion fatigue crack under pitting conditions is, by combining the previous equations with $N = t.\nu$:

$$N_i = \nu\left[t_0 + \left(dCL/C_2\right)^{V_\beta}\right] \tag{4}$$

The dependence of N_i on the applied stress range $\Delta\sigma_0$, calculated according to the previous equation is schematically represented in Figure 2a. Also, under pitting conditions no corrosion fatigue limit exists. For $N_i \le \nu.t_0$, the influence of corrosion on the fatigue crack initiation life disappears. Then, the life time is determined by the air fatigue behaviour. Figure 2b shows an example for which the proposed calculation of N_i seems quite appropriate.

Nevertheless, the main problem is related to the fact that the coefficients C and β of the pit kinetics are often not constant during cycling : it is a clear example of a cooperative effect between plasticity and electrochemistry which needs finer analyses.

Figure 3 illustrates the fatigue crack initiation from a pit in a fcc Fe-Mn-Cr alloy cyclically deformed at low strain rate in a Cl⁻ solution. In many multiphase engineering materials, the presence of a constituent particle favors pitting and crack initiation.

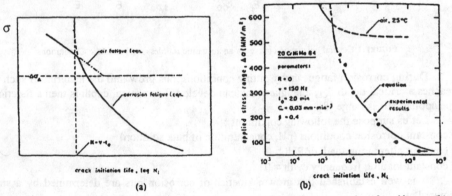

FIGURE 2 Corrosion fatigue crack initiation by pitting corrosion : (a) Schematic representation of a σ-N curve. (b) Comparison of experimentally and theoretically derived fatigue lives for the 20 Cr Ni Mo alloy in 30 g/l NaCl solutions [8].

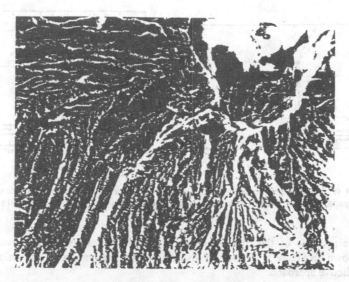

FIGURE 3 Crack initiation from a pit for a Fe-17 Mn - 13 Cr alloy during CF in a 110°C Cl⁻ solution at a plastic strain amplitude $\Delta\varepsilon p/2 = 4 \times 10^{-3}$ and a strain rate $\dot{\varepsilon} = 10^{-5} \text{ s}^{-1}$.

(ii)   In zone 2, $Ep < E < E_r$, pits are repassivated. If the plastic strain amplitude is too small for localised depassivation (by slip band emergence), pits will not grow and the CF behaviour is then close to that in air. On the other hand, pits can grow if mechanical depassivation occurs and the CF behaviour is then close to that of zone 1.

(iii)   In the passive region 3, a competition between the kinetics of depassivation by slip and that of repassivation takes place. Thus the influence of the plastic strain amplitude and strain rate is quite obvious. In the same mechanical conditions as in (i), the quantity of matter dissolved per cycle in the depassivated slip bands can be expressed, using the Faraday's law :

$$dN = \frac{M}{nF\rho} \, 2 \int_0^{\gamma/\nu} i \, (repassivation) \, dt \tag{5}$$

where i (repassivation) = i max exp(-γt) with γ taken as a constant. Then $N_i = N$ (for $dN = d_c$). If d_c is taken as the grain size φ for instance, then :

$$N_i = \frac{\phi}{dN} = \phi \, \frac{nF\rho}{M} \, \frac{\gamma}{2i_{max}} \, [1 - \exp(\gamma/\nu)]^{-1} \tag{6}$$

A schematic representation of the previous equation is given on Figure 4a and corresponding experimental result are shown on Figure 4b.

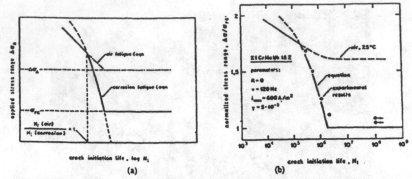

FIGURE 4 (a) Schematic representation of a σ-N curve for CF crack initiation under passive conditions. (b) Comparison of experimentally and theoretically derived fatigue lives for the X1Cr MoNb 182 alloy in 30 g/l NaCl at 80°C [8].

Nevertheless, one of the main problem is that the repassivation law evolves during cycling as shown on Figure 5. This result also emphasises the synergy in CF which leads to a complex predictive approach.

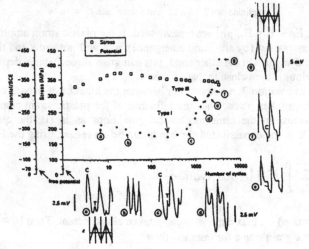

FIGURE 5 Simultaneous evolution of the cyclic stress σ, the average potential and the shape of the cyclic potential transients for a 316 L alloy in 30 g/l Na Cl ($\Delta\epsilon p/2 = 10^{-3}$, $\dot{\epsilon} = 10^{-2} s^{-1}$).

(iv) In zone 4, generalised dissolution occurs, which is generally quite dangerous for materials even without stress !

(v) In zone 5, cathodic reactions are favoured. If the reduction of hydrogen occurs, we can have :

$$2H^+ + 2e^- \rightarrow H_2$$
$$\rightarrow 2H_{adsorbed} \rightarrow H_2$$
$$H_{ads} \rightarrow H_{absorbed}$$

which can induce hydrogen after diffusion and transport by dislocations leading generally to macroscopic brittle fracture under stress !

The electrochemical approach presented here has many limitations. First of all, the kinetics of the electrochemical reactions are closely dependent on the cyclic plasticity and the number of cycles. Thus, predictive laws are very complex. Moreover these laws use the local current densities which are very difficult to model. Finally this approach does not take really into account the local corrosion-deformation interactions (CDI) and the effects of the corrosive solution on the deformation mode. Indeed, such synergetic effects between corrosion and deformation can be of prior importance : the following examples emphasize the role of corrosion deformation interactions in crack initiation processes.

Influence of cyclic plasticity on electrochemical reactions

The evolution of dissolution current density transients during cycling of ferritic and austenitic stainless steels in NaCl solutions is shown in Figure 6, where curves $J_T = f(N)$ are plotted. J_T is the peak current density related to the depassivation process due to cyclic plasticity, and N the number of cycles. During the first cycles the amount of dissolution increases, particularly at high strain rate. One of the ferritic steel exhibits twinning, which induces a more marked depassivation during cycling, compared to the behaviour of the second ferritic steel which deforms by pencil glide.

Figure 6 clearly illustrates the influence of the deformation mode on the electro-chemical reactions and the evolutions of such electrochemical transients as a function of N (i.e. the localization of the plastic deformation with a decrease of J_T after the first cycles, and then the formation of microcracks with a new increase of J_T till fracture due to a more difficult repassivation).

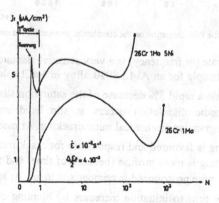

FIGURE 6 $J_T = f(N)$ curves for stainless steels during CF in a 3.5 % NaCl solution.

290

PSB and intense slip bands are very prone to preferential dissolution, not only for passivated alloys but also in conditions of generalized dissolution as shown on Figure 7 for copper single crystals in NaClO₄ solution [9].

As soon as the PSB form, the anodic current increases although the applied plastic strain remains constant. This effect is not only related to the localization of the cyclic plasticity but also to the influence of the dislocation microstructure of PSB on the free enthalpy of dissolution (- ΔG) and the energy of activation (ΔG^*).

Moreover cyclic plasticity has been also shown to often promote localized pitting well below the pitting potential without stress [10]. For the ferritic Fe-26 Cr-1Mo stainless steel in 3.5 % NaCl solution, a high strain rate $\dot{\varepsilon}$ promotes strain localization at grain boundaries, which induces an intergranular pitting for an applied -100 mV/SCE.

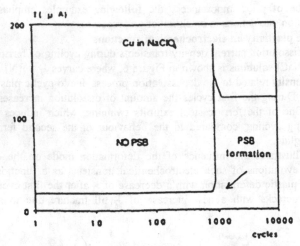

FIGURE 7 Influence of the PSB formation on the dissolution amount for Cu single crystals in NaClO₄ [9].

The applied strain rate (or frequency) is a very sensitive parameter for CF damage. Figure 8 gives an interesting example for an Al-Li 8090 alloy in NaCl solutions. N_i is defined as the number of cycles to obtain a rapid 3% decrease of the saturation stress. At high strain rate ($\dot{\varepsilon} > 5 \times 10^{-3} s^{-1}$), the anodic dissolution occurs at slip band emergence and induces an enhancement of the transgranular mechanical microcracking. At medium strain rate ($5 \times 10^{-5} s^{-1} < \dot{\varepsilon} < 5 \times 10^{-3} s^{-1}$), pitting is favoured and responsible for crack initiation. So when the plastic strain rate decreases, pitting is more profuse (because of time) and the reduction in the fatigue life to crack initiation is more pronounced in comparison to air. At low strain rate ($5 \times 10^{-6} s^{-1} < \dot{\varepsilon} < 10^{-5} s^{-1}$), the fatigue time to initiation increases by blunting of the mechanically formed microcracks because of generalised pitting which acts as general corrosion. At very low strain rate ($\dot{\varepsilon} < 5 \times 10^{-6}\,\mathrm{s}^{-1}$) CF crack initiation occurs by intergranular stress corrosion. The rapid occurrence of SCC induces a marked decrease of N_i.

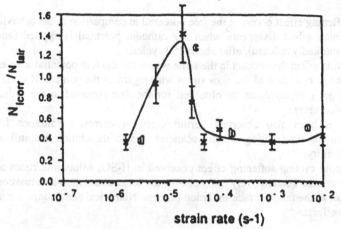

FIGURE 8 Influence of strain rate on Al-Li 8090 alloy fatigue life to crack initiation in a 3.5 % NaCl solution for $\dfrac{\Delta\varepsilon p}{2} = 4\times10^{-3}$ at free potential.

Softening effect due to anodic dissolution

CF tests on smooth specimens were performed at room temperature on a 316 L austenitic stainless steel in a 0.5 N H_2SO_4 solution at different electrochemical potentials and for a prescribed plastic strain amplitude of 4×10^{-3} ($\dot{\varepsilon} = 10^{-2}\,\mathrm{s}^{-1}$). The depassivation-repassivation process occurs in a very regular way, well before any microcracks can form. It is of particular interest to follow the evolution of the maximum flow stress in the corrosive solution at free

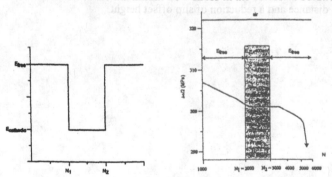

FIGURE 9 Evolution of the peak stress $\Delta\sigma/2$ during cycling in a 0.5 N H_2SO_4 solution at free potential, for $\dfrac{\Delta\varepsilon p}{2} = 4\times10^{-3}$ and $\dot{\varepsilon} = 10^{-2}\,\mathrm{s}^{-1}$, compared to the air behaviour.

potential and at imposed cathodic potential, and to compare this evolution with that observed in air (Figure 9). It clearly appears that :

(i) a cyclic softening effect occurs at the free potential in comparison to the behaviour in air.
(ii) this softening effect disappears when the cathodic potential is applied (and the anodic dissolution is markedly reduced), after about 150 cycles.
(iii) the softening effect then occurs in the same way when the free potential is re-established.
(iv) a delay in the evolution of the flow stress with regard to the number of cycles for which a potential change is imposed can be observed for the free potential to the cathodic potential change (and vice-versa).

This effect has been also observed during creep in corrosive solutions for copper. It corresponds to the time during which vacancies due to dissolution are still acting on the dislocation mobility.

The macroscopic cycling softening effect observed in H_2SO_4 solution at room temperature is very relevant to take quantitatively into account the local dissolution-deformation interactions which will lead to the fatigue crack initiation process. Numerical simulations are under study to quantify such effects.

Influence of corrosion on PSB configurations

Electrochemical control of corrosion has been shown to significantly affect the morphology of surface deformation. A modification of the number of persistant slip bands (PSB) and of the slip offset height in PSB has been observed for Ni single crystals [4] in a 0.5 N H_2SO_4 and in copper single crystals [4,9] according to the applied potential, in comparison to air. It is easy to understand that such influences on PSB distribution will affect the crack initiation conditions. Figure 10 shows histograms of the PSB distribution produced on monocrystalline nickel in 0.5 N H_2SO_4 at a constant strain amplitude.

Experiments conducted at the corrosion potential and at + 160 mV/SCE result in a reduction of the inter PSB distance and a reduction of slip offset height.

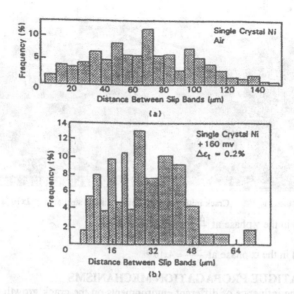

FIGURE 10 Histograms of distance between PSB clusters, from the replicate Ni single crystal specimens fatigued for 1000 cycles under total strain control $\Delta\varepsilon_t$. Shear strain = 0.12 %, (a) air and (b) + 160 mV/SCE in 0.5 N H_2SO_4 [4].

An example of mechanical and electrochemical coupling effects : the CF crack initiation mechanisms of a two - phase stainless steel in NaCl solutions

Mechanical and electrochemical coupling effects are generally the key to understand the crack initiation mechanisms in multiphase alloys. This is clearly illustrated for a duplex α/γ stainless steels (without nitrogen) in a 3.5% NaCl solution at pH 2 and free potential [11]. At low plastic strain amplitude, the softer γ phase is depassivated but this phase is cathodically protected by the non-plastically deformed α phase. This coupling effect reduces the dissolution of the γ phase and delays CF damage, which is not the case at higher strain amplitude when the α phase is also depassivated by slip band emergence.

Observations of the crack initiation sites by scanning electron microscopy show that at low plastic strain amplitudes ($\Delta\varepsilon_p/2 < 10^{-3}$) for which the fatigue resistance of the α-γ alloy is close to that of the γ alloy, cracks nucleate only in the austenitic phase (Figure 11a) but, at higher strain amplitudes ($\Delta\varepsilon_p/2 > 10^{-3}$), the first cracks nucleate principally in the ferritic phase (Figure 11b). The excellent CF resistance of duplex stainless steels (for $\Delta\varepsilon_p/2 < 10^{-3}$) can then be understood through the electrochemical and mechanical coupling effects on crack initiation processes.

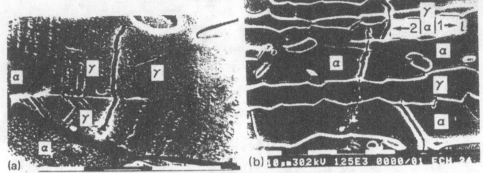

FIGURE 11   Crack initiation in a duplex stainless steel at $\dot{\varepsilon} = 2\times10^{-3}\,s^{-1}$.

(a) in the γ phase at $\dfrac{\Delta\varepsilon p}{2} = 3\times10^{-4}$.

(b) in the α phase at $\dfrac{\Delta\varepsilon p}{2} = 4\times10^{-3}$.

CORROSION FATIGUE PROPAGATION MECHANISMS

Figure 12 shows the influence of different environments on the crack growth rate of a classical industrial steel [4]. The aim of this section is to analyse the possible mechanisms at crack tip leading to crack advance.

The determination of the crack-tip environment is not easy due to the numerous electrochemical reactions and the associated mass transport and thermodynamic criteria which govern this environment. The nature of the solution (composition, pH, species, corrosion products which can induce roughness...) can be very different at the crack tip and in the bulk solution. The mathematical modelization of the crack tip environment is therefore possible. Figure 13 schematizes the possible electrochemical reactions at the crack tip. Both anodic dissolution effects and coupled hydrogen reaction most be often taken into account.

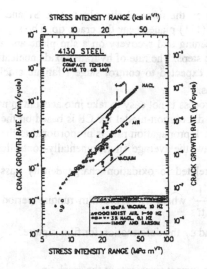

FIGURE 12 Influence of environment on the crack propagation velocity for a 4130 steel [4].

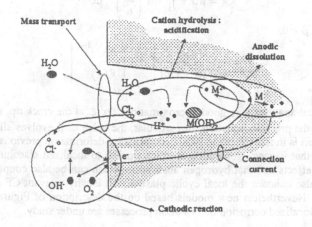

FIGURE 13 Possible electrochemical reactions at a corrosion fatigue crack tip.

Anodic Dissolution Effects

Anodic dissolution has been shown to occur preferentially in slip bands at the very near crack tip, which leads to many consequences on the CF behaviour as well as in CF as shown on Figure 14. The restricted slip reversibility model for stage I fatigue crack propagation has been proposed. The average crack growth increment per cycle, $\Delta\chi$, is determined by factors controlling slip reversibility on the forward slip planes (*i.e.* S1 and S4). For fcc materials, it involves alternating slip on (111) planes near the crack tip. The slip reversibility is controlled by the degree of work hardening and recovery on a slip plane and the presence of corrosion products (oxides) on the slip steps. The rate of hydrated oxide nucleation plays a critical role in this model and therefore is expected to control stage I fatigue crack growth of an austenitic stainless steel in Cl⁻ solutions.

This localized dissolution process is not easy to take into account in numerical modelisations of crack propagation. The slip dissolution-model for CF is based on the fact that for many alloys in different solutions the crack propagation rate is proportional to the oxidation kinetics. Thus, by invoking the Faraday's law, the average environmentally-controlled crack propagation rates $\overline{V_t}$ for passive alloys is related to oxidation charge density passed between film rupture events, Q_f : $\overline{V_t} = \frac{M}{n\rho F} Q_f \frac{1}{t_f}$ where t_f is the film rupture period. Thus $\overline{V_t} = \frac{M}{n\rho F} Q_f \frac{\dot{\varepsilon}}{\varepsilon_f}$

where $\dot{\varepsilon}$ is the strain rate and ε_f the deformation for film rupture (about 10^{-3}).

If we take a classical law for current transients at the crack tip,

$$Q_f = i_0 t_0 + \int_{t_0}^{t_f} i_0 \left(\frac{t}{t_0} \right)^{-\beta} dt,$$

Then, for $t_f > t_0$, $\beta > 0$,

$$\overline{V_t} = \frac{M i_0 t_0 \dot{\varepsilon}}{n\rho F (\beta - 1) \varepsilon_f} \left[\beta - \left(\frac{\dot{\varepsilon} t_0}{\dot{\varepsilon}_f} \right)^{\beta - 1} \right]$$

Even if mechanical analyses give good approximations for $\dot{\varepsilon}$ at the crack tip, many problems still remain with the previous equation. In particular, the value of β evolves all along cycling. But the main effect is in fact related to localized corrosion-deformation previously described. It has been shown that vacancy generation at crack tip due to localized dissolution can induce cyclic softening effects and that hydrogen absorption which can be also coupled to localized dissolution can also enhance the local cyclic plasticity. It is why relevant CF predictive laws are still missing. Nevertheless new models based on the description of Figure 14 but taking into account the localized corrosion deformation processes are under study.

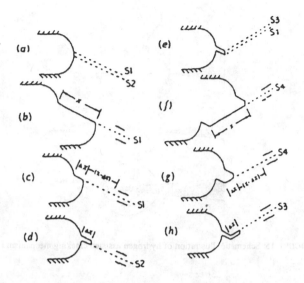

FIGURE 14 Schematic diagram of stage I corrosion fatigue crack propagation mechanism [18].

Hydrogen Effects

Hydrogen assisted cracking is often invoked, particularly for bcc materials but also together with anodic dissolution for fcc alloys. Figure 15 schematizes a hydrogen assisted cracking event. Interactions between a discretized dislocation array and the crack tip under an applied stress produce a maximum stress field from behind the tip. When the hydrogen concentration reaches a critical value, a microcrack is nucleated because either the local cohesive strength is reduced, dislocation motion is blocked in the hydrogen-enriched zone, or both. The microcrack arrests about 1 μm ahead of the original location of the tip and these processes then repeat leading to discontinuous microcracking.

Other mechanisms have been proposed, particularly the hydrogen-induced plasticity model for precipitates containing materials such as Al-Zn-Mg alloys. This model is described in Figure 16 [5].

Absorbed hydrogen atoms weaken interatomic bonds at crack tip and thereby facilitates the injection of dislocations (alternate slip) from crack tip. Crack growth occurs by alternate slip at crack tips which promotes the coalescence of cracks with small voids nucleated just ahead of the cracks. In comparison to the behaviour in neutral environments, the CF crack growth resistance decreases as the proportion of dislocation injection to dislocation egress increases. More closely spaced void nuclei and lower void nucleation strains should also decrease the resistance to crack growth in CF. This mechanism is proposed for Al-Zn-Mg alloys and is highly supported by observations that environmentally assisted cracking can occur at high crack velocities in materials with low hydrogen diffusivities and that the characteristics of cracking at high and low velocities are similar.

298

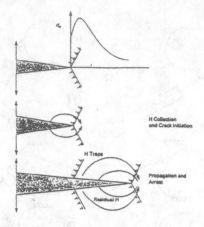

FIGURE 15 Schematic illustration of hydrogen assisted cracking mechanism [12].

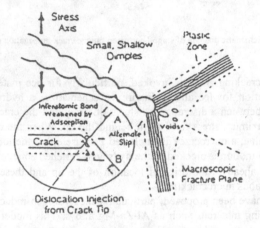

FIGURE 16 Schematization of the hydrogen-induced plasticity model for CF cleavage like cracking [5].

Finally, it must be said that very often anodic dissolution and hydrogen effects must be both taken into account.

Corrosion Fatigue Crack Growth

Figure 17 summarizes the influence of corrosion on the fatigue crack growth characteristics. To the purely mechanistic mode I crack propagation behaviour under vacuum (curve 1), must be added : (i) the effect of closure and mode II on the near-threshold (curve 2), (ii) the hydrogen assisted crack propagation behaviour (curve 3) and (iii) the influence of absorption and diffusion of hydrogen at low loading R ratio (curve 4) and at higher R ratio (curve 5).

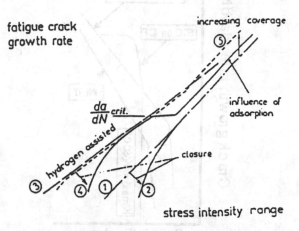

FIGURE 17 Schematization of the different CF crack growth mechanisms [12].

In fact, one must distinguish between long crack growth and short crack growth.
For long crack growth, using the linear elastic fracture mechanics (LEFM), typical characteristics of the corrosion fatigue behaviour are given on Figure 18 as a function of the stress intensity factor ΔK.

300

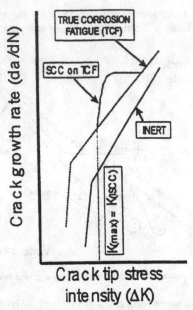

FIGURE 18 Schematic of fatigue and corrosion fatigue crack growth rate as a function of crack tip stress intensity.

The models then propose that the rate of crack growth corresponds either to the sum of pure mechanical fatigue and the rate of stress corrosion cracking or to the fastest available mechanisms among those previously described [13].

However, it has been shown that short cracks propagate at stress intensity factors well below the long crack threshold, as shown on Figure 19 for carbon steel in seawater.

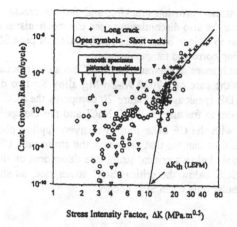

FIGURE 19 Comparison of long and short CF crack growth rates for carbon steels (σ_g = 500 Mpa) in artificial water, 0.2 Hz, R = 0.1 [14].

The application of microstructural fracture mechanisms (MFM) has been successfully used to predict the growth of short cracks during fatigue in air [15]. These models have been adapted to characterize and predict the uniaxial and multiaxial corrosion fatigue loading [16].

The two following equations provide the basis of the Brown–Hobson model [17] for which the fatigue crack growth rate is expressed as a function of an equivalent strain term γ_{eq} for stage I and stage II cracks respectively.

$$\frac{da}{dN} = B_I \, \gamma_{eq}^{\beta_I} \left(d_i - a \right) \qquad \text{(stage I), shear crack growth}$$

$$\frac{da}{dN} = B_{II} \, \gamma_{eq}^{\beta_{II}} \left(a - D \right) \qquad \text{(stage II), tensile crack growth}$$

where a is the crack length, d a microstructural parameter as grain size, i the number of grains through which the crack has traversed, B_I, B_{II} and β_I, β_{II} are constants depending upon the material/environment system and D is the long crack threshold.

INTERACTIONS WITH STRESS CORROSION CRACKING (SCC)

It is well known that SCC can interfer with corrosion fatigue during crack propagation (Figure 18). But this interaction also occurs for crack intiation and crack propagation near the threshold as illustrated in Figure 3 for a fcc Fe-Mn-Cr alloy cyclically deformed at low strain rate in a Cl⁻ solution and on Fig. 8 for Al-Li alloys in 30 g/l NaCl. On the fracture surface of the Fe-Cr-Mn alloy, a crack initiates from a pit (which is favoured by fatigue processes) by SCC with a cleavage-like process which has been described in details elsewhere [18]. Then fatigue and corrosion fatigue take place with classical striations. Finally, it must be noted that the

302

distinction between stress corrosion cracking and fatigue cracking is not always easy to be made because SCC can be also discontinuous (see the mechanisms in Ref. 18) with crack arrest markings very similar to fatigue ones. Thus combination of pure SCC and fatigue mechanisms is often possible during corrosion fatigue damage.

CF is often much more dangerous than SCC because it occurs whatever the strain rate. This is illustrated in the case of a 7020 Al-Zn-Mg alloy for two different heat treatments, T4 (underaged) and TGDR (peak aged). Figure 20 compares the SCC behaviour through the ratio between the elongation to fracture in 30 g/l NaCl and the elongation in air for tensile tests at imposed strain rate with the CF behaviour at a given applied plastic strain $\Delta\varepsilon p/2 = 10^{-3}$ for different strain rates. One can see that, whatever the strain rate, CF always occurs unlike SCC which takes places only below a critical strain rate dependent on the heat treatment. Moreover, CF is affected by SCC below the critical SCC strain rate, as shown by the fact that crack initiation is intergranular only when SCC occurs.

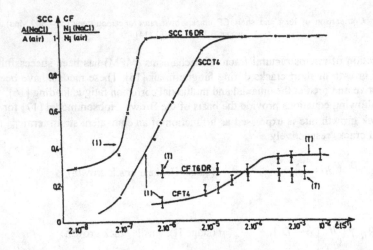

FIGURE 20 Comparison between SCC and CF (at $\Delta\varepsilon p/2 = 10^{-3}$) behaviour of a 7020 T4 and T6DR alloy at free potential in a 30 g/l NaCl solution. Crack initiation can be transgranular (T)

on intergranular (I) according to the strain rate $\dot{\varepsilon}$.

Nevertheless, cyclic loading is not always deleterious in regards to SCC. The example of cracking in alloy 600 for pressurised water reactor (PWR) illustrates such behaviour. Intergranular SCC is known to occur in the alloy 600/PWR system [19] when tests are performed on CT specimens for constant loading conditions with a stress intensity factor $K_I = 30$ MPa√m. The effect of cycling loading at a load ration R = 0.7 with a triangular wave form at $K_{max} = 30$ MPa√m is shown on Figure 21 according to the imposed frequency.

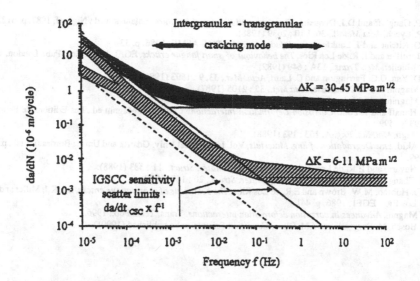

FIGURE 21 Crack growth rates versus frequency during CF and SCC on CT alloy 600 specimens in PWR environment [19].

An increase in frequency induces a decrease of the crack growth rate and a transition from intergranular to transgranular cracking mode, untill air cyclic behaviour becomes predominant.

This last example clearly shows that cycling can completely change the crack propagation mode when stress corrosion crack propagation is intergranular.

CONCLUDING REMARKS

The analysis of CF micromechanisms related to anodic dissolution and hydrogen effects is under progress but needs to be more quantitative through the localized corrosion deformation interactions. The trends for future researches are mainly related to :

(1) the modelling of crack tip chemistry,
(2) the quantitative analysis of corrosion-deformation interactions at CF crack tip (scale of 1 ∙m) according to the electrochemical conditions,
(3) a comparison between CF and SCC based on a detailed analysis of micromechanisms near the fatigue threshold,
(4) developments of numerical simulations at mesoscopic scales.

These researches are needed to propose predictive laws for CF damage based on physico-chemical controlling factors.

REFERENCES
1. C. Laird and D.J. Duquette, *Corrosion Fatigue*, NACE ed., New York, 1972, p. 82.
2. C. Patel, *Corrosion Science 21* : 145 (1977).
3. R.P. Wei and G.W. Simmons, *Int. J. Fract.*, 17 : 235 (1981)

304

4. R.P. Gangloff and D.J. Duquette, *Chemistry and Physics of Fracture*, Latanision ed., Nighoff, 1987, p. 612.
5. S.P. Lynch, *Acta Metall.*, 36, L10 : 2639 (1988).
6. R.O. Ritchie and J. Lankford, *Small fatigue cracks*, TMS-ASME, 1986, p. 33.
7. K.J. Miller and E.R. de Los Rios, *The behaviour of short fatigue cracks*, EGF1, Mech. Eng. Pub., London, 1986.
8. M. Mueller, *Met. Trans.*, 13A : 649 (1982).
9. B.D. Yan, G.C. Farrington and C. Laird, *Acta. Met.*, 33, 9 : 1593 (1985).
10. T. Magnin and L. Coudreuse, *Acta Met.*, 35 : 2105 (1987).
11. T. Magnin and J.M. Lardon, *Mater. Sci. Eng. A*, 104 : 21 (1988).
12. G. Henaff and J. Petit, *Corrosion Deformation Interactions CDI'92*, T. Magnin ed., Les Editions de Physique, 1993, p. 599.
13. L. Hagn, *Mat. Sci. Eng. A*, 103 : 193 (1988).
14. R. Akid, *Env. Degradation of Eng. Materials*, Vol. I, Technical Univ. Gdansk and Univ. : Bordeaux 1 ed., p. 112, 1999.
15. A. Navarro and E.R. de Los Rios, *Fat. Fract. Eng. Mater. Struct.*, 11 : 383 (1988).
16. W. Zhang and R. Akid, *Fat. Fract. Eng. Mater. Struct.*, 20 : 167 (1997).
17. P.D. Hobson, M.W. Brown and E.R. de Los Rios, *The behaviour of short fatigue cracks*, Ed. K.J. Miller and E.R. de Los Rios, EGF1, 1986, p. 441.
18. T. Magnin, *Advances in corrosion deformation interactions*, Trans. Tech. Pub., 1996.
19. C. Bosch, M. Puiggalli and J.M. Olive, *Corrosion Science*, to be published (1999).

STRESS CORROSION OF GLASS

R. GY
Saint-Gobain Recherche
Aubervilliers, France

1. Introduction

It has been known for a long time that silicate glasses are sensitive to static fatigue : the duration of the application of the loading has an effect on the strength of glass. The " long term " strength is different from the " short term " strength. In the present paper a short review of some practical manifestations of this phenomenon is done. Its interpretation is then overviewed: the role of ambient water is well established. It is a stress-activated chemical reaction of water from the environment with the silicon-oxygen bonds. It takes place very efficiently in the highly strained material at the tip of a surface crack. This enables its sub-critical growth to be explained. In the last part, questions and issues are presented, which, to the author's opinion, still need to be investigated.

2. Stress-assisted corrosion of glass: examples from the field

The practical manifestations of the stress-assisted corrosion of glass are almost ubiquitous, as is water, and some examples from different application areas are given.

2.1. ARCHITECTURAL GLASS

Most of the times, design recommendations for architectural glazing come from the accumulation of experience gained over many years by glaziers and designers in the field and are not very much based on scientific considerations. They do, however, distinguish between the transient loads (wind bursts on vertical glazing...) and the more permanent ones (glass' own weight, accumulation of snow on inclinated glazing,) For instance, Saint-Gobain's recommended design stress for standard silico-soda-lime glass, as-float, is 20 MPa for a vertical window, whereas, the same glass in a safely-designed roof window would not undergo a tensile stress larger than 10 MPa. In an aquarium, the same glass would not be allowed to undergo more than 6 MPa.

305

E. Bouchaud et al. (eds.), Physical Aspects of Fracture, 305–320.
© 2001 *Kluwer Academic Publishers. Printed in the Netherlands.*

2.2. CAR GLASS

May be one of the most widely known manifestation of the effect of stress corrosion of glass is the one which is experienced by a car driver after a small gravel has impacted the laminated windshield. Generally, there are very small cracks close to the impact point. But one of these cracks may then grow, in days or months after the impact, slowly and continuously under the driver's eye, up to the point where it has become so long that the windscreen has to be replaced. This is a macroscopic demonstration of the sub-critical crack growth process.

2.3. FLAT GLASS CUTTING

Flat glass cutting is done by breaking after scribing with a diamond or a sharp tungsten carbide wheel. During scribing, a blind crack is created on the surface. The deposition of a thin oil film during the scribing process is effective in lowering the required load in the subsequent breaking process, and thus in increasing the quality of the cut edge. The role of stress corrosion in the glass cutting process is further explained below, after Swain [1].

2.4. BOTTLES

A glass bottle is subjected to a permanent loading, namely internal pressure, when it contains a carbonated drink: soda, bier, and champagne. The sudden explosion for no apparent reason, of champagne bottles stored in the cellar, has indeed been reported and attributed to the sub-critical growth of surface defects in the glass items of lesser quality.

2.5. EVACUATED GLASSES

In every conventional TV set, the cathode ray tube is an evacuated glass body that is subjected to the permanent atmospheric pressure on the outside. To preclude any risk of implosion, the design of glass thickness has to take the stress corrosion into account. This issue will be made much more serious in the case of thin flat evacuated glass devices currently under development, like the ones that will be included in the new flat panel displays like the Field Emission Displays, or in flats lamps, or in high performance insulating glazing (vacuum glazing) [2].

2.6. INSULATION FIBERS

Glass wool for insulation application is a very low-density product that has to be highly compressed before transportation in order to reduce the delivery cost. The compression is removed just before use in the buildings. The original thickness is not fully recovered yet, and a part of the irreversibility is attributed to the stress corrosion and delayed fracture of the highly bent fibers in the compressed product.

2.7. REINFORCEMENT FIBERS

Glass fibers for reinforcement of plastics have to be light, stiff, and strong. Their tensile strength, in the pristine state, can be very high, up to more than 3000 MPa. It has been shown [3] however that it is sensitive to the nature of the atmospheric environment below the bushing. This is attributed to stress corrosion of hot glass during the drawing process. The tensile stress experienced by the hot glass fiber during the drawing can be as high as 300 MPa. Though this stress is not applied for a long time, stress corrosion takes place because glass is hot between the bushing and the sizing roller.

2.8. OPTICAL FIBERS

The development of silica-based optical wave-guides has motivated a lot research on coatings impervious to water in order to try to prevent the stress corrosion from taking place under the fiber's service conditions of stress and moisture and to improve their long term reliability. Metallic, carbon, SiON coatings have shown the best efficiency [4]. Polymer coatings which normally are applied to protect the optical fiber against mechanical damage do not prevent from fatigue as they are permeable to water [5].

2.9. SPECIAL CASE: TEMPERED GLASS

In a thermally tempered glass plate, the residual stress field is compressive close to the surfaces and to the edges and tensile in the core. Delayed spontaneous fracture of tempered architectural glazing has been reported, but it is not an effect of the stress-assisted corrosion of glass. It is attributed to an internal defect: equatorial cracks surrounding a nickel sulfide inclusion within the glass become unstable as the metastable nickel sulfide undergoes a slow transformation at room temperature into a lower density phase. The increase of volume of the inclusion drives the slow growth of the cracks up to the point where they become unstable in the residual tensile stress field in the core of the tempered glass and this triggers the catastrophic fracture of the tempered plate [6]. A modification of the manufacturing process of tempered glass ("Heat Soaking") allows this phase transformation to not take place during service and thus efficiently guarantees the product against such spontaneous failures. Moreover, since a properly tempered glass has a compressive residual stress everywhere in the surfaces, stress-assisted corrosion by ambient water cannot take place, provided that the glazing has been designed in such a way that the service tensile stress does not exceed the residual compression. Tempered glazing designed in this way should be considered as immunized against aging due to static fatigue.

308

3. Sub-critical crack growth

3.1 CRACK SPEED VERSUS STRESS INTENSITY FACTOR (SIF).

Static fatigue of glass is interpreted as a stress-activated sub-critical crack growth. It is well known that silicate glasses are brittle. Their toughness (K_{IC} , critical value of the SIF) is very low, generally ranging between 0.6 MPa.m$^{1/2}$ and 1 MPa.m$^{1/2}$. But the actual situation is even worse: for glass in ambient conditions, the tip of a surface crack undergoing a SIF half the critical value, actually is slowly running. Figure 1 shows the typical general features of the speed v of a running crack as a function of the strain energy release rate. Alternatively, the speed v is sometimes plotted versus the SIF. Around and above K_{IC} the crack speed does not depend on the environment. It increases sharply with SIF but levels off at some characteristic speed (around 1500 m/s for soda-lime glass). In a narrow range around K_{IC} the crack speed ranges between 10^{-3} m/s and 1 m/s (region III). The slope of the curve is very steep. In absence of any water, this curve would extrapolate linearly to lower crack speed. But actually, for lower speeds, the curve strongly depends on the environment. In Region I, starting from very a low speed (down to 10^{-10} m/s), the crack speed v generally has an empirical power law dependence on the SIF, $v = A(K_I)^n$, n, the slope of the curve, being called the fatigue parameter. The smaller is n, the larger is the susceptibility to fatigue. The range of variation of n is very broad, from 12 to 50, depending on many parameters as explained below and in section 4. Region I and region III are connected together by region II in which the speed of the crack does not depend very much on the SIF, but does depend on the amount of water in the environment. Below some low value of SIF, which is called the " fatigue limit " or " threshold SIF ", no crack growth can be measured. This is sometimes also called " region 0 ".

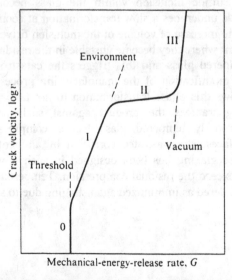

Figure 1. Velocity of a running crack versus Strain energy release rate: general features. (from Lawn [24])

3.2. INFLUENT PARAMETERS

3.2.1. *Effect of amount of water*
The effect of an increasing amount of water in the environment (in the atmosphere or in another inert liquid medium) is a shift of region I towards lower SIF, without changing the slope n, and a shift of region II towards higher speeds (Figure 2). Actually, it can be shown [7], that the absolute amount of water does not actually matter, but rather the ratio of actual partial pressure of water to that at saturation (humidity ratio). More generally speaking, the chemical activity of water is the control parameter. This is important because it means that a fluid that cannot dissolve water, but only a very small quantity, should not be considered as inert from the stress corrosion point of view, especially if it is not too far from saturation [8].

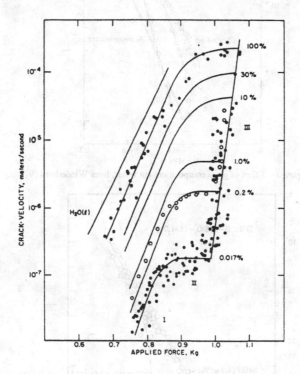

Figure 2. Effect of an increasing amount of water in the environment. (from Wiederhorn [7])

3.2.2. *Effect of temperature*
The main effect of an increase of temperature in liquid water, below 100°C is also a shift of the curve towards lower SIF [9].

3.2.3. *Effect of glass composition*
The effect of glass composition is complex, see Figure 3. The location of region I on the SIF scale and the fatigue parameter (slope) can be different depending on the glass

310

composition. It is worth mentioning that there generally is a correlation of the fatigue susceptibility with the alkali content of the glass.

3.2.4. *Effect of pH*

For a given SIF, the crack speed in aqueous media increases with the pH, see for example [10]. The slope n in region I decreases as pH increases in silica glass. This is illustrated on Figure 4. There is also a very strong effect of the pH on the fatigue limit.

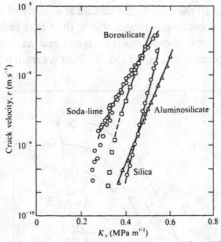

Figure 3. Effect of glass composition. (in water, from Wiederhorn [9])

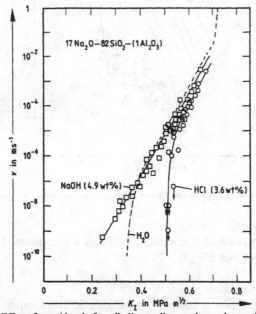

Figure 4. Effect of an acid and of an alkaline medium on the crack growth. (from Gehrke [10])

3.3. INTERPRETATION

The first theory to account for this phenomenology is the Charles & Hillig model [11], in which the chemical reaction rate of water with silica is a thermally activated process. The activation energy depends on the local stress and on the radius of curvature at the tip. The activation energy is actually reduced by a term of the order of the product G.a, G being the strain energy release rate and a being an "activation area". In experimental (v-G) diagrams, the fit of region I allows an activation area a of the order of 1 Å^2 to be estimated, which can be considered satisfactorily from the physical point of view, for the chemical reaction of an individual molecule of water with a strained Si-O bond at the tip of a crack. This theory also accounts for the existence of a limit to fatigue. The limit is explained by the rounding or blunting of the crack tip at low speed. Wiederhorn showed that region II can be explained by the fact that at some higher speed level, the kinetics of the chemical reaction is not controlled anymore by the activation of the chemical process, but by the supply rate of one of the reactant: water. It takes time for a water molecule to travel from the environment to the tip of the crack. As the tip is moving faster and faster, a shortage in the supply of water at the tip takes place.

A more detailed model of the chemical interaction between the adsorbed H_2O and the Si-O bond was given by Michalske and Freiman [12], explaining why water is so efficient in the stress-activated corrosion of glass whereas the chemical durability of silica glass in pure water is so good. They showed that any molecule able to donate a proton on one side, and with a lone pair of electron on the other side, is able to react with the strained Si-O bond at the tip of a crack, provided that it is small enough to be transported to the tip [13]. The fact that constrained Si-O bonds have their chemical reactivity with water enhanced according to the intensity of the strain was directly evidenced [14], in hydrolysis experiments for different siloxane ring structures (cyclosiloxanes).

The interaction of Si-O bond with the hydroxide ion OH⁻ instead of molecular water explains the greater susceptibility to fatigue in basic solutions [15].

3.4. APPLICATIONS

Actually models do not predict a power law for the crack speed as a function of the SIF, in region I: they generally give an exponential law. However, it is difficult in practice to clearly distinguish a power law with a large exponent from an exponential law. The practical interest of the empirical fit to a power law is that it allows an easy computation of engineering lifetime predictions for static or dynamic stress conditions, involving only the fatigue parameter and an idea of the initial flaw size, or "inert" strength. Actually, the applied stress as function of time to failure has a slope equal to −1/n, in logarithmic scales.

The " dynamic fatigue " curves are obtained by plotting the strength of test-samples as a function of the stressing rate on logarithmic scales. It can be shown that, if the test-samples are free from residual stresses, the slope in this diagram is $1/(n+1)$, and the fatigue parameter n is commonly evaluated from strength measurement at different stressing rates.

Numerical procedures according for example to the one sketched in the flow chart on Figure 5 allow the computation of time to failure to be done for any v-K_1 curve, and any loading history.

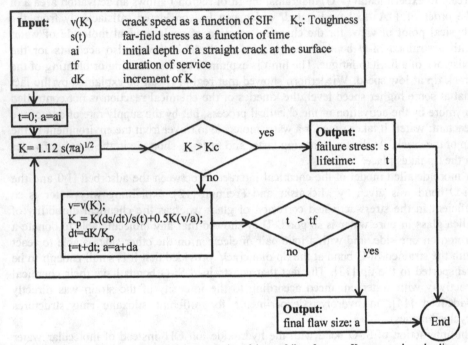

Figure 5. Numerical procedure for the computation of time to failure for any v-K_1 curve, and any loading.

4. Questions & Issues for further investigations

4.1. EXPERIMENTAL METHODS

Some difficulties with the experimental methods are worth mentioning:

In the static fatigue test, the glass under investigation is loaded, and the time to failure is monitored as a function of the steady state load. This can be very close to the real engineering situation, but can be very time consuming too.

The "dynamic fatigue" is a quicker and convenient way of evaluating the susceptibility to fatigue. The result it gives however is very much dependant on the nature of the initial flaws on the test sample. This raises the following interesting question: do all the different kinds of flaws age in the same way? The answer is no, and this will be further discussed below. In other methods, the growth of a large application of acoustic waves that result in a slight periodic modulation of the orientation of the plane of the crack) as a function of a known applied SIF. Different geometrical configuration can be used: the Double Torsion, the Double Cleavage Drilled Compression etc... This might seem a more direct way of obtaining the K_I-v diagram, however, from the engineering point of view, the interesting flaws in real glasses are not these large cracks, and moreover, a recent investigation done in Saint-Gobain Recherche has pointed out the strong influence of the quality of annealing of the glass on the location of the curve on the SIF scale (Figure 6). The low residual stress field (below 1 MPa) within the plate indeed has a non-negligible contribution to the SIF for such a large crack and this may result in an increase of the measured crack speed by a factor of ten.

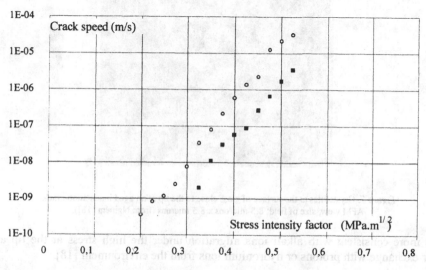

Figure 6. Influence of the quality of annealing on the v-K_I curve obtained with DT test-samples. (Circles: as-float; residual core stress: 0.75 MPa. Squares: special annealing, residual core stress 0.25 MPa).

4.2. THE THRESHOLD SIF

At the threshold SIF, the crack does not grow anymore, or its growth is so slow that it is hardly measurable. But, the existence of an intrinsic threshold SIF can be questioned. It appears to be very much dependent on the environment, on the pH, for a liquid aqueous medium and on glass composition. It is more easily evidenced, with alkali containing glasses and in neutral or acidic environment. In alkali containing glasses, there is also a hysteresis effect: a crack, which is first aged at the threshold SIF, will not resume its propagation immediately on reloading. A thorough investigation of this phenomenon [16] strongly supports the hypothesis that alkali are leached out of the glass, and that this change in the chemical composition at the tip of the crack is responsible for the fatigue limit rather than a geometrical change

314

(blunting). Though, direct evidence of crack tip blunting in pure silica glass was given [17] by TEM observation of the tip of a crack in a very thin film of silica glass aged in water. Now, in alkali containing glass, crack blunting is not believed to be responsible for the fatigue limit. AFM observations of aged indentation cracks do not bring any evidence of blunting. Sodium containing crystallites were actually found on the surface of glass close to the tip of the indentation crack.

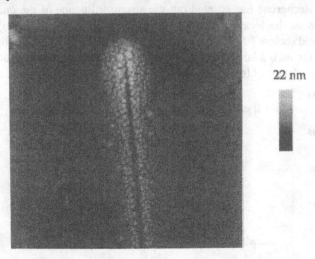

22 nm

Figure 7. Crystallites at the surface of glass close to the tip of an aged indentation crack.
AFM view; size of field: 8.5 microns x 8.5 microns (from Nghiêm [18]).

It is more consistent with alkali ions migration under the high stress at the tip and their exchange with protons or hydronium ions from the environment [18].

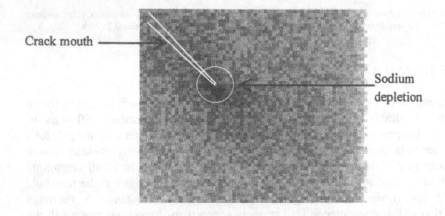

Crack mouth

Sodium
depletion

Figure 8. EDX cartography of sodium close to the tip of an aged indentation crack.
Top view, size of field: 20 microns x 20 microns (from Nghiêm [18]).

The hysteresis effect, (and by the way the arrest lines that can sometimes be seen on multi steps fracture surfaces) is convincingly explained by renucleation of the aged crack in a plane different from the original one, as if the path of the crack has to turn around the area just in front of the former crack tip. A direct evidence of the non-coplanar repropagation of an indentation crack has been obtained with an AFM observation [19].

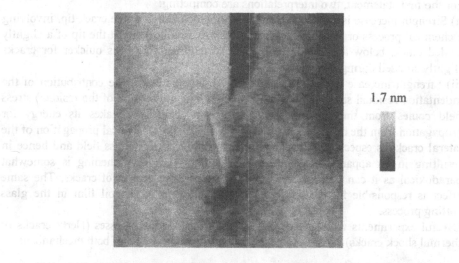

1.7 nm

Figure 9. Direct AFM evidence of the non-coplanar repropagation of an aged indentation crack
Size of field: 2 microns x 2microns. (from Hénaux [19]).

4.3. CRACK HEALING. CAN AGING MAKE A BETTER GLASS?

In glass, a crack is considered " healed " when it is closed, more or less aged, and when its reopening and repropagation requires an increase of the applied load and thus the supply of some strain energy to the cracked body. This is reported for silicate glasses with large cracks, in double torsion loading configuration [20]. This experiment shows that crack closure takes place before complete unloading of the crack, actually below the threshold SIF. A crack closure speed of 10^{-7} m/s for a loading corresponding to a strain energy release rate G = 0.17 J/m^2 is reported for soda-lime silicate glass in air. The repropagation energy release rate (i.e. the quality of healing) was found to increase with the aging time in air, and with increasing temperature. The essential role of water in the healing process is demonstrated by the fact that no healing can be observed for a crack that was first propagated in an organic inert liquid. The idea that healing of cracks can take place in real glass products, and that aging can make a better glass is rather common. To make it short, the observations can be summarized as follow:

- An increase of the strength of glass after wet aging is reported only for items with abrasion or sharp indentation defects. This increase is more effective in high alkali glass [21].
- Glass fibers never take advantage from aging, wet or dry. Instead a decrease of strength is always reported after any aging. The case of fibers is further discussed below in a specific section.

For the first statement, two interpretations are competing:

(i) Strength increase is due to rounding, blunting or healing at the crack tip, involving a chemical process or alkali ion exchange with hydronium ions at the tip of a slightly loaded crack, below the fatigue limit). The strengthening rate is quicker for cracks slightly stressed during aging [21].

(ii) Strength increase is due to the progressive attenuation of the contribution of the indentation residual stress field to the total SIF. The attenuation of the residual stress field comes from the sub-critical crack growth itself as it takes its energy for propagation from the residual stress field [22], [23]. The sub-critical propagation of the lateral cracks is especially efficient in releasing the residual stress field and hence in resulting in an apparent strengthening by aging. This strengthening is somewhat paradoxical as it can be considered as produced by the growth of cracks. The same effect is responsible for the above-mentioned efficiency of an oil film in the glass cutting process.

Careful experiments with surface cracks free from residual stresses (Hertz cracks or thermal shock cracks) should enable to sort out the relative part of both mechanisms.

4.4. DIFFERENT FATIGUE BEHAVIOR FOR DIFFERENT KINDS OF SURFACE FLAWS

All the above interpretations assume that there is a preexisting crack in the glass surface and that everything is taking place at the tip of this crack. But the "natural" flaws do not necessarily fall into that ideal category. In many cases, the "natural" flaws on the glass surface come from concentrated mechanical contacts with a sharp body. In this case, there are always some irreversible and inhomogeneous deformations of the glass below the contacting area, and the corresponding residual stresses in the close vicinity of the deformed zone, but, it is known that well-developed indentation cracking does not take place below some threshold contact load [24]. Such a situation is further referred to as a sub-threshold flaw.

For post-threshold cracks, the role of residual stresses has already been emphasized: they are responsible of some strengthening after aging. They are also responsible for a larger susceptibility to fatigue [23] and for reduced lifetime, compared to cracks free from residual stresses. This is illustrated on Figure 10.

For sub-threshold flaws, the susceptibility to fatigue is also larger than that of well-developed cracks free from residual stresses. The strength of glass with sub-threshold flaws is higher than in the post-threshold case, but also much more scattered. During steady-state loading, crack "pop-in" can take place spontaneously [25], and reduce the strength significantly. This is very important, from the engineering point of view, since the lifetime of glass products with initial sub-threshold flaws, predicted in the conventional way is likely to be too optimistic if it does not take into account the possibility of delayed crack pop-in.

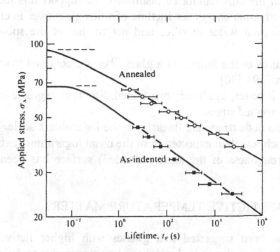

Figure 10. Increased susceptibility to static fatigue due to indentation residual stresses. Vickers-indented soda-lime glass (load: 5 N), tested in water (from Lawn [24]).

4.5. THE CASE OF HIGH STRENGTH GLASS FIBERS.

The fatigue behavior of high strength thin glass fibers is far from being clearly understood. The strength of "pristine" glass fibers is very high, and hardly scattered. It exhibits a lot of the different features of fatigue of glasses with surface cracks: spontaneous delayed fracture under steady-state loading, strength increase at higher stressing rates or after application of impervious coatings, as already mentioned, and in liquid nitrogen. It is hard however to assign the responsibility of the tensile rupture of such a thin fiber to a surface crack. For a measured room temperature tensile strength of 3000 MPa, the size of such a crack at the moment of the rupture would be of the order of 200Å. In liquid nitrogen, the strength is roughly twice larger, indicating that the original surface crack depth would be only 50 Å. Can this still be called a crack? Moreover, the fact that the fracture originates from the surface, in this case has not been directly evidenced: high speed images recording of the fracture process shows that the fiber is fully pulverized by the release of the stored elastic energy, making hopeless any attempt to obtain an insight into this question from an examination of the fracture surfaces. On the other hand, brittle fracture of glass fibers at the same tensile stress level, but at high temperature, above the glass transition, is known. At that temperature, stress relaxation prevents the elastic energy from accumulating within the test sample and enables the analysis of fracture surfaces. The fracture clearly originates from within the bulk [26]. Since molecular water apparently can enter very quickly within highly strained silica glass [27], the possibility that in the case of high strength glass fibers, stress corrosion in ambient conditions does not necessarily take place at the surface, but within the bulk cannot be dismissed. To support this idea, Guillemet pointed out that the activation energy for lifetime of silica glass fiber is close to that of the diffusivity of molecular water in silica and not to that of the sub-critical crack growth [28].

Other distinctive features of the fatigue of a glass fiber are obtained from silica light-guides fibers literature [29], [30]:

- Aging never makes it better, as already mentioned. Strength degradation takes place during aging with no applied stress.
- The fatigue curves do not extrapolate linearly on the logarithmic scale. The lifetimes at low stresses are much less than expected from the usual logarithmic extrapolation.

An increase in the roughness of the aged (corroded) surface has been proposed to explain these features.

4.6. DOES THE GLASS FICTIVE TEMPERATURE MATTER?

More recently, it has been suggested that glasses with higher fictive temperature (cooled at a higher rate from the melted state) are less susceptible to fatigue [31]. But, since this effect was shown on samples with abrasion flaws, the effect of fictive temperature might be indirect: lowering the fictive temperature is known to increase the brittleness of the glass [32]. This is associated with an increase in the density, and also likely an increase of the hardness. Correspondingly the intensity of the residual stress field associated with indentations flaws, which is already known to strongly affect the stress corrosion behavior is expected to be increased for glasses cooled at very low rate or carefully annealed. However, the situation can be even more complicated because the presence of water also has a role in the rate of the structural relaxation [31]. This also raises the question of a possible effect of minute amounts of internal water within the bulk glass on the stress-corrosion behavior.

4.7. MISCELLANEOUS QUESTIONS

- How is molecular water transported to the tip of a crack? How much molecular water is adsorbed on the fracture surface? Is there a liquid meniscus at the tip of a crack in air? A positive answer to the last question along with the already known pH effect might give a better understanding of the role of the glass chemical composition (especially the role of the easily leached alkali ions) on its fatigue behavior in ambient conditions. Such a liquid meniscus, in equilibrium with adsorbed water, has been evidenced [33] between two silica rods in close contact in a very dry environment.
- In a previous section, the contribution of the indentation residual stress field to the fatigue behavior was emphasized, but another question is: has environmental water a role on the indentation damaging process itself? The effect of water on the hardness is

known [34]. The threshold load for radial crack pop-in is very much lowered in water at 80°C, compared to ambient conditions [35]. Repeated sub-threshold indentation loading in the ambient condition is also known to trigger the crack pop-in [36]. For soda-lime silica glass, indentation loads sustained before cracking, are much larger in liquid nitrogen than in ambient conditions [37].
- Is there a special effect of cyclic loads? For SIF close to the threshold level, an enhancement of the fatigue effect was found, by cycling the loading [38].

5. Conclusions

The importance of water for crack propagation in glasses has been shown clearly and theoretical models have been proposed which account for stress corrosion in a semi-quantitative matter. Consequences of this effect have been listed both for applications and for more fundamental questions, which to our opinion would still deserve further analysis and careful investigations. Among them, the role of water on mechanical properties other than crack growth, like room temperature non-elastic deformation, viscosity, relaxation, diffusion/ion exchange coupled to stress, is of particular interest. The characterization of these effects requires experimental techniques, which have to resolve microscopic scales.

Acknowledgement: The critical reading of the manuscript by S. Roux and suggestions from C. Guillemet are acknowledged.

References:

1. Swain, M.V., Metras, J.C. (1982) The breaking of scored glass. *Glass Technology* Vol. 23, N°2, 120-124.
2. Collins, R.E. and Fisher-Cripps, A.C. (1991) Design of support pillar arrays in flat evacuated windows, *Aust. J. Phys.* 44, 545-63.
3. Burgman, J. A. and Hunia, E.M. (1970) The effect of fibre diameter, environmental moisture, and cooling time during fibre formation on the strength of E glass fibres, *Glass technology*, Vol. 11 N°6, 1970, 147-152.
4. Kurkjian, C.R. et al. (1993) Strength, degradation and coating of silica lightguides, *J. Am. Ceram. Soc.* 76 [5], 1106-12.
5. Wei, T., Skutnik, B.J. (1988) Effect of coating on fatigue behavior of optical fiber, *J. Non-Cryst. Sol.* 102, 100-105.
6. Swain, M. V. (1981) Nickel Sulfide inclusions in glass : an example of microcracking induced by a volumetric expanding phase change, *J. Mat. Science* 16, 151-158.
7. Wiederhorn, S.M. (1967) Influence of water vapor on crack propagation in soda-lime glass, *J. Am. Ceram. Soc.* 50, N°8, 407-414.
8. Freiman, S. W. (1985) Environmentally enhanced crack growth in glasses, in *Strength of Inorganic Glass*, C. R. Kurkjian ed., Plenum Press, New York.
9. Wiederhorn, S. M. and Bolz, L. H. (1970) Stress corrosion and static fatigue of glass, *J. Am. Ceram. Soc.* 53, N°10, 543-548.
10. Gehrke, E., Ullner, C. and Hähnert, M. (1990) Effect of corrosive media on crack growth of model glasses and commercial silicate glasses, *Glastech. Ber.* 63 Nr.9
11. Charles, R. J. and Hillig, W. B. (1962) The kinetics of glass failure by stress corrosion, *Symp. on Mechanical strength of glass and ways of improving it.* USCV, Charleroi, Belgium.
12. Michalske, T.A. and Freiman, S.W. (1983) A molecular mechanism for stress corrosion in vitreous silica, *J. Am. Ceram. Soc.* 66, N°4, 284-288.

13. Michalske, T.A. and Bunker, B.C. (1987) Steric effects in stress corrosion fracture of glass, *J. Am. Ceram. Soc.* 70, N°10, 780-84.

14. Michalske, T.A. and Bunker, B.C. (1993) A Chemical kinetics model for glass fracture, *J. Am. Ceram. Soc.* 76, N°10, 2613-18.

15. White, G.S., Freiman, S.W., Wiederhorn, S.M. and Coyle, T.D. (1987) Effects of counterions on crack growth in vitreous silica, *J. Am. Ceram. Soc.* 70, N°12, 891-95.

16. Gehrke, E., Ullner, Ch. and Hähnert, M. (1991) Fatigue limit and crack arrest in alkali-containing silicate glasses, *J. Mat. Science* 26, 5445-5455.

17. Bando, Y., Ito, S. and Tomozawa, M. (1984) Direct observation of crack tip geometry of SiO_2 glass by high-resolution electron microscopy, *J. Am. Ceram. Soc.* 67, N°3, C36-C37.

18. Nghiêm, B. (1998) Fracture du verre et hétérogénéités à l'échelle submicronique, *Thèse*, Université Paris-VI.

19. Hénaux, S. and Creuzet, F. (1997) Kinetics fracture of glass at the nanometer scale, *J. Mat. Science. Lett.* 16, 1008-1011.

20. Stavrinidis, B., Holloway, D.G. (1983) Crack healing in glass, *Phys. and Chem. of Glasses* Vol.24, N°1.

21. Han, W.-T. and Tomozawa, M. (1989) Mechanism of mechanical strength increase of soda-lime glass by aging, *J. Am. Ceram. Soc.* 72, N°10, 1837-43.

22. Marshall, D.B. and Lawn, B.R. (1985) Surface flaws in glass, in *Strength of Inorganic Glass*, C. R. Kurkjian ed., Plenum Press, New York.

23. Lawn, B.R. Marshall, D.B., and Dabbs, T.P. (1985) Fatigue strength of glass: a controlled flaw study, in *Strength of Inorganic Glass*, C. R. Kurkjian ed., Plenum Press, New York.

24. Lawn, B.R. (1993) *Fracture of Brittle Solids - Second Edition*, Cambridge University Press.

25. Ritter, J.E. Ray, C.A. and Jakus K. (1986) Dynamic fatigue of soda-lime glass with sub-threshold flaws, *Collected Papers*, XIV Intl. Congr. On Glass.

26. Gy, R. and Guillemet, C. (1992) Characterization of a mode of rupture of glass at 610°C, in *The Physics of Non-Crystalline Solids* Ed. L.D. Pye, W.C. La Course, H.J. Stevens, Taylor & Francis.

27. Tomozawa, M. and Han, W.-T. (1991) Water entry into silica glass during slow crack growth, *J. Am. Ceram. Soc.* 74, N°10, 2573-76.

28. Guillemet, C. (1995) Fracture et plasticité des verres, *La Revue de Métallurgie-CIT/Science et Génie des Matériaux*. Février 1995.

29. Matthewson, M.J. and Kurkjian, C.R. (1988) Environmental effects of static fatigue of silica optical fiber, *J. Am. Ceram. Soc.* 71, N°3, 177-83.

30. Kurkjian, C.R., Simpkins, P.G. and Inniss, D. (1993) Strength, degradation and coating of silica lightguides, *J. Am. Ceram. Soc.* 76, N°5, 1106-12.

31. Li, H., Agarwal, A. and Tomozawa, M. (1995) Effect of fictive temperature on dynamic fatigue behavior of silica and soda-lime glasses, *J. Am. Ceram. Soc.* 78, N°5, 1393-96.

32. Ito, S. Sehgal, J. and Deutschbein S. (2000) Fictive temperature and fracture behavior of glass, *Glass in the new Millenium* - ICG 2000, Amsterdam.

33. Barthel, E., Lin, X.Y. and Loubet, J.-L. (1996) Adhesion energy measurements in the presence of adsorbed liquid using a rigid surface force apparatus, *J. of Colloid and Interface Sci.* 177, 401-406.

34. Keulen, N.M. (1993) Indentation creep of hydrated soda-lime silicate glass determined by nanoindentation, *J. Am. Ceram. Soc.* 76, N°4, 904-12.

35. Keulen, N.M. and Dissel, M. (1993) Temperature dependence of indentation cracking in soda-lime silicate glass, *Glass Technology* Vol. 34, N°5, 200-205.

36. Banerjee, R., Sarkar, B. K. (1997) Crack initiation by indentation fatigue in lead alkali and soda-lime glass, *J. Am. Ceram. Soc.* 80, N°10, 2722-24.

37. Kurkjian, C.R., Kammlott, G.W. (1995) Indentation behavior of soda-lime silica glass, fused silica and single-crystal quartz at liquid nitrogen temperature, *J. Am. Ceram. Soc.* 78, N°3, 737-44.

38. Dill, S.J., Bennison, S.J. and Dauskardt, R.H. (1997) Sub-critical crack-growth behavior of borosilicate glass under cyclic loads: evidence of a mechanical fatigue effect, *J. Am. Ceram. Soc.* 80, N°3, 773-76.

DYNAMICS OF FRACTURE

EXPERIMENTAL CHALLENGES IN THE INVESTIGATION OF DYNAMIC FRACTURE OF BRITTLE MATERIALS

K. RAVI-CHANDAR
Center for Mechanics of Solids, Structures and Materials
Department of Aerospace Engineering and Engineering Mechanics
The University of Texas at Austin
Austin, TX 78712-1085
e-mail: kravi@mail.utexas.edu

1. Introduction

Experimental investigation of dynamic fracture of brittle materials has a long history, beginning with the early pioneering experiments of Schardin [1] and continuing currently with more sophisticated experimental tools. In examining the dynamic fracture of brittle materials, it is important to perform carefully controlled experiments, where a well-controlled loading is applied and the resulting response is evaluated through proper diagnostic tools. The interpretation of the experiments further requires a theoretical foundation from which the results could be understood. In this paper, we describe the basic experimental schemes that have been used along with a discussion of some of the conclusions and unresolved issues.

2. Theoretical Background

The foundation for a theoretical description of dynamic fracture in brittle materials rests in continuum mechanics; the brief description provided here is intended to capture the essence of the elastodynamic theory of brittle fracture. The complete theory is described very well in the monograph by Freund [2]. Consider a homogeneous, isotropic, linearly elastic medium containing a mathematically sharp crack that moves rectilinearly at an instantaneous speed v. Fracture processes that rupture the material and cause the extension of the crack are assumed to operate at a scale much smaller than that resolved in continuum mechanics. Therefore, the only description of the fracture process that is required in the continuum theory is the total dissipation that occurs within the fracture process zone at the point of the crack tip. In other words, the dynamics of the processes within the process zone itself are not considered to be important in controlling the dynamics of the overall growth of the crack. The resulting theory is therefore scale independent and applied equally well at all length scales. It has been noted that dynamic

323

E. Bouchaud et al. (eds.), Physical Aspects of Fracture, 323–342.

growth of cracks within a molecular dynamics simulation exhibit response similar to the continuum model [3,4].

The Cartesian components of the stress-field near the tip of the crack can be described as follows:

$$\sigma_{\alpha\beta}(r,\theta) = \frac{K_I^{dyn}(t,\upsilon)}{\sqrt{2\pi r}} \Sigma_{\alpha\beta}^I(\theta;\upsilon) \tag{1}$$

where r, θ are polar coordinates centered at the crack tip and $\Sigma_{\alpha\beta}^I(\theta;\upsilon)$ are nondimensional functions indicating the angular variation of the stress field and its dependence on the crack speed, υ. $K_I^{dyn}(t,\upsilon)$ is the *dynamic stress intensity factor* that characterizes the strength of the square-root singular elastodynamic stress field. We note that only loading symmetric about the crack line resulting in the opening mode or mode I crack loading is considered. The stress intensity factor is actually a reinterpretation of the global loading in terms of the near crack tip conditions and can therefore be taken to be the driving force. Enforcing the energy balance equation – thereby extending Griffith theory of fracture to include the influence of material inertia – results in a completion of the dynamic fracture problem and provides the crack tip equation of motion [2]:

$$\frac{1-v^2}{E} A(\upsilon)\left[K_I^{dyn}(t,\upsilon)\right]^2 = \Gamma, \tag{2}$$

where $A(\upsilon) = \dfrac{\upsilon^2 \alpha_d}{(1-v)c_s^2 D}$, $\alpha_d = \sqrt{1-\upsilon^2\Big/c_d^2}$, $\alpha_s = \sqrt{1-\upsilon^2\Big/c_s^2}$,

$D = 4\alpha_d\alpha_s - (1+\alpha_s^2)^2$, and

c_d, c_s are the dilatational and shear wave speeds respectively

E is the modulus of elasticity and v is the Poisson's ratio.

Γ is the fracture energy and is a material property to be determined experimentally; note that Γ may depend on the crack position due to a spatial distribution of heterogeneities and on the speed if the material exhibits strain rate dependence. If Γ is known, Eq. (2) is a unique relationship between the driving force, $K_I^{dyn}(t,\upsilon)$, and the crack speed. This equation also suggests that $K_I^{dyn}(t,\upsilon) \to 0$ as $\upsilon \to c_R$, the Rayleigh wave speed, indicating that this is the limiting crack speed. The crack tip equation of motion in Eq. (2) can be written equivalently in terms of the stress intensity factor as follows:

$$K_I^{dyn}(t,\upsilon) = K_{ID}(\upsilon), \tag{3}$$

where $K_{ID}(\upsilon)$ is the *dynamic fracture resistance* of the material and depends on the crack speed. The motivation for such a formulation is to include the influence of other dissipative processes into the crack tip equation of motion, still within the limitations of linear elastody-namics. The fracture resistance, $K_{ID}(\upsilon)$, might increase rapidly with crack speed due to inherent rate dependence within the fracture process zone and thereby result in a limiting speed that is smaller than the Rayleigh wave speed. The experimental challenge has really been in the characterization of the dynamic fracture toughness and its dependence on the crack speed through careful measurements.

3. Experimental Methods in Dynamic Fracture

Over the last century, a number of special methods have been developed for the generation of dynamic loading in order to explore material behavior over a range of loading rates. A number of diagnostic tools have also been developed to examine the response of the materials to such high strain rate loading conditions. In this section, we provide a brief overview of methods of loading and diagnosis that have been used to explore dynamic fracture examining the sensitivity of each method.

3.1. METHODS OF GENERATING DYNAMIC LOADING

Numerous methods have been used to generate reproducible loading for the purpose of examining dynamic crack growth, ranging from static or quasi-static loading of blunt cracks to impact with projectiles at speeds ranging from 5 m/s to 100 ms/. Furthermore, novel methods such as electromagnetic loading and explosive loading have been developed to generate high intensity, short duration pulse loading on cracks. Since the stress and strain state in the near crack tip region that dominates the fracture process zone is proportional to the dynamic stress intensity factor, the rate of loading or the strain rate near the crack tip must be proportional to the rate of increase of the dynamic stress intensity factor: $\dot{K}_I^{dyn}(t,\upsilon)$. In experiments, this rate has been varied in the range from zero in the quasi-static loading situations to about 10^8 MPa $\sqrt{m}$ s^{-1} in the projectile impact experiments. However, the rate of change of the dynamic stress intensity factor after the crack begins to grow is determined primarily by the crack growth rate itself and therefore by the material, and is typically on the order of 10^8 MPa $\sqrt{m}$ s^{-1}. A short description of each loading method and the appropriate reference is provided below.

3.1.1. *Static Loading of Blunt Cracks*
This method of loading has by far been the most popular of all loading methods owing to its simplicity; specialized equipment is not necessary to generate dynamic crack growth. Almost every imaginable configuration of quasi-static loading geometry has been used by investigators – double cantilever beam [5], compact tension specimen [6], single-edge-notched plate [7-9], large strip geometry [10], center-cracked panel [11], and many others. Typically, pre-cracked specimens are loaded in a standard testing machine to initiate crack growth from blunt crack

tips. The bluntness of the crack tip will introduce transients that impede analysis of the initiation event. Therefore, the method is suitable only when the interest focuses on crack propagation or crack arrest.

3.1.2. *Dropweight Tower*

The dropweight tower is a simple device where a mass (typically in the range of a few tenths of a kilogram to a few tens of kilograms) is raised to some height (< 10 m, but typically 2 to 3 m) and dropped on to the test specimen. The impact speeds achieved are on the order of 5 – 10 m/s resulting in a reasonably modest rate of loading. The loading duration is usually quite large (on the order of 10 ms) and commercial instruments are readily available to implement this loading scheme. Typical loading rates achieved in the dropweight towers is in the range of $\dot{K}_I^{dyn}(t,\upsilon) = 10^4$ MPa $\sqrt{m}$ s^{-1}. Modifications of the standard Charpy test apparatus have also been used to obtain loading rates in this range. It is possible to design many different specimen geometries to be tested in the dropweight tower, but the most common specimen is the edge-notched specimen in a three point bend configuration, although a bending stress field may not be developed in the specimen under the dynamic loading conditions in the duration of interest. Since the loads are applied rapidly, a naturally cracked specimen can be used; therefore the use of dropweight towers for dynamic fracture toughness testing has now become standard practice [12].

3.1.3. *Projectile Impact*

Loading generated by the impact of a projectile onto a specimen can yield loading rates in the range of $\dot{K}_I^{dyn}(t,\upsilon) = 10^5$ MPa $\sqrt{m}$ s^{-1}. In these impact experiments, a projectile (weighing anywhere from a few grams to a few kilograms) is launched from the barrel of a gun at speeds in the range of a 10 m/s to 1 km/s, but typically in the range of 100m/s. The loading duration is governed by the length of the projectile and is typically much shorter than in the dropweight tower – about 10 μs to 200 μs. These devices are not commercially available and have to be fabricated specially for each implementation. Because the projectile velocity can be controlled to within close tolerances, this method of loading is quite repeatable. At the high projectile speeds, $\dot{K}_I^{dyn}(t,\upsilon) = 10^8$ MPa$\sqrt{m}$s^{-1} has been achieved for a very short duration (less than 1 μs) [13]. Some variations of this scheme also appear in the literature. For example, the impact from the projectile occurs in a one-dimensional waveguide first, and this guided wave then imposes the impact load on the specimen. Monitoring the motion of the waveguide results in a precise determination of the impact conditions imposed on the specimen [14]. It should be noted that the by impacting the projectile on many different specimen geometrical configurations different loading rates and constraints can be imposed on the cracks; many such examples can be found in the literature [see for example, 15].

3.1.4. *Explosives*

Lead azide and pentaerythritol tetranitate (PETN) explosives have been used to apply high strain rate loading in the examination of dynamic fracture. There are safety issues involved in the handling and operation of these explosives and are therefore not that commonly used. The loading rates obtained are comparable to that obtained with moderate speed impact of projectiles, ~ $\dot{K}_I^{dyn}(t,\upsilon) = 10^5$ MPa $\sqrt{m}$ s^{-1} [16]; the duration is typically in the range of 10 to 200 μs.

3.1.5. Electromagnetic Loading

This loading scheme is based on electromagnetic interaction between two current carrying conductors. A folded copper strip is inserted into a crack and the space between them is filled with a mylar insulating strip. A current pulse of a predetermined rise time, amplitude and duration is derived from discharging a capacitor bank. The interaction of the current in each leg of the folded conductor with the magnetic field of the other conductor forces the conductors and thus the crack faces apart. If the specimen is large, this loading configuration is equivalent to an infinite plate, with a pressurized semi-infinite crack for the duration of the current pulse. Crack surface pressures of about 15 MPa, loading duration of about 150 μs and a loading rate in the range of $\dot{K}_I^{dyn}(t,\upsilon) = 10^5$ MPa $\sqrt{m}$ s^{-1} have been attained using this method [17]. Since synchronization of the loading scheme with the diagnostic schemes is readily accomplished, this loading method is well suited for studies of crack initiation as well as continued growth. However, specialized equipment is necessary and is not commercially available, thereby limiting the use of the method.

3.2. DIAGNOSTIC METHODS IN DYNAMIC FRACTURE

In addition to the ability to apply a well-characterized, reproducible loading on the cracks, it is necessary to use diagnostic methods for observing the response of the cracks to the applied loading. The response must be characterized in terms of the crack position, speed, υ, and the dynamic stress intensity factor, $K_I^{dyn}(t,\upsilon)$. Therefore, experimental investigations into dynamic fracture must incorporate diagnostic methods not only to determine the crack position and speed, but also a measure of the crack tip stress or deformation field. In this section, we describe some of the common methods used to perform these measurements.

3.2.1. Measurement of Crack Position and Speed

High Speed Photography – Beginning with Schardin's own invention, the multiple-spark camera, the *high-speed camera* has been a very powerful tool in observing crack extension with time. Many other types of high speed imaging systems, based on different operating principles have since been developed. Field [18] has provided a review of the many methods commonly used. Today, there are many high-speed cameras commercially available that provide a spatial resolution of 0.2 mm and a temporal resolution in the range of 1 to 10 μs. Such cameras have enabled the imaging of cracks growing at speeds that are a significant fraction of the Rayleigh surface wave speed.

Potential Drop Method – The *electrical resistance grid technique* in various manifestations has also been used to determine the crack speed; in this method a number of electrical wires are laid across the path of the crack. As the crack propagates, it severs the wires sequentially and provides an electrical signal that can then be used to determine the crack position and speed with time [19-23]. Carlsson *et al.*, [6] demonstrated that if a film is used instead of a grid, a much higher resolution can be obtained. In this implementation, a high-frequency alternating current is passed through the specimen; the impedance of the specimen is determined by the film thickness, crack length and crack opening. Many investigators have used this method to determine the speed of running cracks [10, 24, 25]. According to Fineberg *et al.*, [25] this method provides for an evaluation of the crack position to within 200 μm and the crack speed to within 10 m/s, really not much more accurate than high speed photographic techniques. The main difference from high speed photography is that the sampling rate is much higher – instead of the 100k samples per second obtained in high speed photographic techniques, they are able to obtain 10M samples per second – and thus resulting in a higher temporal resolution of the measurements.

A major source of concern in all of the methods of crack position and crack speed determination lies in the spatial and temporal averaging that arises in the measurement scheme that influences the accuracy of the measurements. While the resolution that can be obtained varies with the particular design of the camera, typically, a temporal resolution of about 1 μs and a spatial resolution of about 200 μm are easily obtained in high-speed cameras. While other devices might be capable of greater spatio-temporal resolution, the high-speed camera has also the added advantage that it provides a clearly identifiable crack tip in addition to the possibility of determining the crack tip stress field parameters through other methods such as photoelasticity or caustics discussed below.

Crack Surface Modulations – Wallner [26] showed that the interaction of the propagating crack with shear waves emanating from the fracture can lead to characteristic undulations on the fracture surface; he suggested that these lines – now called *Wallner lines* – can be used to determine the crack speed quite accurately. A number of investigators have used this idea to determine the crack growth speed [see 21, 27-29]. Recently, however, these markings have been reinterpreted as a crack front wave that possess soliton-like characteristics [30, 31]. This idea of Wallner lines was generalized by Kerkhof [32], who demonstrated that by imposing a small-amplitude high-frequency stress wave – typically generated by an ultrasonic transducer – ripple marks can be created on the fracture surface; these ripple marks can be analyzed *post-mortem* and used to determine the crack speed. The accuracy of the spatial measurement can be very high since one relies on a *post-mortem* examination of the fracture surface modulation at high magnifications. Through suitable choice of the ultrasonic transducer, crack speeds in the range of 0.02 m/s to 2000 m/s could be measured by this technique called *stress wave fractography*; a recent review of this technique can be found in Richter and Kerkhof, [33].

3.2.2. *Measurement of Crack Tip Field Parameters*

In providing an overview of the diagnostic methods used for measuring crack tip stress and/or deformation field parameters, it is useful to group the methods into two categories: point-wise measurement techniques and full-field measurement techniques. Strain gages and the method of caustics fall under the category of pointwise measurements because these methods rely on measurement of the field quantities at a single point where as photoelasticity, interferometry and coherent gradient sensing rely on information gathered over some region near the crack tip. Each one of these techniques is sensitive to one or more specific components of the crack tip field and therefore the reliability in establishing the crack tip condition using these techniques is dependent on the particular technique. One major consideration in implementing any of these methods of measurement is that in the region of measurement, the stress, strain and/or displacement field must be described by the singularity dominated elastic analysis – i.e., a K-dominant field must be established in order to interpret the experimental measurements appropriately in terms of the dynamic stress intensity factor. There has been a significant amount of work on resolving these issues and we will refer to the major contributions in the following.

Strain gages – The crack tip strain field corresponding to the stress field in Eq. (1) is given by:

$$\varepsilon_{\alpha\beta}(r,\theta) = \frac{K_I^{dyn}(t,\upsilon)}{E\sqrt{2\pi r}}\Phi_{\alpha\beta}^I(\theta;\upsilon,v) \tag{4}$$

where $\Phi_{\alpha\beta}^I(\theta;\upsilon,v)$ is a known function of θ, υ and v. Strain gages are placed at some position (r,θ) to measure one or more components of the strain tensor. Practical considerations on- the type, dimensions, and sensitivity of the strain gage are extremely important and are addressed in many standard textbooks. This measurement can then be used in Eq. (4) to determine the dynamic stress intensity factor, $K_I^{dyn}(t,\upsilon)$. This method was used very successfully by Kinra and Bowers [34], Shukla and Dally, [35] and others. Since this method provides a measurement only at one point, the method can provide an estimate of $K_I^{dyn}(t,\upsilon)$ only over a small time interval during which the gage is positioned within the K-dominant zone of the crack tip field (see [36] for a discussion).

Photoelasticity – The development of the stress field near a dynamically growing crack tip stress has been examined using different optical techniques. Most of the polymers used in dynamic fracture investigations – polymethylmethacrylate, Homalite and polycarbonate for example – exhibit stress-induced optical birefringence to different degrees. Thus, placing the specimen between crossed circular polarizers reveals isochromatic fringes that are contours constant in-plane shear stress. Assuming that the fringe pattern formation is governed by the K-dominant stress field near the crack tip, the geometry of the fringe pattern can be expressed as follows:

$$\frac{Nf_\sigma}{h} = (\sigma_1 - \sigma_2) = f\left(r,\theta;K_I^{dyn}(t,\upsilon)\right) \tag{5}$$

where N is the fringe order, f_σ is the fringe sensitivity, h is the specimen thickness, and σ_1, σ_2 are the principal stress components. $f\left(r,\theta;K_I^{dyn}(t,\upsilon)\right)$ is determined from the linear elastic crack tip stress field; in this representation, only the singular term is indicated whereas in actual applications higher order terms in the crack tip stress field are also taken into account in interpreting the experimental fringe pattern. Using a high-speed camera, the time evolution of the fringe patterns can be obtained. The dynamic stress intensity factor $K_I^{dyn}(t,\upsilon)$ can be obtained at each instant in time by using a least-squares matching of the experimentally measured fringe pattern – described by a collection of (N, r, θ) – with simulations based on Eq. (5). Many investigators working over a long period of time have contributed to the development of this method of determining the stress intensity factor. Details may be found in the Handbook of Experimental Mechanics [37]. Although most investigators have assumed that three-dimensional effects are important at distances less than about one half of the plate thickness (using observations of Rosakis and Ravi-Chandar [38]), K-dominance can be observed in photoelastic experiments at much smaller distances from the crack tip.

The Method of Caustics – The method of caustics was discovered accidentally by Schardin [1], when the high-speed camera he used to photograph running cracks was slightly out of focus! The principle of formation is very simple: consider a parallel beam of light incident on the surface of a specimen containing a crack. Far away from the crack tip, these rays pass through in a transparent specimen or reflect from the mirrored surface of an opaque specimen and maintain their parallel propagation. On the other hand, in the region near the crack tip, the specimen exhibits a concave surface due to the Poisson contraction; therefore the light rays deviate from parallelism and under suitable conditions form a dark region (called the shadowspot) on a screen where there are no light rays at all, surrounded by a bright curve (the caustic curve). Using the crack tip strain field in Eq. (4) to calculate the deformation and hence the optical ray deviation, the dimensions of the caustic curve can be related to the dynamic stress intensity factor as:

$$K_I^{dyn}(t,\upsilon) = \frac{2\sqrt{2\pi}}{2chz_0 F(\upsilon)}\left(\frac{D(t)}{3.17}\right)^{3/2} \tag{6}$$

where $D(t)$ is the transverse diameter of the caustic curve, z_0 is the distance between the specimen and the screen, h is the specimen thickness, c is the optical coefficient that depends on whether the method is used in reflection from one surface or in transmission in transparent specimens, and $F(\upsilon)$ is a known velocity dependent function. Thus, from time resolved measurements of the transverse diameter of the caustic curve, the dynamic stress intensity factor can be determined. For a complete discussion of the method see the article by Rosakis [39]. Three dimensionality of the deformation and stress wave effects place significant limitations on the use of the method. Rosakis and Ravi-Chandar [38] determined that if the caustic measurements correspond to distances from the crack tip that are larger than one half of the plate thickness, the analysis using Eq. (6) provides a very good estimate of the stress intensity factor for stationary cracks. Ravi-Chandar and Knauss [40] examined the corresponding dynamic problem and concluded that due to stress wave effects, measurements at such distances may not be in regions of K-dominance, thereby introducing large errors in the estimates of the dynamic stress intensity factor.

Coherent Gradient Sensing – This method is a variation of the method of caustics; when parallel light rays reflect or transmit through a plate containing a crack, the emerging rays deviate from parallelism as discussed above. If these rays are made to pass through two diffraction gratings and then subsequently imaged such that the two different orders of diffraction are superposed on the image plane, an interference is observed due to the phase shift between the two diffracted beams. This fringe pattern is related to the deviation from parallelism of the initial rays and therefore to the deformation of the specimen surface. One major improvement over the method of caustics is that a full field image of the gradients of the surface deformation is obtained. Analysis of the data is very similar to that used in photoelasticity, where the observed interference pattern is compared with an analytically calculated pattern and extracting the dynamic stress intensity factor as the best fitting parameter. For a complete description of the technique, see the article by Rosakis [39].

Interferometry –Classical interferometry [41], and high resolution moire [42] have also been applied to dynamic fracture problems. In classical interferometry, coherent light reflected from the deformed specimen surface is compared with light reflected from a reference flat surface resulting in optical interference fringes that are interpreted in terms of the out-of-plane displacement of the specimen. In high resolution moire interferometry, a fine grating is coated on to the specimen and its deformations are then compared with a reference grid to extract the components of the in-plane displacement components. Both of these methods indeed pose a challenging task in dynamic situations and very few successful attempts have been reported in the literature. The crack tip displacement corresponding to the asymptotic stress field in Eq. (1) is given by:

$$u_i(r,\theta) = \frac{K_I^{dyn}(t,\upsilon)}{E\sqrt{2\pi r}} \Psi_i^I(\theta;\upsilon,v) \tag{7}$$

where nd $\Psi_i^I(\theta;\upsilon,v)$ is a known function of θ, υ and v. The fringe patterns observed in the interferometry experiments are contours of constant displacement component: u_3 in the case of classical interferometry, and u_1 or u_2 in the case of moire interferometry. Therefore fitting the observed fringe patterns to a prediction based on Eq. (7) using a least-square error method, the dynamic stress intensity factor can be determined, in much the same way as for the methods of photoelasticity and coherent gradient sensing. A major difficulty in using these methods arises from the fact that while the measurements themselves are very accurate, the application of the plane elastodynamic fields in the interpretation of the crack tip field parameters introduces significant errors, particularly when one approaches close to the crack tip.

4. Some Observations on the Dynamic Fracture in Nominally Brittle Materials

As a result of the spurt of activity in the latter part of the 20th century, there has been a substantial amount of progress in the understanding of dynamic fracture. The continuum description of dynamic fracture is fully captured by Eq. (2) or Eq. (3). There are a number of predic-

332

tions that arise from either one of these equations. These concern the initiation, arrest, and growth of cracks under dynamic loading. In this section, we will examine these predictions and the corresponding experimental observations and then to outline the areas where open issues still exist.

4.1. CRACK INITIATION AND ARREST

From Eq. (3), taking the limit as $v \to 0$, we can show that crack propagation must begin and end when $K_I^{dyn}(t,0) = K_{ID}(0)$. $K_{ID}(0)$ is the threshold value for crack growth and is typically called the *critical stress intensity factor* or the *dynamic fracture toughness*. However, even in brittle polyesters, it turns out that the stress intensity factor at initiation is always higher than the stress intensity factor at crack arrest [43]. For engineering design purposes, it is necessary to establish this dynamic fracture toughness and compare it to the fracture toughness established from quasistatic tests even if the limiting case of Eq. (3) does not lead to a unique value for initiation and arrest. Typically, one might perform repeated experiments to identify the smallest value of the dynamic stress intensity factor at which crack growth begins; however, experiments on crack initiation are quite sensitive to the preparation of the initial crack. In almost all experiments, an abrupt initiation at a fairly high speed is observed. Furthermore, even in nominally similar cracks – that is cracks that were generated under nearly identical conditions – the onset of crack growth exhibits a significant rate dependence in materials that are quite rate insensitive! This behavior is shown in Figure 1, where

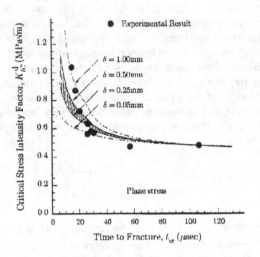

Figure 1. Variation of the critical stress intensity factor at initiation with the time to fracture. Experimental results are from Ref. [43]. The plane stress model is from Ref [44] and the figure is reproduced from Ref [44] with permission. δ is the distance at which the critical stress is evaluated.

the variation of the stress intensity factor at initiation is plotted as a function of the time to initiate crack growth [43]. Clearly, at high loading rates (inversely proportional to the time to fracture) the critical stress intensity factor at initiation increases dramatically; it was also found that the crack speed immediately upon initiation jumps to a much larger value at the higher loading rates. Liu *et al.*, [44] analyzed the inertial effects on crack initiation and suggested that if a failure criterion based on the stress attained at some critical distance from the crack tip is adopted, then the inertia of the material surrounding the crack tip is sufficient to account for the observed rate dependence in the brittle polyester, Homalite 100. Predictions from their model are overlaid on the experimental results in Figure 1. Owen *et al.*, [45] demonstrated a similar rate dependence of crack initiation in a 2024-T3 aluminum alloy.

For a nominally brittle material, assuming that $K_{ID}(v)$ is a well-defined curve, crack arrest must occur at $K_{ID}(0)$ as well. However, the initiation and arrest of a crack under dynamic loading exhibit a hysteretic behavior as illustrated in Figure 2. The crack always exhibits an abrupt jump to a large speed at crack initiation, whereas it decelerates gradually upon unloading; furthermore there are indications that there might not be a jump in speed at arrest. Marder and Gross [46] argue that this hysteretic behavior is due to the absence of steady-state solutions in a lattice model at slow speeds; therefore a

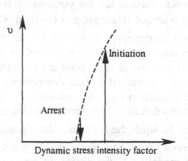

Figure 2. Hysteresis in the crack initiation and arrest behavior

crack at rest continues to be trapped at rest beyond the minimum critical stress intensity factor while a decelerating crack continues to be trapped in the growing state even at stress intensities well below the initiation level. While the reasons for this behavior are not quite completely settled, for many engineering applications, it is important to identify an upper bound for the stress intensity factor as a way of ensuring safe operating conditions. This is accomplished simply by performing a number of experiments to identify the lowest value of the dynamic stress intensity factor at which a running crack comes to rest and label this value the *crack arrest toughness*, K_{Ia}. Standard procedures have been developed for characterization of $K_{ID}(0)$.

4.2. CRACK PROPAGATION

Between initiation and arrest, cracks can grow at very high speeds. Once again, Eq. (2) or Eq. (3) provides an instantaneous relationship between the dynamic stress intensity factor and the crack speed. While Eq. (2) provides this relationship from a more fundamental perspective – i.e., given the fracture energy Γ, it determines the crack speed – Eq. (3) incorporates the idea that the resistance of cracks might itself depend on the crack speed. This idea is easily justified for a material that exhibit plastic flow, but is more difficult to rationalize for nominally brittle materials. Many investigators have performed careful experiments to determine the variation of the dynamic stress intensity factor with the crack speed. For crack speeds that are small relative to the Rayleigh wave speed, the description provided by either Eq. (2) or Eq.(3) appears to work quite well. On the other hand, these investigations have also demonstrated that the relationship between the instantaneous crack tip stress intensity factor and the crack speed may depend on (i) the geometry of the specimen used to characterize it [5, 47], (ii) the crack tip state influenced by the crack parallel stress, called the T-stress effect [48], (iii) the rate of load-

334

ing [49, 50], and (iv) the lack of dominance of the dynamic stress intensity factor [40, 51]. Of course, the errors in measurement of the stress intensity factor and the crack speed discussed in Section 3 add to the problems in interpreting the experimental data, but are not the main cause of the lack of a unique evaluation of the parameters in Eq. (2). While one must still pursue the fundamental reasons for the variation of the dynamic fracture resistance with crack speed, adopting a unique relationship of the form $K_{ID}(v)$ is quite useful for engineering purposes. In addition, experiments indicate that the fast fracture surface exhibits a progressive roughening [52], the limiting crack speed is well below the Rayleigh wave speed in nominally brittle materials [49] and that at some point cracks branch into multiple cracks and grow equally rapidly [49, 53]. Of course, all of these observations indicate a departure from the ideal elastodynamic model discussed in Section 2 and point to the need to take into account the details of the fracture process that have been ignored in formulating the energy balance equation in Section 2. In other words, while the continuum description in terms of $K_I^{dyn}(t,v)$ may provide a description of the driving force experienced by the crack, in order to determine the response of the crack, a detailed modeling of the dissipative processes in the fracture process zone is essential. This problem is only now beginning to be tackled. While we restrict our attention in this section to the issue of the limiting crack speed, a discussion of this issue will bring to focus almost all aspects of the dynamic fracture problem.

A compilation of the limiting speeds from different sources may be found in Ravi-Chandar and Knauss, [49]. Schardin [1] performed an illuminating investigation and demonstrated that the limiting crack speed is not a fixed fraction of the characteristic wave speeds. He measured the limiting crack speed in a 29 inorganic glasses obtained by systematically varying the com-

position and thereby the Rayleigh wave speed. This investigation showed that the limiting crack speed, while a constant for each material, was not the same fraction of the characteristic wave speeds for all the materials examined; his data are plotted in Figure 3 and clearly, the liming speed varied in the range from 0.38 c_R to 0.68 c_R. Schardin suggested that the limiting crack speed be considered a new physical constant, perhaps related to other physical parameters that govern the fracture process. There are two distinct classes of explanations for the existence of a limiting speed less than the Rayleigh wave speed. The first

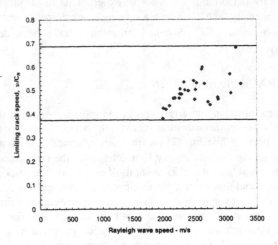

Figure 3. Variation of the limiting crack speed with the Rayleigh wave speed. Data from Schardin [1].

class relies on the fact that beyond a certain crack speed, alternate crack paths off the initial crack plane become possible – this is really a generalization of the Yoffe model [54] which suggests that the rearrangement of the maximum stress near the crack tip when the crack speed reaches about 60% of the Rayleigh wave speed is responsible for crack branching and therefore limiting the crack speed. The limiting speed arguments of Fineberg et al., [25], Abraham et al.,[4], Nakanao et al.,[5], and Xu and Needleman [55] fall into this class. The second class of explanations is based on the evolution of the crack tip dissipation processes that occur over a finite process zone involving microcracking or other damage mechanisms. The works of Kerkhof [32], Ravi-Chandar and Knauss [43,49,52,53], Johnson [56], and Gao [57] fall into this class. There is a fundamental difference between these two classes of models that needs some consideration: the Yoffe instability idea depends centrally on the inertial rearrangement of the stress field and is therefore a function of the crack speed relative to characteristic wave speeds in the model. On the other hand, in the process zone based models, the change in the material behavior near the crack tip is behind the limiting crack speed. Hence, in this class of models the limiting speed will depend not on the characteristic wave speeds, but on characteristic material properties within the process zone or the kinetics of the evolution of the process zone. Both of these depend primarily on the energy flux into the process zone. The strong dependence of the limiting speed on the material composition, the absence of a correlation between the limiting speed and the wave speeds, and the observation of low limiting speeds well below the Yoffe threshold, together suggest that the Yoffe type instability is not at the root of the limiting crack speed. The process zone dynamics must play a crucial role.

The roughness of the fracture surface is an indication that the fracture processes occur over a size scale at least of the scale of the surface roughness. Evaluation of the surface roughness provides evidence regarding their formation and hence important input into the debate between the two classes of models. Once again, the most important question regarding the generation of the dynamic fracture surface roughness is whether it is driven by the crack speed or by the stress field (characterized by the dynamic stress intensity factor). Most observers state that the surface becomes rough at high crack speeds, but exact correlations between the surface roughness and crack speed are seldom found in the literature. The most compelling case for roughness triggered by crack speed is provided by Fineberg et al., [10]. In their experiments, crack speed was determined using an electrical resistive film method discussed in Section 3.2. Their crack speed measurements show that large-amplitude, high-frequency oscillations set in beyond a critical crack speed which they interpret as a dynamical instability associated with the crack path instability; the surface roughness is attributed to this crack path instability. However, there are a number of other observations that indicate that surface roughness is not correlated to the velocity, but the stress field.

Levengood, [58], Shand [59], Johnson and Holloway[60], and Abdel-Latif et al., [61] among many others have shown that on the fracture surface of inorganic glasses, if one considers the transition from "mirror" to "mist", $\sigma_f \sqrt{l}$ remains a constant, where σ_f is the macroscopic stress at initiation of fracture and l is the crack length at the transition. Similarly, if one considers the transition from "mist" to "hackle", "hackle" to "branching", $\sigma_f \sqrt{l}$ remains constant, where l is now the appropriate transition length (see Mecholsky, [62], for a tabulation of these constants for a number of different glasses). While these observations strongly indicate that the crack tip stress must govern roughness generation, the actual stress state at the crack tip has not been considered in these estimates. The quasi-static failure stress has been used, but the actual crack tip stress depends on the crack speed and thus the crack speed might enter into the picture. However, there are many experiments, where the crack speed has been measured; Congleton and Petch [27] evaluated crack speed from Wallner lines and determined that at the onset of crack branching, $\sigma_f \sqrt{l}$ was a constant, but the crack speed varied from 0.35 c_s to 0.57 c_s. Kerkhof [32] measured the crack speed using ultrasonic fractography and showed that "the surface of the crack, after having reached this final velocity, shows a fine roughness in the beginning of this phase with, later, rough hackles and typical branching". In other words, the crack surface roughness continues to evolve even though the crack speed has attained a constant value at the limiting speed implying that the roughness must be correlated to the stress field and not the crack speed. Perhaps the most dramatic demonstration of the stress dependence of roughness was shown by Ravi-Chandar and Knauss, [52]. In their experiments, they observed that the crack speed remained constant while the crack tip loading increase substantially; correspondingly, the crack surface roughness increased and was correlated to the stress field. Arakawa and Takahashi [63] made profilometric measurements of the crack surface roughness in Homalite-100 and attempted to correlate the measurements to the stress intensity factor and the crack speed. Their results also point to a correlation with the stress intensity factor rather than the crack speed. These observations imply a significant history dependence on the fracture surface development and would be quite consistent with a model of dynamic fracture based on the development of a fracture process zone, but not based on crack speed induced instabilities. Recently, the roughening of the fracture surface has been examined from the point of view of self-affine scaling; Bouchaud [64] has shown that regardless of the material, and the manner in which the surface was created, the surface exhibits a self-affine scaling with a scaling exponent that appears to be universal. Such observations, while not yet completely validated, provide an entry point into the modeling of fracture process kinetics in detail.

The dynamic fracture process zone evolution in polymethylmethacrylate (PMMA) is examined in this section. Dynamic crack growth was induced in large PMMA specimens by an electromagnetic loading device. This loading corresponds to a pressure loaded semi-infinite crack in an infinite medium for about 150 μs. The crack speed was determined, through high speed photography, to be nearly constant at about $0.5 C_R$; since the photographs were taken at a rate of 100 kHz, fluctuations at frequencies greater than 50 kHz cannot be identified. The dynamic stress intensity factor, $K_I^{dyn}(t,v)$, was measured using the method of caustics and found to increase slowly along the crack from 1.03 MPa$\sqrt{m}$ at crack initiation to 1.2 MPa$\sqrt{m}$ as the crack extended about 25 mm. The details of these variations can be found in Ravi-Chandar and Yang [65]. An examination of the fracture surface of this specimen reveals a dramatic variation in the surface features that provide insight into the mechanism of dynamic fracture. Figure 4a shows the fracture surface from a PMMA specimen during the early stages of crack growth. This region exhibits a rather flat surface tiled with conic markings. Figure 4b shows a high magnification view of one of the conic markings. The focus of the conic is clearly visible and radial lines emanate from the focus; Smekal [66] suggested that in the enhanced stress field of the main crack, inhomogenieties triggered the initiation of a secondary (micro) crack ahead of the primary crack front. The secondary fracture may not be in the same plane as the main crack front and when these two fronts intersect in space and time, the ligament separating the two cracks breaks up leaving a conic marking on the fracture surface. While the origin of these microcrack nuclei is hence still uncertain, a naive estimate of the stress levels that may cause cavitation can be obtained rather easily as shown later.

First, an estimate of the conditions of the formation of these conic markings and their foci must be obtained by an evaluation of their geometry and the stress state. The density and spacing of the conic markings increase along the crack path from about 350 per mm^2 to about 1500 - 2600 per mm^2 prior to the appearance of the periodic bands on the fracture surface. Note that the speed of the *main crack* – which is now the ensemble average of all the microcracks – is nearly constant along the crack path and only the stress amplitude increases. These observations suggest that the nucleation of a microcrack is a stress-induced event; at larger stress levels, more microcracks are nucleated. The focal length is also an indication of how far ahead of the main crack the microcrack was nucleated; this nucleation distance d_n is equal to twice the focal length. From measurements of the conic markings in Figure 4, we observe that the average spacing between nuclei decreases from 53 μm to 20 μm as the stress level increases. Furthermore, the average nucleation distance d_n increases from about 5.5 μm to 8.5 μm over this same crack path. *Thus as the stress level increases, more flaws are nucleated and they are nucleated farther ahead of the main crack. These flaws grow, interact, and coalesce leading to the overall growth of the "ensemble crack". This appears to be the primary mechanism of crack growth in PMMA.* Ravi-Chandar and Yang, [65] also explore the fracture surface features in other polymers and show some similarity between PMMA and other polymers.

The dynamic stress intensity factor at crack initiation was 1.03 MPa√m and gradually increased to about 1.2 MPa√m as the density of flaws increased from about 350 per mm^2 to about 2600 per mm^2. The yield stress (determined from quasi-static experiments) for PMMA is approximately 100 MPa and is likely to be higher at the high strain rates encountered at the dynamically growing crack tip; in an effort to estimate the size of the plastic zone, assume that PMMA obeys a power law hardening behavior of the type $\varepsilon = \sigma^n$, with $n = 3$. The size of the plastic zone under plane strain conditions appropriate to the interior of the specimen is then given by

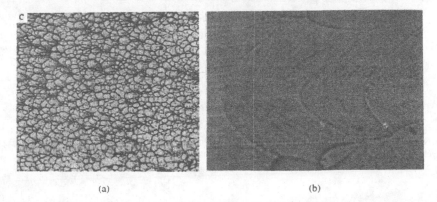

(a)   (b)

Figure 4. Surface morphology of the PMMA fracture surface. (a) and (b) are optical micrographs at different magnifications. The horizontal dimension in (a) is 770 μm, and (b) is 50 μm.

$$r_p = \frac{1}{6\pi} \left(\frac{n-1}{n+1} \right) \left(\frac{K_I}{\sigma_y} \right)^2 . \qquad (8)$$

Hence, the plastic zone size increases from about 2.8 μm to 3.8 μm as the stress intensity factor increases from 1.03 to 1.2 MPa$\sqrt{}$m. Note that the plastic zone is very small, and in particular smaller than the average nucleation distance d_n which increases from about 5.5 μm to 8.5 μm over this same crack path; thus conditions of small-scale yielding seem to be appropriate. Assuming then that the linear elastodynamic stress field estimates are appropriate at these distances, the normal stress component σ_{22} at the nucleation of the flaws can be determined and is shown in Figure 5; note that the stress level is much higher than the craze nucleation threshold commonly observed in PMMA. The triaxial stress state at the nucleation distance of 5.5 μm, corresponding to a stress intensity factor of 1.03 MPa$\sqrt{}$m, and a crack velocity of about $0.25C_R$ can then be calculated as: $\sigma_{11} = 189$ MPa, $\sigma_{22} = 175$ MPa and $\sigma_{33} = 125$ MPa; similarly at the nucleation distance of 8.5 μm, corresponding to a stress intensity factor of 1.2 MPa$\sqrt{}$m, and a velocity of about $0.25C_R$, the triaxial stress state is given by: $\sigma_{11} = 172$ MPa, $\sigma_{22} = 164$ MPa and $\sigma_{33} = 118$ MPa; note that these stresses are much higher that typical values quoted for crazing in PMMA under uniaxial tension, which is in the range of 60 to 100 MPa. It appears that at such high triaxial tensile stress levels, extensive cavitation occurs and these cavities act as the nuclei for further development of the dynamic fracture process. The energy dissipation in these nucleation and growth processes and the dynamics of evolution of this cluster of interacting micropores and microcracks is quite complex; this interaction is what leads to crack surface roughness, periodicity in fracture surface roughness, attempted and successful branching and is

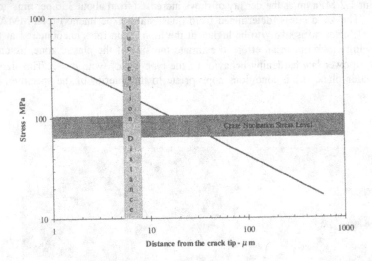

Figure 5. Variation of the stress normal to the crack tip according to linear elasticity; the craze nucleation stress level and the observed nucleation distance range are also shown.

instrumental in limiting of the crack speed to levels much below the Rayleigh wave speed allowable by the continuum. It is, however, quite difficult to incorporate this interaction into models of dynamic fracture, derived purely from continuum considerations.

The main conclusions from fracture surface observations are that the fracture surface is not smooth, down to the nanometer scale and that the surface roughness changes are not abrupt, but appear continuously. In the "mirror" region, the fracture process zone is small and continuum fracture mechanics theory provides an adequate description of fast fracture. In the later stages, the fracture process zone becomes large, with a large number of independent microcracks growing simultaneously. The interaction of microcracks provides a structure to the fracture process zone and the dynamics of evolution of the process zone must be considered in formulating a dynamic fracture theory; neither the continuum theory, nor the discrete models are currently able to do this satisfactorily.

The key elements of this microcrack dominated fracture process model can be summarized as follows:

- Propagation of the crack occurs by the nucleation, growth and coalescence of microcracks.
- The number of microcracks that are activated into growing is a function of the stress intensity factor, the size of the fracture process zone, the distribution of voids.
- The size of the fracture process zone increases as the stress intensity factor increases. As the fracture process zone grows in size, the number of potential microcracks contributing to the fracture process zone increases. Thus the probability of initiating flaws into growth increases, thereby increasing the energy that is dissipated in the fracture process zone.
- Also, as the number of initiated microcracks increases, the interaction of these microcracks becomes important in determining crack propagation behavior.

This microcrack model for the growth of cracks in nominally brittle materials provides a comprehensive mechanistic view of what happens during fast fracture, not only in opening mode fracture, but also in the shearing mode. The experimental observations point to a regime of slow crack speeds where these fracture processes develop autonomously in a small zone and hence the framework of continuum mechanics is appropriate. In this regime, rather than developing a quantitative characterization of the fracture processes, a global description of the dissipation in the fracture process zone obtained from calibration experiments is adequate. On the other hand, at higher crack speeds, microcrack nucleation, growth and coalescence dominates the fracture processes. The dissipation within the process zone can increase quite rapidly and this is what sets the limit on the crack speed and contributes to crack branching. The remaining challenge to develop a quantitative model is incorporating the microcracking in the fracture process zone. The microcracking observations also point to the need for caution in generalizing results from atomistic simulations to the larger scales; the presence of discrete fracture events at the microscale suggests that in proceeding from the atomistic to the continuum scale, the microscale events have to be modeled realistically.

5. References

1. Schardin, H., 1959, "Velocity effects in fracture", in *Fracture*, Edited by Averbach *et al.*, John Wiley, 297-330.
2. Freund, L.B., 1990, *Dynamic Fracture Mechanics*, Cambridge University Press.

3. Nakano, A., Kalia, R.K. and Vashishta, P., 1995, "Dynamics and morphology of brittle cracks: A molecular-dynamics study of silicon nitride", *Physical Review Letters*, **75**, 3138-3141.

4. Abraham, F.F., Brodbeck, D., Rudge, W.E., and Xu, X., 1997, "A molecular dynamics investigation of rapid fracture mechanics", *Journal of the Mechanics and Physics of Solids*, **45**, 1595-1619.

5. Kalthoff, J.F., Beinert, J., Winkler, S., and Klemm, W., 1980, "Experimental analysis of dynamic effects in different crack arrest test specimens", **ASTM STP 711 – Crack Arrest Methodology and Application**, American Society for Testing and Materials, Philedelphia, 109-127.

6. Carlsson, J., Dahlberg, L., and Nilsson, F., 1973, "Experimental studies of the unstable phase of crack propagation in metals and polymers", in **Dynamic Crack Propagation**, (G.C. Sih Ed.), Noordhoff International Publishing, Leyden, 165-181.

7. Kobayashi, A.S., Wade, B.G., and Bradley, W.B., 1973, "Fracture dynamics of Homalite-100", in Deformation and Fracture of High Polymers, H.H. Hausch et al., Editors, Plenum Press, New York, 487-500.

8. Brickstad, B. 1983, "A FEM analysis of crack arrest experiments", *International Journal of Fracture*, **21**, 177-194.

9. Taudou, C., Potti, S., and Ravi-Chandar, K., 1992, "On the dominance of the dynamic crack tip stress field under high rate loading", *International Journal of Fracture*, **56**, 41-59.

10. Fineberg, J., Gross, S.P., Marder, M., and Swinney, H.L., 1991, "Instability in dynamic fracture", *Physical Review Letters*, **67**, p.457.

11. Beebe, W.M., Ph.D Thesis, California Institute of Technology, Pasadena, CA, 1966.

12. ASM Handbook, 1994, Volume 11, Materials Park, Ohio.

13. Ravichandran, G., and Clifton, R.J., 1989, "Dynamic fracture under plane wave loading", *International Journal of Fracture*, **40**, 157-201.

14. Costin, L.S., Duffy, J., and Freund, L.B., 1977, "Fracture initiation in metals under stress wave loading", in *Fast Fracture and Crack Arrest*, ASTM STP 627, 301-308.

15. Maigre, H., and Rittel, D., 1995, " Dynamic fracture detection using the force-displacement reciprocity: Application to the compact compression specimen", *International Journal of Fracture*, **73**, 67-79.

16. Shukla, A., and Rossmanith, H.P., 1995, "Dynamic photoelastic investigation of wave propagation and energy transfer across contacts", *Journal of Strain Analysis*, **21**, 213-218.

17. Ravi-Chandar, K., and Knauss, W.G., 1982, "Dynamic crack tip stresses under stress wave loading - A comparison of theory and experiment", *International Journal of Fracture*, **20**, 209-222.

18. Field, J.E., 1970, "Brittle fracture: its study and application", *Contemporary Physics*, **12**, 1-31.

19. Dulaney, E.N., and Brace, W.F., 1960, "Velocity behavior of a growing crack", *Journal of Applied Physics*, **31**, 2233-2236.

20. Cotterell, B., 1965, "Velocity effects in fracture propagation", *Applied Materials Research*, **4**, 227-232.

21. Cotterell, B., 1968, "Fracture propagation in organic glasses", *International Journal of Fracture Mechanics*, **4**, p.209.

22. Anthony, S.R., Chubb, J.P., and Congleton, J., 1970, "The crack branching velocity", *Philosophical Magazine*, **22**, 1201-1261.

23. Paxson, T.L., and Lucas, R.A., 1973, "An investigation of the velocity characteristics of a fixed boundary fracture model", in **Dynamic Crack Propagation**, (G.C. Sih Ed.), Noordhoff International Publishing, Leyden, 415-426.

24. Stalder, B., Beguelin, P., and Kausch, H.H., 1983, "A simple velocity gauge for measuring crack growth", *International Journal of Fracture*, **22**, R47-R54.

25. Fineberg, J., Gross, S.P., Marder, M., and Swinney, H.L., 1992, "Instability in the propagation of fast cracks, *Physical Review*, **B45**, 5146-5154.

26. Wallner, H., 1938, "Linienstrukturen an bruchflächen", *Z. Physik*, **114**, 368-370.

27. Congleton, J., and Petch, N.J., 1967, "Crack-branching", *Philosophical Magazine*, **16**, 749-760.

28. Hull, D., 1997, "Influence of stress intensity and crack speed on fracture surface topography: mirror to mist transition", *Journal of Materials Science*, **31**, 1829-1841.

29. Hull, D., 1997, "Influence of stress intensity and crack speed on fracture surface topography: mirror to mist to macroscopic bifurcation", *Journal of Materials Science*, **31**, 4483-4492.

30. Morrissey, J.W., and Rice, J.R., 1998, "Crack front waves", *Journal of the Mechanics and Physics of Solids*, **46**, 467-487.

31. Sharon, E., Cohen, G., and Fineberg, J., 2000, "Crack front waves: Localized solitary waves in dynamic fracture", preprint.

32. Kerhkof, F., 1973, "Wave fractographic investigation of brittle fracture dynamics", in **Dynamic Crack Propagation**, (G.C. Sih Ed.), Noordhoff International Publishing, Leyden, 3-35.

33. Richter, H.G., and Kerkhof, F, 1994, "Stress wave fractography", in **Fractography in Glass**, (Edited by R.C. Bradt and R.E. Tressler), Plenum Press, New York, 75-109.

34. Kinra, V.K., and Bowers, C.L., 1981, "Brittle fracture of plates in tension. Stress field near the crack", *International Journal of Solids and Structures*, **17**, 175-182.

35. Khanna, S.K., and Shukla, A., 1995, "On the use of strain gauges in dynamic fracture mechanics", *Engineering Fracture Mechanics*, **51**, 933-948.

36. Ravi-Chandar, K., "A note on the dynamic stress field near a propagating crack", *International Journal of Solids and Structures*, **19**, (1983), 839-841.

37. Kobayashi, A.S., Editor, The Handbook of Experimental Mechanics, Prentice Hall,

38. Rosakis, A.J. 1993, "Two optical techniques sensitive to gradients of optical path difference: The method of caustics and the coherent gradient sensor", in **Experimental Techniques in Fracture**, J.S. Epstein, Editor, VCH Publishers, Inc, New York, p.327-425.

39. Rosakis, A.J., and Ravi-Chandar, K., 1986, "On crack tip stress state: An experimental evaluation of three-dimensional effects", *International Journal of Solids and Structures*, **22**, 121-134.

40. Ravi-Chandar, K., and Knauss, W.G., 1987," On the characterization of the transient stress field near the tip of a crack", *Journal of Applied Mechanics*, **54**, 72-78.

41. Pfaff, R.D., Washabaugh, P.D., and Knauss, W.G., 1995, "An interpretation of Twyman-Green interferograms from static and dynamic fracture experiments", *International Journal of Solids and Structures*, **32**, 939-955.

42. Dadkah, M.S., and Epstein, J.S., 1993, "Moire interferometry in fracture research", in **Experimental Techniques in Fracture**, J.S. Epstein, Editor, VCH Publishers, Inc, New York, p.427-508.

43. Ravi-Chandar, K., and Knauss, W.G., 1984, "An experimental investigation into dynamic fracture - I. Crack initiation and crack arrest", *International Journal of Fracture*, **25**, 247-262.

44. Liu, C., Knauss, W.G., and Rosakis, A.J., 1998, "Loading rates and the dynamic initiation toughness in brittle solids", *International Journal of Fracture*, **90**, 103-118.

45. Owen, D.M., Zhuang, S, Rosakis, A.J., and Ravichandran, G., 1998, "Experimental determination of dynamic crack initiation and propagation fracture toughness in thin aluminum alloys", *International Journal of Fracture*, **90**, 153-174.

46. Marder, M., and Gross, S.P., 1995, "Origin of crack tip instabilities", *Journal of the Mechanics and Physics of Solids*, **43**, 1-48.

47. Kobayashi, A.S., and Mall, S., 1978, "Dynamic fracture toughness of Homalite-100", *Experimental Mechanics*, **18**, 11-18.

48. Dally, J.W., Fourney, W.L., and Irwin, G.R., 1985, "On the uniqueness of the stress intensity factor-crack velocity relationship", *International Journal of Fracture*, **27**, 159-168.

49. Ravi-Chandar, K., and Knauss, W.G., 1984, "An experimental investigation into dynamic fracture - III. Steady state crack propagation and crack branching", *International Journal of Fracture*, **26**, 141-154.

50. Arakawa, K., and Takahashi, K., 1987, "Dependence of crack acceleration of the dynamic stress intensity factor in polymers", *Experimental Mechanics*, **27**, 195-200

51. Ma, C.C., and Freund, L.B., 1986, "The extent of the stress intensity factor field during crack growth under dynamic loading conditions", ASME *Journal of Applied Mechanics*, **53**, 303-310.

52. Ravi-Chandar, K., and Knauss, W.G., 1984, "An experimental investigation into dynamic fracture - II. Microstructural aspects", *International Journal of Fracture*, **26**, p.65-80.

53. Ravi-Chandar, K., and Knauss, W.G., 1984, "An experimental investigation into dynamic fracture - IV. On the interaction of stress waves with propagating cracks", *International Journal of Fracture*, **26**, 189-200.

54. Yoffe, E., 1951, "The moving Griffith crack", *Philosophical Magazine*, **42**, 739-750. Yoffe, E., 1951, "The moving Griffith crack", *Philosophical Magazine*, **42**, 739-750.

55. Xu, X.-P., and Needleman, A., 1994, "Numerical simulations of fast crack growth in brittle solids", *Journal of the Mechanics and Physics of Solids*, **42**, 1397-1434.

56. Johnson, E., 1992, "Process region changes for rapidly propagating cracks", *International Journal of Fracture*, **55**, 47-63.

57. Gao, H., 1996, "A theory of local limiting speed in dynamic fracture", *Journal of the Mechanics and Physics of Solids*, **44**, 1453-1474.

58. Levengood, W.C., 1958, "Effect of origin flaw characteristics on glass strength", *Journal of Applied Physics*, **29**, 820-826.

342

59. Shand, E.B., 1959, "Breaking stress of glass determined from dimensions of fracture mirrors", *Journal of the Americal Ceramic Society*, **42**, 474-477.
60. Johnson, J.W., and Holloway, D.G., 1966, "On the shape and size of the fracture zones on glass fracture surfaces", *Philosophical Magazine*, **14**, 731-743.
61. Abdel-Latif, A.I.A., Bradt, R.C., and Tressler, R.E., 1977, "Dynamics of fracture mirror boundary formation in glass", *International Journal of Fracture*, **13**, 349-359.
62. Mecholsky, J.J., 1994, "Quantitative fractographic analysis of fracture origins in glass", in **Fractography of Glass**, edited by R.C. Bradt and R.E. Tressler, Plenum Press, New York, 37-73.
63. Arakawa, K., and Takahashi, K., 1991, "Relationship between fracture parameters and surface roughness of brittle polymers", *International Journal of Fracture*, **48**, 103-114.
64. Bouchaud, E., 1997, "Scaling properties of cracks", *Journal of Physics: Condensed Matter*, **9**, 4319-4344
65. Ravi-Chandar, K., and Yang, B., 1997, "On the role of microcracks in the dynamic fracture of brittle materials", *Journal of the Mechanics and Physics of Solids*, **45**, 535-563.
66. Smekal, A., 1953, "Zum Bruchvorgang bei sprodem Stoffverhalten unter ein- and mehrachsigen Beanspruchungen", *Osterr. Ing. Arch*, **7**, 49-70.

EXPERIMENTS IN DYNAMIC FRACTURE

D. RITTEL

Faculty of Mechanical Engineering
Technion, Israel Institute of Technology
32000 Haifa, Israel
email: merittel@tx.technion.ac.il

1. Abstract

This paper summarizes the material which was presented orally at the Workshop on the *Physical Aspects of Fracture*, held in Cargèse. Experimental fracture is a wide subject which combines many aspects of Mechanics and Materials Science. Rather than giving a comprehensive overview of the experimental aspects of dynamic fracture, we have selected a number of topics, which are under current investigation, to get the reader familiar with recent results and current issues in that field.

The first concerns *thermomechanical couplings* in dynamic fracture. We present our experimental investigation of transient thermal effects which accompany dynamic crack initiation in amorphous polymers. Two "model" commercial polymers, "brittle" polymethylmethacrylate (PMMA) and "ductile" polycarbonate (PC) were chosen. Dynamic mode I and mode II conditions were investigated, during which the temperature in the vicinity of the crack-tip was continuously monitored until the onset of crack propagation. For thermoelastic PMMA loaded in mode I, a marked tendency for crack-tip *cooling* was observed. Mode II loading in thermoplastic "ductile" PC caused the more familiar *temperature rise*, associated with shear band formation. *These results show that dynamic crack initiation (and propagation) is not an isothermal process.*

The next topic concerns dynamic *crack propagation* under shear mode II (shear) loading, in homogeneous and heterogeneous bimaterials. Here, the results of the experiments showed that high crack velocities (intersonic) can be reached, which are not attainable under mode I (opening) loading. *These results shed a new light on the propagation phase of dynamic fracture both in terms of material systems and loading mode.*

E. Bouchaud et al. (eds.), Physical Aspects of Fracture, 343–352.

2. Introduction

2.1 THERMOMECHANICAL COUPLING

The conversion of mechanical energy into heat is a long known fact (Taylor and Quinney, 1934), which manifests itself sometimes by spectacular shear localization (adiabatic shear banding), or sometimes by the slower evolution of heat in a cyclically loaded dissipative material. The coupled heat equation, for small temperature changes, is given by (Boley and Weiner, 1960):

$$k\nabla^2 T - \alpha(3\lambda + 2\mu)T_0\dot{\varepsilon}_{ij}^e + \beta\sigma_{ij}\dot{\varepsilon}_{ij}^p = \rho c\dot{T} \tag{1}$$

where k is the heat conductance and α is the thermal expansion coefficient. ρ, c, λ and μ stand for the material's density, heat capacity and Lamé constants respectively. T is the temperature and the strain rates are divided into elastic and plastic. Finally the factor β expresses the fraction of plastic work rate converted into heat. The term T0 denotes a reference temperature. Equation (1) expresses the convertibility of mechanical energy into heat (coupling) by introducing two heat sources (sinks) which are related to elastic volume changes (reversible thermoelastic effect) and irreversible plastic work of deformation.

While the overall situation is described by equation (1), one frequently assumes that thermoelastic effects are negligible and that adiabatic conditions prevail so that the heat equation can be rewritten in a simpler form:

$$\beta\sigma_{ij}\dot{\varepsilon}_{ij}^p = \rho c\dot{T} \tag{2}$$

Equation 2 has been used to study crack-tip adiabatic heating effects with the assumption that the *rate* of conversion of mechanical into thermal energy is a constant, $\beta=0.9$. Recent studies on metals (Mason et al., 1994) and on polymers (Rittel, 1999) have questioned this assumption and shown that β is both strain and strain-rate dependent.

Thermoelastic effects, on the other hand are often neglected since temperature drops associated with *uniform* elastic extension seldom exceeds half a degree. For this case, equation (1) writes:

$$-\alpha(3\lambda + 2\mu)T_0\dot{\varepsilon}_{ij}^e = \rho c\dot{T} \tag{3}$$

While these effects are certainly negligible in the case of a homogeneously deforming elastic solid, the vicinity of a crack experiences a triaxial stress/strain state with a strong spatial singularity. Therefore, one could expect (and observe) a significant amount of crack-tip cooling during the phase of crack initiation.

In this paper, we will illustrate each case by showing recent results about the dynamic crack initiation in a "brittle" PMMA, adiabatic shear banding in a "ductile" PC.

2.2 SHEAR LOADING OF HETEROGENEOUS AND HOMOGENEOUS BIMATERIAL SYSTEMS

Whereas mode I interfacial fracture in bimaterials has been extensively investigated, it appears that relatively little has been done, by comparison, for the case of mode II (shear) loading. Nevertheless, bimaterials, whether made of two different materials (heterogeneous) or of the same material (homogeneous), are likely to experience dynamic loading with a significant shear component. During the last decade, Rosakis and coworkers have investigated the shear fracture of such systems. Sophisticated experimental techniques were used. These include high-speed photography (typically 10^6 frames/s and more) optical methods (e.g. photoelasticity, CGS) to monitor the stresses and displacements in the vicinity of the crack-tip. The bimaterial systems which were investigated included metal-polymers (heterogeneous), polymer-polymer (homogeneous), and ultimately unidirectional composite materials. The specimen was impacted by a gas propelled projectile, in parallel to the crack line. While the theoretical backround describing crack propagation in such sytems can be quite complicated, a very simple and striking observation came out of these experiments. Namely, provided specific impact velocities, cracks were observed to propagate at intersonic velocities, that is between the shear and longitudinal wave speed. Such velocities had not been observed in previous mode I dynamic fracture studies, in which the crack-tip velocity limiting is limited by the Rayleigh wave speed, with the exception of propagation along weak crystal planes (Winkler et al., 1970).

3. Typical experimental results

3.1 DYNAMIC CRACK INITIATION IN "BRITTLE" PMMA: MODE II LOADING

Mode I loading of a stationary crack (whether dynamic or static) causes expansion of the crack-tip material. If the experiment is carried out rapidly such as to induce

346

adiabatic conditions, the temperature of the material should drop as long as the response of the material is linear thermoelastic. Indeed, Fuller et al. (1975) monitored the temperature of propagating cracks in various brittle polymers and they reported a noticeable temperature drop which preceeded a very large temperature rise. Other experiments with glass plates, carried out by Weichert and Schönert (1978), clearly indicate a temperature drop at the onset of crack propagation. However, none of the above mentioned authors addressed the cooling effect in detail.

Rittel (1998-a) investigated dynamic crack initiation by performing stress wave loading experiments and monitoring *simultaneously* the onset of crack propagation and the temperature slightly ahead of the crack-tip. The temperature was monitored by means of small embedded thermocouples, for which it was shown that they can adequately measure transient temperature changes (Rittel, 1998-b ; Rabin and Rittel, 1999). As shown in Figure 1, the temperature drops noticeably prior to the onset of crack initiation. This result is new, compared to that about a propagating crack, in the sense that it shows the *conditions which prevail at the onset of crack initiation.*

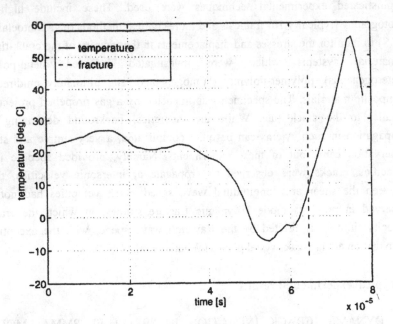

Figure1: Temperature ahead of the stationary crack-tip as a function of time. Dynamic mode I experiment. Commercial PMMA. Time origin coincides with impact of the specimen. The dashed line indicates the onset of crack propagation. A noticeable temperature drop preceded crack initiation. (Rittel, 1998-a).

Thus, the main result is that for PMMA, as an example of linear thermoelastic solid, dynamic crack initiation occurs during a crack-tip cooling phase. The temperature drop is significant (typically 20 degrees and more). While the temperature is measured at a single point with the above mentioned technique, this result nevertheless illustrates the fact that the thermoelastic coupling is not negligible for a (linear) cracked solid. Thus, the crack-tip temperature is different from the ambient temperature. The importance of such an observation lies in the development of adequate *failure criteria* which should reflect this observation. This point will be further addressed in the subsequent section.

3.2 ADIABATIC SHEAR BANDING IN "DUCTILE" PC: MODE II LOADING

Shear impact experiments of cracked (notched) steel plates showed that, with increasing impact velocity, the failure mode shifted from opening into shear failure, by adiabatic shear band formation (Kalthoff, 1988). At low velocities, the crack propagated with a high kink angle value of the order of 70°. At the higher velocities, the crack propagated along its initial line, i.e. at 0°. A similar failure mode transition was investigated for other metallic alloys by Zhou et al. (1996). Similarly, the room failure mode transition was established by Ravi-Chandar (1995) for commercial polycarbonate. All observations were reported to occur at room temperature. Rittel and Levin (1998) confirmed this observation and explained it in general terms of mode-mixity. The interesting point is that of the development of an adiabatic shear band, which once again indicates that the crack-tip temperature is fairly different from the ambient temperature. Two experiments were performed to assess the crack-tip temperature. The first, of an "indirect" nature consisted in side impact (mode II) of cracked plates for which the base temperature was systematically varied between -120°C and +70°. Rittel and Levin (1998) observed the kink angle value reduced from about 70° to less than 40° when the temperature reached -25°C. The fracture surface examination clearly showed the development of a 200μ wide shear band ahead of the crack-tip. Thus, a simple calculation, based on a glass transition temperature of 150°C for this material, indicated a temperature elevation of the order of 175°C. Real time monitoring of the crack-tip temperature was subsequently carried out by Rittel (2000). As shown in Figure 2, the temperature rise is of the order of 70°C, as recorded by an embedded thermocouple, located about 1 mm ahead of the crack-tip. To our knowledge, there is no similar result in the literature, and keeping in mind that the thermocouple measures the tip of the shear band, the "direct" and "indirect" assessment of the temperature rise are of the same order of magnitude. At this stage, it can be concluded, that when plastic deformation takes place (shear loading of a ductile material), a noticeable temperature rise occurs, as expected.

348

However, the response of each material (PMMA and PC) differs, which calls for a *thermomechanical charaterization* of the different polymers, as reported in Rittel (2000). In this case, commercial PMMA was shown to remain essentially isothermal until the later stages of the specimen failure, and this behavior should be contrasted with that of the previously mentioned PC.

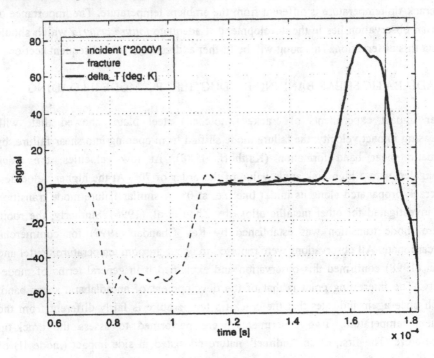

Figure2: Temperature ahead of the adiabatic shear band as a function of time. Dynamic mode II experiment. Commercial PC. The dashed line corresponds to the strain gage signal of the incident bar. The thin solid line indicates the onset of crack propagation (Rittel, 2000).

3.3 SHEAR DOMINATED FRACTURE IN BIMATERIAL SYSTEMS

Three typical bimaterial systems were studied. The first system consists of metal bonded to a brittle polymer (aluminum-Homalite100, steel-PMMA). A detailed description of the optical and impact setup can be found in Rosakis et al. (1998). In these experiments, it was observed that the terminal crack-tip speeds were 1.4 times the shear wave speed of PMMA (1330m/s) and 1.3 times that of Homalite100 (1255 m/s). As the dilatational wave speed of these polymers was not exceeded, crack propagation is intersonic with respect to the polymer. Another related phenomenon is the observation of shock-like wave emanating from the crack-tip, in the intersonic regime.

One question which arises is that of the requirement for the system to be heterogeneous. Subsequent impact experiments were carried out by Rosakis et al. (1999) on homogeneous bimaterials, in which two Homalite 100 (brittle polyester) plates were bonded together. When the impact velocity exceeded 25 m/s, the authors observed a crack propagating along the interface, whose stabilized velocity was √2 the shear wave speed of the material. While this limit bears theoretical implications, it also shows that intersonic crack propagation may develop in constitutively homogeneous bimaterials *as well*. A typical example of real time crack-tip monitoring is given in Figure 3.

The last type of experiments to be reported here was carried out by Coker and Rosakis (2000) on 48 ply thick unidirectional graphite/epoxy composites. Here, both mode I and mode II experiments were carried out. Mode I cracks propagated at speeds which did not exceed the Rayleigh wave speed of the material. By contrast, for shear dominated crack propagation along the fibers, crack speeds of up to 7400 m/s were recorded. Such a high speed was caused by impact applied at 57 m/s. These results shed new light on the dynamic fracture of composite materials.

INITIATION & GROWTH OF AN INTERSONIC SHEAR CRACK

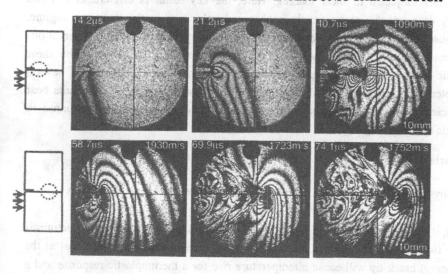

Figure 3: Isochromatic fringe pattern showing the initiation and propagation of an intersonic shear crack along a weak plane in Homalite. The impact velocity was 28m/s.Two Mach waves radiating from the crack tip can be clearly distinguished.Two fields of view (50mm diameter) are shown in each row. The time after impact and crack-tip speed are indicated in each frame (Rosakis et al., 1999).

4. Discussion

We have presented in this paper several recent results pertaining to the conversion of mechanical energy into heat in glassy polymers. These can be extended beyond the particular case of polymeric materials, to investigate the nature of the coupling in general. The examples illustrate the respective effect of the *thermoelastic* and *thermoplastic* coupling which cause transient temperature changes during dynamic crack initiation. Specifically, a significant temperature drop was observed to accompany mode I dynamic crack initiation in the "brittle" material, by contrast with the more usual case of a temperature rise for the "ductile" crack-tip material subjected to shear fracture.

These results illustrate the very fact that dynamic fracture is far from being isothermal and that the local crack-tip temperature is dictated both by the loading mode and the material thermomechanical response. As a result, the crack-tip temperature is different from the ambient temperature at which fracture takes place. This physical observation should be taken into account when investigating and developing dynamic fracture criteria.

Another topic which was addressed is that of intersonic crack propagation in bimaterials subjected to shear loading. Here, the key result is that cracks can and actually do propagate at previously unobserved velocities, in the intersonic regime. Such observations are not restricted to heterogeneous systems, as similar observations were made in homogeneous polymeric bimaterial systems. Furthermore, the shear dynamic fracture of unidirectional composites disclosed crack velocities of several kilometers/second, as a result of a relatively low velocity impact. These results bear both scientific and technological implications as they show a distinct response to dynamic shear loading.

5. Conclusion

Two physical aspects of dynamic fracture have been presented at the Workshop.

Firstly, it has been shown that *dynamic crack initiation* and, more generally fracture, is not an isothermal phenomenon. The conversion of mechanical energy into heat at the (stationary) crack-tip will cause a temperature rise for a thermoplastic response and a temperature drop in the thermoelastic case. Both effects may affect the actual material separation processes at the tip of the crack, and should therefore be accounted when investigating dynamic fracture.

The second aspect relates to the seldom adressed problem of *dynamic crack propagation* in bimaterials (homogeneous and heterogeneous) materials under shear loading. Recent observations have shown that crack velocities may exceed the shear wave speed, thus yielding shock-like phenomena. Intersonic crack propagation is triggered by impacting the material at relatively moderate velocities, and is therefore likely to be encountered in many cases (e.g. composites, bonded structures). Furthermore, such observations bear implications at the larger geological scale (earth crust) at which shear fracture is frequently observed.

Acknowledgement
Prof. A. Rosakis is gratefully acknowledged for discussing and providing the results quoted in this paper.

References

1. Boley, B.A. and Weiner, J.H. (1960), *Theory of thermal stresses*, J. Wiley and Sons, New York, NY.

2. Coker, D. and Rosakis, A.J. (2000), Experimental observations of intersonic crack growth in asymmetricallyloaded unidirectional composites plates, to appear in *Phil Mag. A*.

3. Fuller, K.N.G., Fox, P.G. and Field, J.E. (1975) The temperature rise at the tip of a fast-moving crack in glassy polymers, *Proc. R. Soc.* , **A 341**, 537-557.

4. Kalthoff, J.F. (1988), Shadow optical analysis of dynamic fracture, *Optical Engng*, **27**, 835-840.

5. Mason, J.J., Rosakis, A.J., and Ravichandran, G. (1994) On the strain and strain rate dependence of the fraction of plastic work converted into heat: an experimental study using high speed infrared detectors and the Kolsky bar, *Mechanics of Materials*, **17**, 135-145.

6. Rabin,Y. and Rittel, D. (1999-b) A model for the time response of solid-embedded thermocouples, *Experimental Mechanics,* **39**(1), 132-136.

7. Ravi Chandar, K. (1995) On the failure mode transitions in polycarbonate under dynamic mixed mode loading, *Int. J. Solids and Structures*, **32**(6/7), 925-938.

8. Rittel, D., (1998-a) Experimental investigation of transient thermoelastic effects in dynamic fracture, *Int. J. Solids and Structures*, **35**(22), 2959-2973.

9. Rittel, D., (1998-b) Transient temperature measurement using embedded thermocouples, *Experimental Mechanics,* **38**(2), 73-79.

10. Rittel, D., (1999) "On the conversion of plastic work to heat during high strain rate deformation of glassy polymers", *Mechanics of Materials*, **31**, 131-139.

11. Rittel, D., (2000), Experimental investigation of transient thermoplastic effects in dynamic fracture, *Int. J. Solids and Structures,* **37**(21), 2901-2913.

12. Rittel, D., and Levin, R., (1998) Mode-mixity and dynamic failure mode transitions in polycarbonate, *Mechanics of Materials*, **30**(3), 197-216.

13. Rosakis, A.J., Samudrala, O., Singh, R.P. and Shukla, A. (1998) Intersonic crack propagation in bimaterial systems, *J. Mech. Phys. Solids,* **46**, 1789-1813.

14. Rosakis, A.J., Samudrala, O. and Coker, D. (1999) Cracks faster than shear wave speed, *Science,* **284**, 1337-1340.

15. Taylor, G.I. and Quinney, H. (1934) The latent energy remaining in a metal after cold working, *Proc. R. Soc.*, **A 143**, 307-326.

16. Weichert, R. and Schönert, K. (1978) Heat generation at the tip of a moving crack, *J. Mech. Phys. Solids,* **26**, 151-161.

17. Winkler, S., Shockey, D.A. and Curran, D.R. (1970) Crack propagation at supersonic velocities, *Int. J. Fract.*, **6**(2), 151-158.

18. Zhou, M., Rosakis, A.J. and Ravichandran, G. (1996) Dynamically propagating shear bands in impact-loaded prenotched plates. I- Experimental investigations of temperature signatures and propagation speed, *J. Mech. Phys. Solids,* Vol. **44**(6), 981-1006.

PROPAGATION OF AN INTERFACIAL CRACK FRONT IN A HETEROGENEOUS MEDIUM: EXPERIMENTAL OBSERVATIONS

Jean Schmittbuhl

Laboratoire de Géologie, UMR 8538, Ecole Normale Supŕieure, 24 rue Lhomond, 75232 Paris Cedex 05, France

Jean.Schmittbuhl@ens.fr

Arnaud Delaplace

Laboratoire de Mécanique et Technologie, Ecole Normale Supérieure de Cachan, Avenue du Président Wilson, F-9 Cachan, France

Knut Jørgen Måløy

Fysisk Institutt, Universitetet i Oslo, P.O. Boks 1048 Blindern, 0316 Oslo S, Norway

Keywords: inter-facial fracture, front roughness, depinning, local dynamics

Abstract   We study experimentally the propagation of an inter-facial crack through a weak plane of a transparent Plexiglas block. The toughness is controlled artificially by a sand blasting procedure and fluctuates locally in space like an uncorrelated random noise. The block is fractured in mode I at low speed ($10^{-7} - 10^{-4}$ m/s). The crack front is observed optically with a microscope and a high resolution digital camera for pinned fronts or a fast digital camera for local dynamics of fronts. During the propagation, the front is pinned by micro-regions of high toughness and becomes rough. Roughness of the crack front is analyzed in terms of self-affinity. The roughness exponent of trapped fronts is shown to be 0.63 ± 0.05. During propagation, bursts of the front are described as an interface growth process. The dynamical exponent of a Family-Vicsek scaling is measured to: $z \approx 1.15$. Velocity fluctuations are measured in the space-time domain.

353

E. Bouchaud et al. (eds.), Physical Aspects of Fracture, 353–369.
© 2001 *Kluwer Academic Publishers. Printed in the Netherlands.*

1. INTRODUCTION

The propagation of rupture front through heterogeneous solids is a central question for numerous mechanical problems. One of the main evidence of the fluctuating front propagation is the roughness of fracture surfaces. As first mentioned by Mandelbrot *et al* [15], the geometry of fractured surfaces exhibits scaling invariance. Numerous works have shown the self-affine properties of crack surfaces [2, 4, 5, 14, 28]. The roughness exponent is found to be very robust over different materials, different fracture modes [2] and a broad range of length scales. This range can be extended if earthquake faults are considered as fractured surfaces [4, 25]. The estimate of the roughness exponent is close to 0.80. A possible "universal" self-affine crack geometry of heterogeneous material has been proposed by Bouchaud *et al* [2].

Studies of fracture surfaces and models of crack propagation are generally linked on the basis of the following assumption: the morphology of the crack surface is inherited from the geometry of the crack front [3]. Subsequently understandings of the mechanisms of the crack front geometry during its propagation becomes a key question.

Even when the rupture front is inter-facial, i.e. restricted to move within a plane, the dynamics results from interactions between non local elastic coupling, inertia, wave effects, and influence of quenched heterogeneities at different scales. The front is trapped temporarily by local asperities that may be related to local material heterogeneities or residual stresses. Depinning from these asperities involves local instabilities. When the elastic coupling is small, the dynamics is controlled by individual instabilities. On the contrary for strong elastic coupling, the dynamics is controlled by the whole front. Moreover during asperity depinning, dynamical coupling can develop from waves and result in stress overshoots elsewhere along the front.

In recent years, there have been theoretical developments within the context of crack theory that include an explicit three dimensional modeling and describe the long range elastic interactions through the bulk for the determination of the stress intensity factor variations along the crack front. In the framework of first-order perturbation theory a relationship between the stress intensity factor for tensile crack (Mode I) and small deviations from straightness of the edge of a semi-infinite plane crack, extending at constant speed in an infinite elastic body, was first proposed [11] using weight functions [21] in the approximation of scalar elasticity. This work has been extended to dynamic crack growth both for scalar and vectorial elasticity [10, 17, 19, 22, 32] as well as for non planar crack growth under mixed loading [18].

The relevance of a specific crack model is difficult to estimate owing to the lack of experimental observations. A direct observation of the crack front is usually impossible and inverse descriptions of the crack front are generally observed with a low spatial resolution. Daguier *et al* [6] proposed a first experimental roughness characterization of a crack front propagation in a randomly heterogeneous medium. Using ink, they casted the crack tip at a given stage of the propagation. From the three-dimensional observation of the crack front, they extracted the self-affine exponent describing the in-plane roughness of the crack: $\zeta = 0.60$. Mower and Argon [16] were able to capture directly the propagation of a crack in a transparent epoxy containing defaults as Nylon rods. However, in this case, defaults are not randomly distributed.

The aim of this paper is to review previous works on direct observations of an inter-facial crack front along a weak heterogeneous plane at rest [7, 26] and present recent preliminary analyses of the dynamics of the crack front using a fast digital camera.

The paper is organized as follow: we introduce in section (2) the experimental setup. In section (3), we summarize the main results for pinned fronts at rest; in sections (4), we present the local dynamics of the crack using the framework of interface growth. Finally in section (5) we discuss our observations and compare them to numerical results.

2. EXPERIMENTAL SETUP

2.1. A WEAK PLANE WITH RANDOM TOUGHNESSES

Samples are made of transparent polymethylmethacrylate (PMMA). Each PMMA sample is obtained from the assembling of two plates. The above plate is $32cm \times 14cm$, and $1cm$ thick (drawn in gray in Figure 1). The second one (in black in Figure 1) is $34cm$ long, $12cm$ wide, and $0.4cm$ thick.

Each plate is sand-blasted on one side. Sand-blasted surfaces are blown with a air jet to remove dust. One important consequence of the sand-blasting is that the transparency of the PMMA is lost. Light scatters from introduced micro-structures.

The plates are then annealed together in an oven under a normal homogeneous pressure using a press. The press is made of two parallel aluminum plates and loaded with eight clamps. The annealing procedure lasts 30 min at a temperature of $205\,^\circ C$.

After annealing both plates together in the oven, the new-formed block recovers its transparency. New polymer chains are formed through the rough interface and air bubbles are extracted with the pressure load.

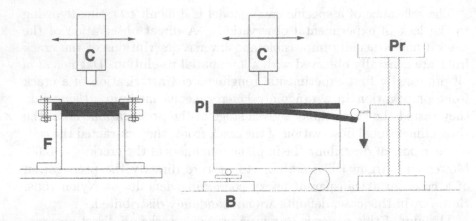

Figure 1 Pictures of the experimental setup: mechanical press (Pr), plexiglas plate
(Pl), camera (C), balance (B), rigid frame (F).

Difference of the refraction index along the interfaces disappears. When
both plates are moved apart during the fracture propagation, the polymer chains within the weak plane are broken. The difference of refraction
indexes across the interface is recovered and the fracture surface appears
opaque due to the light scattering from the micro-structures. The transition between the transparent and opaque regions corresponds to the
crack front (see Figure 2).

The annealing process (specially its uniformity) was checked by visu-
al inspection with a microscope. In order to check that the plates were
sufficiently annealed, a colored nigrosine water solution was injected between the two plates at a given position of the crack front before the
plates were completely separated. Water wets the Plexiglas and spontaneously invades any free space between the plates. No transport of
water nigrosine solution was observed across the fracture front, indicating a continuous soldering of the plates without air bubbles trapped in
the annealed part.

Sand-blasting roughens the surfaces of the Plexiglas plates. The magnitude of the roughness is less than a few micrometers. The random position of the defaults was checked optically using a microscope. Optical
signatures of the flaws were observed to be smaller than a cutoff of about
$50 \mu m$ corresponding to the bead size. This length scale is much smaller
than the long range correlations observed in this experiment which is
of the order of several cm. The height fluctuations introduced by the
sand-blasting procedure introduce fluctuations in the toughness during
the annealing process. We expect the toughness to be uncorrelated on

length scales larger than the cutoff length. However the correlations in the toughness up to this length scale is an open question.

We obtain at the final stage one block with a weak annealing plane where the crack will propagate. The crack plane is referred as (x, y): the y-axis is along the crack propagation direction and the x-axis is parallel to the mean crack front.

2.2. MECHANICAL SETUP

The mechanical setup is sketched in Figure 1. The thick plate of the PMMA block is clamped to a stiff aluminum frame. A normal displacement is imposed to the thin plate with a rod parallel to the crack front. By using an imposed displacement, we obtain a *stable* crack propagation in mode I. Oil is added to reduce friction between the thin plate and the rod. The rod is moved vertically by a continuous motor which allows a quasi-static loading. Typically the imposed displacement rate is in the range $10^{-7} - 10^{-4}$ m/s.

For a subset of experiments, the Plexiglas block is lying on a balance with a high dynamic resolution (1 mg - 1 kg) which provides a force measurement during the crack propagation. The load-displacement history was then obtained during experiments as the measurement of the force F acting on the plate with respect to the imposed displacement δ. Assuming an elastic rheology of the Plexiglas plate, a straightforward expansion leads to:

$$F(\delta, \langle h \rangle_x) \propto E \frac{\delta}{\langle h \rangle_x^{\,3}} \tag{1.1}$$

where E is the Young modulus and $\langle h \rangle_x$ the average position of the front along the x axis. The elastic modulus has been measured as: $E = 3.3 GPa$. The viscous behavior of the Plexiglas was obtained by relaxation tests. Departure from a linear relationship between force F and displacement δ corresponds to the onset of the crack propagation.

The fracture propagates within the weak plane of the Plexiglas block. During the propagation, the front is pinned by local regions of high toughness and becomes rough. The geometry of the crack front is observed with a microscope. When static positions of the fronts are analyzed, a digital camera, set on the top of the microscope provides, high resolution images: 1536×1024 pixels. Each pixel covers a region of $2.6 \mu m \times 2.6 \mu m$. A sample picture is shown in Figure 2. In the image, which is inverted, the sintered part is seen as light while the dark region represents the fracture aperture. The front is defined as the contrast boundary between the dark and the light areas and is found by using a specifically developed software.

Figure 2 Example of a crack front picture using the Kodak DSC420 camera. The resolution is 1536 × 1024. The image covers an area of 6.5mm×4mm. The crack is propagating from bottom to top. The cracked area is in dark gray. The front obtained from image treatment is superimposed as a black line.

To observe dynamical propagation of the front, a fast digital camera was used: Kodak Motion Korder Analyzer camera which takes up to 1000 images per second with a 512x240 pixel resolution.

3.   ROUGHNESS OF PINNED FRONTS AT REST

Measurements have been performed after a complete stop of the press when the crack front is at rest. The microscope is translated along the front (x direction) and up to 20 pictures of the same front are taken. Neighboring pictures overlap one another over one third of the picture width (i.e. over $1.3mm$). Fronts are extracted from each picture and average at overlapping positions. An example of three neighbor fronts is presented in Figure 3a. The assembling of 20 pictures provides a high resolution of the crack front (up to 2^{14} data points) over $45mm$. Figure 3b shows one reconstructed large scale front (i.e. a macro-front). A complete description of the assembling procedure can be found in Ref. [7].

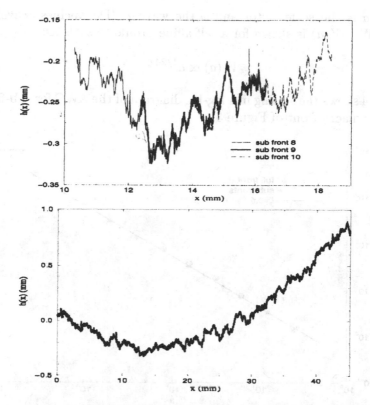

Figure 3 a) (upper picture) Fronts obtained from different pictures taken along the fracture front are assembled to reconstruct a large scale macro-front. Overlap between pictures is of one third of the picture. b) (lower picture) Example of a macro-front obtained from the assembling of 20 pictures. The total extend along the crack front is 45 mm with a resolution of more 17000 points (2.6μm/pixel) and plotted as filled black disks. Nine of such front were obtained.

The roughness of the crack front is analyzed in terms of self-affinity [9]. A self-affine profile $h(x)$ is statistically invariant under the scale transformation

$$(x, h) \rightarrow (\lambda x, \lambda^\zeta h) \tag{1.2}$$

where ζ is the roughness exponent. Numerous techniques exist to probe the self-affine behavior and estimate the roughness exponent [1, 24, 30]. We present here analyses based on a recent technique that uses wavelet transform: The Average Wavelet Coefficient (AWC) technique [31]. It consists of a wavelet transform of the front and an average over the translation factor b:

$$W(a) = \langle W(a, b) \rangle_b = \left\langle \frac{1}{\sqrt{a}} \int h(x) \phi\left(\frac{x - b}{a}\right) dx \right\rangle_b \tag{1.3}$$

where a is the scale factor and ϕ the wavelet (Debauchies' wavelets). The AWC $W(a)$ is shown for a self-affine profile to scale as:

$$W(a) \propto a^{1/2+\zeta} \tag{1.4}$$

Figure 4 shows the scaling in a log-log diagram of the AWC for sub-fronts and the macro-front of Figure 3.

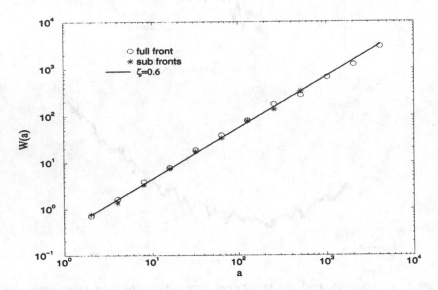

Figure 4 Roughness analysis of the front is measured using the average wavelet coefficient (AWC) technique. The average spectrum of sub fronts obtained from each constitutive pictures is superimposed using *. The fit is: $W(a) \propto a^{1.1}$ which is very consistent with an roughness exponent of $\zeta \approx 0.6$. The scale factor is in unit of pixels $(2.6\mu m)$.

A power law behavior fits nicely the data both for the entire macro-front and the sub-fronts which translates the self-affine property of the crack front at rest. The slope of the fit in Figure 4 provides an estimate of the roughness exponent: $\zeta \approx 0.6$. In Ref. [7] a full analysis of nine independent macro-fronts has been performed using four different techniques: the variable bandwidth method, the multiple return probability method, the power spectrum analysis and the AWC method. All methods are consistent, and the final estimate of the roughness exponent is: $\zeta = 0.63 \pm 0.03$.

Doing the full analyses of the self-affinity property of crack fronts, one observes that the scaling shows an upper cut-off that might be related to the thickness of the Plexiglas plate. Further studies are on going for a careful check of the influence of the Plexiglas plate thickness.

4. LOCAL DYNAMICS OF THE CRACK

When the press is moved with a constant speed of $0.107mm/s$ a fracture front propagates along the weak plane with an average speed over the camera window: $v_l \approx 68\mu m/s$.

Images during the crack propagation are acquired at a rate of 500 per second and cover a region of $5.12mm \times 2.4mm$. Figure 5 shows 6 fronts with a constant time delay between pictures of $0.4sec$ (the time delay between two consecutive images is $2ms$). The figure clearly shows the burst-like propagation at large scale (i.e. at the picture scale). The inset in Figure 5 presents the relative evolution of the crack front with respect to the loading: $\delta\tilde{h}(t) = v_l t - \langle h(x,t) \rangle_x$. When $\delta\tilde{h}(t)$ is positive, the crack front is going slower than the loading and can be considered as pinned. This situation corresponds to fronts a, b and c which are very close. When $\delta\tilde{h}(t)$ decreases, a crack burst develops and the front accelerates to recover the loading speed (fronts d and e). The burst activity exist at all scales. They exist also at small scales: see for instance differences between front b and c on the right side of Figure 5. In that case the burst extends only over $0.3mm$.

Figure 6 shows three different space (horizontal) time (vertical) diagrams. In Figure 6a the grey levels represent the front position $h(x,t)$, where light represents far developed front positions and dark represents low front positions. The inhomogeneous and abrupt front movements are seen in this figure but is better illustrated in Figure 6b where the grey level represents the difference $\delta h = h(x,t) - h(x,t=0)$, where $h(x,t=0)$ is the first recorded front. The fracture propagated a sufficient long time before recording started to be sure that the front was outside the time-dependent initial regime. This figure further illustrates how the long range correlations build up dynamically in sudden irregular jumps. The jumps are also seen in Figure 6c where the grey level representation of the local velocity fluctuations is plotted. In this figure the local front velocity $(h(x,t) - h(x,t=0))/dt$ has been calculated by subtracting the front positions between two fronts separated a time interval $dt = 0.1s$. The grey level representation of the fronts appear as almost black or white. The reason for this is the sharp exponential distribution observed for the velocity fluctuations.

The dynamic scaling of the fracture front has been measured by calculating the wavelet spectrum $W(a,t)$ of the height difference $\delta h(x,t)$, where a is the scale parameter and t is the time. A Family Vicsek scaling [8] of the wavelet spectrum $W(a,t)$ can be written as

$$W(a,t) = t^{(1/2+\zeta)/z} G(a/t^{1/z}), \qquad (1.5)$$

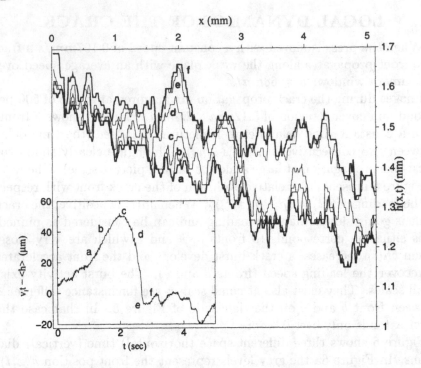

Figure 5 Six fronts positions (a,b,c,d,e,f) around a peak of curve $v_l t - <h> = f(t)$ (shown in the inset). Time difference between consecutive front positions is constant and equal to 0.4 sec.

where $G(x)$ is constant for $x >> 1$ and $G(x) \propto x^{1/2+\zeta}$ for $x << 1$. The dynamic exponent z gives the scaling with time of the horizontal correlation length $\xi \approx 0.3t^{1/z}$ (the prefactor is estimated for ξ in m and time in sec). In Fig. 7a, the wavelet spectrum of the relative height $h(x,t) - h(x,t=0)$ is shown at different times from $0.03s$ to $7.11s$. At small times, no spatial correlation exists and the spectrum appears flat. For larger times, however, the correlations become apparent, and a cross-over to a power law behavior is observed for small a. For large times data are consistent with a power law: $W(a) \propto a^{1/2+\zeta}$ with $\zeta = 0.6$. In Fig. 7b, the scaling function G is plotted as function of $a/t^{1/z}$. This data collapse provide an estimate of the dynamic exponent $z = 1.15 \pm 0.1$. The experimental data were further checked for possible anomalous scaling behavior [13] but no such behavior was observed.

The estimate of the dynamic exponent z is checked by measuring the roughness evolution of the front with time (see Figure 8). The roughness is estimates as:

$$\sigma(t) = \sqrt{\langle \delta h^2(x,t)\rangle_x - \overline{\langle \delta h(x,t)\rangle_x^2}} \tag{1.6}$$

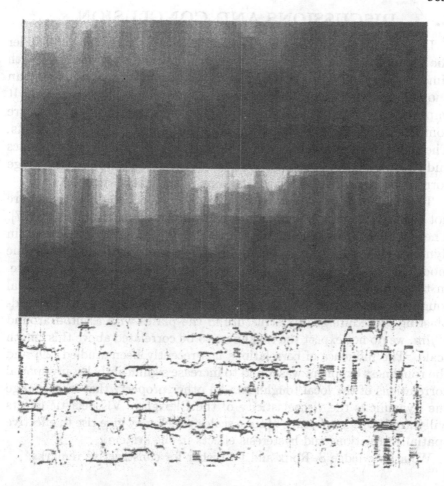

Figure 6 Three representations of the space time diagram: Horizontal direction corresponds to space (the full range of the picture is 5.12 mm), Upwards vertical direction describes time and covers 8.7 sec. The upper graph is a gray scale of the interface position $h(x,t)$. Light grays mean large displacement of the interface. The middle picture shows using similar coding the difference $\delta h(x,t) = h(x,t) - h(x,t=0)$. The bottom graph represents the local velocity of the front: $v(x,t) = (h(x,t+dt) - h(x,t))/dt$ where the time lag is: $dt = 0.1sec$. Dark dots correspond to high speeds.

A Family-Vicsek scaling leads to a roughness evolution: $\sigma(t) \approx t^{\zeta/z}$. Size effects on this relation were not explored to check possible departure from this relationship [12]. In Figure 8 data are reasonably fitted by the power law with previous estimates of ζ and z.

5.   DISCUSSIONS AND CONCLUSION

Dynamic exponent has been studied in three dimensions [13] under the assumption that the average fracture position is proportional with time (constant speed). On the basis of this assumption, it was found an anomalous scaling that differs from the simple Family Vicsek scaling. It important to point out that this experiment was a fast unstable fracture compared with the slow stable fracture performed in our experiments. The speed of a fracture may play an important role both for the dynamics and the structure of the fracture, due to dynamical effects and long range correlations owing to wave propagations in the elastic medium.

The roughness and dynamic exponents obtained in our experiment are not consistent with any simulations or theoretical models [3, 20, 23, 27]. These models are quasi static with no "real" time and do not contain dynamic effects which may be important even at moderate speed. The models do neither contain any long range correlations in the local noise. In the experiments it is not clear if there are correlations in the local toughness introduced by the sand blasting. However since the sand-blasting procedure gives a structure in the plates with a cutoff around $50\mu m$, we do not expect the toughness to be correlated above this length scale. The influence of correlations has recently been studied [29] and the roughness exponent was found to increase significantly by any spatial correlations of the local toughness. An other property that may change the dynamic is the viscoelasticity of the Plexiglas. Viscoelastic effects will introduce a large range of time scales which may give rise to effective spatial correlations and hysteresis effects in the system.

We acknowledge S. Roux and D. Fisher for constructive remarks.

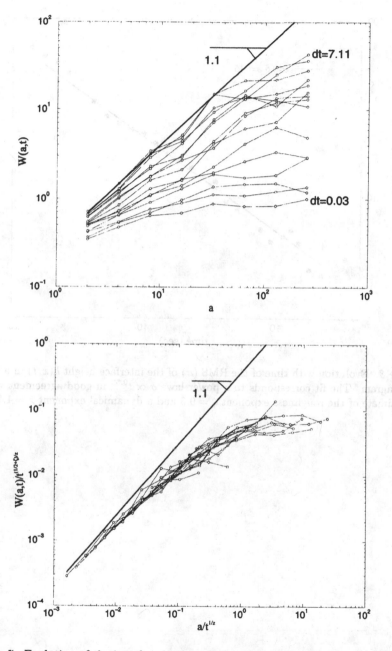

Figure 7 Evolution of the interface in the wavelet domain. In the upper graph, the wavelet spectrum of the relative height $h(x, dt) - h(x, 0)$ is shown for various time differences dt (along a geometrical series). The upper graph shows the superimposition of the AWC spectra of the interface for different times. At a small time difference (e.g. $dt = 0.03$ sec) no spatial correlations emerge and the spectrum is flat. At long time, the scaling: $W(a) \propto a^{1/2 + \zeta}$ with $\zeta = 0.6$ is recovered. The lower diagram shows the collapse as: $W(a, t)/t^{1/2 + \zeta} = G(a/t^{1/z})$. The collapse provides an estimate of the dynamical exponent: $z \approx 1.15$. a is measured in pixel size: $10\mu m$, time is in frame number: $2ms$ between frames.

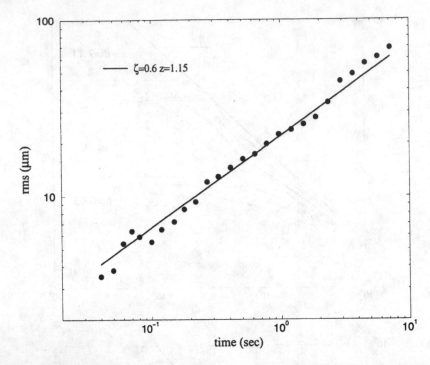

Figure 8 Evolution with time of the RMS (σ) of the interface height $h(x,t)$ in a log-log diagram. The fit corresponds to a power law: $\sigma \propto t^{0.52}$ in good agreement with an estimate of the roughness exponent $\zeta = 0.6$ and a dynamical exponent $z = 1.15$.

References

[1] Barabási, A. and H. Stanley: 1995, *Fractal concepts in surface growth*. Cambridge: Cambridge University Press.

[2] Bouchaud, E., G. Lapasset, and J. Planès: 1990, 'Fractal dimension of fractured surfaces: a universal value ?'. *Europhys. Lett.* **13**, 73–79.

[3] Bouchaud, J. P., E. Bouchaud, G. Lapasset, and J. Planès: 1993, 'Models of fractal cracks'. *Phys. Rev. Lett.* **71**, 2240.

[4] Brown, S. R. and C. H. Scholz: 1985, 'Broad bandwidth study of the topography of natural rock surfaces'. *J. Geophys. Res.* **90**, 12575–12582.

[5] Cox, B. L. and J. S. Y. Wang: 1993, 'Fractal surfaces: measurement and application in earth sciences'. *Fractals* **1**, 87–115.

[6] Daguier, P., E. Bouchaud, and G. Lapasset: 1995, 'Roughness of a crack front pinned by microstructural obstacles'. *Europhys. Lett.* **30**(7), 367–372.

[7] Delaplace, A., J. Schmittbuhl, and K. Måløy: 1999, 'High resolution description of a crack front in a heterogeneous plexiglas block'. *Phys. Rev. E* **60**, 1337–1343.

[8] Family, F. and T. Vicsek: 1991, *Dynamic of fractal surfaces*. Singapore: World Sci. Pub.

[9] Feder, J.: 1988, *Fractals*. New York: Plenum.

[10] Fisher, D.: 1998, 'Collective transport in random media: from superconductors to earthquakes'. *Physics Reports* **301**, 113–150.

[11] Gao, H. and J. R. Rice: 1989, 'A First-Order Perturbation Analysis of Crack Trapping by Arrays of Obstacles'. *ASME Journal of Applied Mechanics* **56**, 828–836.

[12] Krishnamurthy, S., A. Tanguy, and S. Roux: 2000, 'Dynamic exponent in extremal models of pinning'. *Eur. Phys. J. B* **15**, 149–153.

[13] Lòpez, J. and J. Schmittbuhl: 1998, 'Anomalous scaling of fracture surfaces'. *Phys. Rev. E* **57**, 6405–6408.

[14] Måløy, K. J., A. Hansen, E. L. Hinrichsen, and S. Roux: 1992, 'Experimental measurements of the roughness of brittle cracks'. *Phys. Rev. Lett.* **68**, 213–215.

[15] Mandelbrot, B. B., D. E. Passoja, and A. J. Paullay: 1984, 'Fractal character of fracture surfaces of metals'. *Nature* **308**, 721–722.

[16] Mower, T. M. and A. Argon: 1995, 'Experimental investigations of crack trapping in brittle heterogeneous solids'. *Mechanics of Materials* **19**, 343–364.

[17] Perrin, G. and J. Rice: 1994, 'Disordering of dynamic planar crack front in a model elastic medium of randomly variable toughness'. *J. Mech. Phys. Solids* **42**(6), 1047–1064.

[18] Ramanathan, S., D. Ertas, and D. S. Fisher: 1997, 'Quasi-static crack propagation in heterogeneous media'. *Phys. Rev. Lett.* **79**, 873–876.

[19] Ramanathan, S. and D. Fisher: 1997, 'Dynamics and instabilities of planar tensile cracks in heterogeneous media'. *Phys. Rev. Lett.* **79**, 877–880.

[20] Ramanathan, S. and D. Fisher: 1998, 'Onset of propagation of planar cracks in heterogeneous media'. *Phys. Rev. B* **58**, 6026–6046.

[21] Rice, J. R.: 1985, 'First-Order Variation in Elastic Fields Due to Variation in Location of a Planar Crack Front'. *J. Appl. Mech.* **52**, 571–579.

[22] Rice, J. R., Y. Ben-Zion, and K. Kim: 1994, 'Three-dimensional perturbation solution for a dynamic planar crack moving unsteadily in a model elastic solid'. *J. Mech. Phys. Solids* **42**, 813–843.

[23] Roux, S. and A. Hansen: 1994, 'Interface roughning and pinning'. *J. Phys. I Fr.* **4**, 515–538.

[24] Sahimi, M.: 1998, 'Non-linear and non-local transport process in heterogeneous media: from long-range correlated percolation to fracture and material breakdown'. *Phys. Reports* **306**, 213–395.

[25] Schmittbuhl, J., S. Gentier, and S. Roux: 1993, 'Field measurements of the roughness of fault surfaces'. *Geophys. Res. Lett.* **20**, 639–641.

[26] Schmittbuhl, J. and K. Måløy: 1997, 'Direct observation of a self-affince crack propagation'. *Phys. Rev. Lett.* **78**, 3888–3891.

[27] Schmittbuhl, J., S. Roux, J. P. Vilotte, and K. J. Måløy: 1995a, 'Interfacial crack pinning: effect of non local interactions'. *Phys. Rev. Lett.* **74**, 1787–1790.

[28] Schmittbuhl, J., F. Schmitt, and C. H. Scholz: 1995b, 'Scaling invariance of crack surfaces'. *J. Geophys. Res.* **100**, 5953–5973.

[29] Schmittbuhl, J. and J. P. Vilotte: 1999, 'Interfacial crack front wandering : influence of a correlated quenched noise'. *Physica A* **270**, 42–56.

[30] Schmittbuhl, J., J. P. Vilotte, and S. Roux: 1995c, 'Reliability of self-affine measurements'. *Phys. Rev. E* **51**, 131–147.

[31] Simonsen, I., A. Hansen, and O. M. Nes: 1998, 'Using wavelet transforms for hurst exponent determination'. *Phys. Rev. E* **58**, 2779.

[32] Willis, J. and A. Movchan: 1995, 'Dynamic weight-functions for a moving crack: 1. Mode-I loading'. *J. Mech. Phys. Solids* **43**, 319–341.

Index